THE OXFORD ILLUSTRATED
HISTORY OF THE
Royal Navy

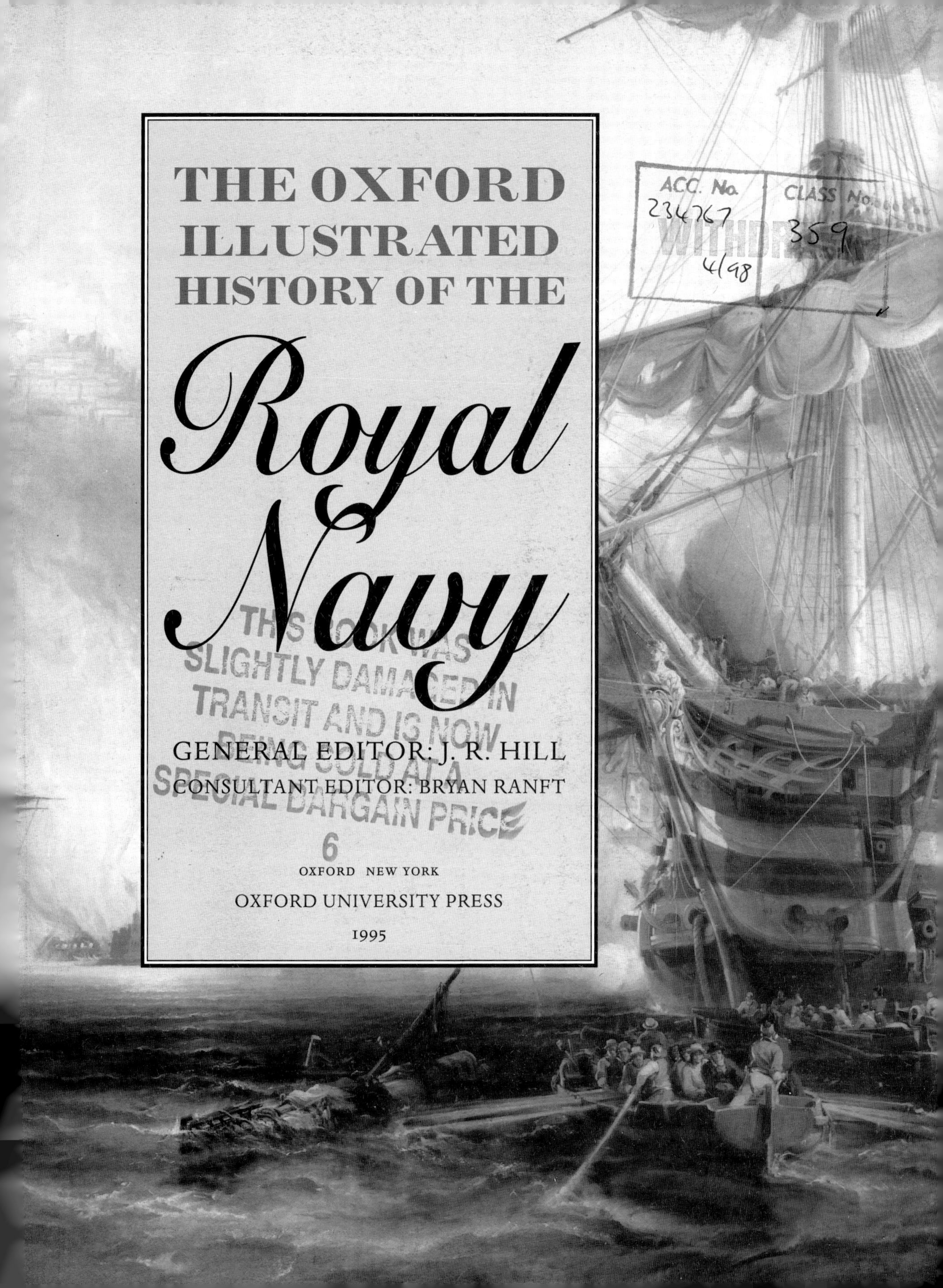

THE OXFORD ILLUSTRATED HISTORY OF THE

Royal Navy

GENERAL EDITOR: J. R. HILL

CONSULTANT EDITOR: BRYAN RANFT

6

OXFORD NEW YORK

OXFORD UNIVERSITY PRESS

1995

Oxford University Press, Walton Street, Oxford OX2 6DP

Oxford New York
Athens Auckland Bangkok Bombay
Calcutta Cape Town Dar es Salaam Delhi
Florence Hong Kong Istanbul Karachi
Kuala Lumpur Madras Madrid Melbourne
Mexico City Nairobi Paris Singapore
Taipei Tokyo Toronto
and associated companies in
Berlin Ibadan

Oxford is a trade mark of Oxford University Press

British Library Cataloguing in Publication Data
Data available

Library of Congress Cataloging in Publication Data
The Oxford illustrated history of the Royal Navy/general editor,
J.R. Hill; consultant editor, Bryan Ranft.
p.   cm.   Includes bibliographical references (p.   ) and index.
1. Great Britain. Royal Navy—History. 2. Great Britain—History, Naval.
I. Hill, J. R.   II. Ranft, Bryan.
359'. 00941—dc20   VA454.094 1995   95–13627

ISBN 0-19-211675-4

1 3 5 7 9 10 8 6 4 2

Printed in Great Britain
on acid-free paper by
Butler & Tanner Ltd.
Frome, Somerset

Title-page illustration: Bombardment of Algiers, 1816, by
George Chambers Senior (1803–40).

CONTENTS

LIST OF COLOUR PLATES

LIST OF MAPS AND ACTION STATIONS

LIST OF CONTRIBUTORS

Daniel A. Baugh is Professor of Modern British History at Cornell University. His principal works are *British Naval Administration in the Age of Walpole* (1965) and *Naval Administration, 1715–1750* (1977). He has written on government and society in England, 1660–1830, especially on relief of the poor. During the past decade his articles and essays have been chiefly concerned with the maritime and financial aspects of British defence policy from the sixteenth century to the early twentieth century.

David K. Brown joined the Royal Corps of Naval Constructors in 1949, retiring as Deputy Chief Naval Architect in 1988. He has written books, technical papers, and articles on the past, present, and future of warship design. Books include: *A Century of Naval Construction, Before the Ironclad, The Future British Surface Fleet, Paddle Warships*, and a biography of William Froude. He is Vice President of the Royal Institution of Naval Architects and a Research Fellow of Bath and London Universities.

J. D. Davies teaches history at Bedford Modern School, having previously taught in Cornwall and Sussex. His doctoral thesis formed the basis of his book, *Gentlemen and Tarpaulins: The Officers and Men of the Restoration Navy* (Oxford, 1991), and he has also written on political, administrative, and strategic aspects of the naval history of the seventeenth century. He won the Julian Corbett essay prize for naval history in 1986. A member of the council of the Navy Records Society, he contributed to the society's centenary volume and is also editing Pepys's later Admiralty letterbooks for it.

Norman Friedman is a naval historian and strategic analyst. His books include: *The Postwar Naval Revolution, British Carrier Airpower, Naval Radar, Modern Warship Design and Development, The US Maritime Strategy*, and various editions of *The Naval Institute Guide to World Naval Weapons Systems*. Recently he edited (and contributed to) *Navies in the Nuclear Age*, a volume of Conway's *Encyclopedia of the Ship*.

James Goldrick, a commander in the Royal Australian Navy, was born in 1958 and joined the RAN in 1974. He holds BA and M.Litt degrees from Australian universities and is an anti-submarine specialist. He has had Sea Service with the RAN, including command of HMAS *Cessnock*, and an exchange with the Royal Navy. After a year as an International Research Fellow at the US Naval War College, he is currently in charge of the RAN's Principal Warfare officer training and its tactical development cell. His publications include *The King's Ships Were at Sea* and a variety of edited books, as well as many articles on contemporary and historical naval subjects.

Eric J. Grove was a civilian lecturer at the Royal Naval College, Dartmouth, from 1971 to 1984, including a year as an exchange professor at the US Naval Academy Annapolis, 1980–81. Leaving Dartmouth as Deputy Head of Strategic Studies, he became a freelance historian, defence analyst, writer, broadcaster and lecturer, also teaching at the Royal Naval College Greenwich and Cambridge University, and working as a Research Fellow at Southampton University. Since 1993 he has been a lecturer in international politics and

Deputy Director of the Centre for Security Studies at the University of Hull. He has written prolifically on twentieth century naval history and contemporary maritime strategy.

John B. Hattendorf D.Phil. (Oxon.), F.R.Hist.S., is the Ernest J. King Professor of Maritime History and Director of the Advanced Research Department at the US Naval War College, Newport, Rhode Island. His books include *England in the War of the Spanish Succession* (1987), and he is the co-editor of *British Naval Documents 1204–1960* (1990).

J. R. Hill served for over 40 years in the Royal Navy in sea and Whitehall appointments, ending his career in 1983 as a Rear-Admiral. He has since been Under-Treasurer of the Middle Temple, retiring in 1994, and since 1980 has written six books on maritime warfare, strategy, and arms control. He has lectured and given conference papers, worldwide, on similar subjects. He is editor of *The Naval Review* and Chairman of the Society for Nautical Research.

Andrew Lambert F.R.Hist.S. is a senior lecturer in War Studies at King's College, London, and a member of Council of both the Society for Navy Records and The International Commission for Maritime History. His books include: *The Crimean War: British Grand Strategy against Russia 1853–1856* (1990), *The Last Sailing Battlefleet: Maintaining Naval Mastery 1815–1850* (1992), and *Battleships in Transition: The Creation of the Steam Battlefleet* (1984).

David Loades is Professor of History at the University of Wales, Bangor, and chairman of the publications committee of the Navy Records Society. His books include: *Politics and the Nation, 1450–1660* (1974; 1992), *The Reign of Mary Tudor* (1979; 1991), *The Tudor Court* (1986; 1993), and *The Tudor Navy* (1992). He has also contributed part ii (1485–1603) to *British Naval Documents, 1204–1960* (NRS, 1993). In 1988–9 he was a Visiting Fellow at All Souls College, Oxford.

Rear-Admiral Roger Morris retired from the Royal Navy in 1990 after a career spent almost entirely in the Hydrographic Service, ending as Hydrographer of the Navy from 1985. He has served on the Councils of the Hakluyt Society and the Navy Records Society, and as Chairman of Council of the Society for Nautical Research. He has just completed writing the History of the Hydrographic Service from 1919 to 1970.

Susan Rose is Lecturer in History at the Roehampton Institute of Higher Education and a course tutor and consultant for the Open University. She has a long-standing connection with the Society for Nautical Research, of which she is currently a Vice-President. Her publications include, for the Navy Records Society, *The Navy of the Lancastrian Kings* and, with Nicholas Rodger, the medieval section of *British Naval Documents 1204–1960*. She has also written on the French Revolution and, for the Open University, on the history of London.

Geoffrey Till is Professor and Head of the Department of History and International Affairs at the Royal Naval College, Greenwich. He also teaches in the Department of War Studies at King's College, London, and before that was in the Department of Management and Systems at the City University. In addition to many articles on various aspects of defence, he is the author of a number of books including *Air Power and the Royal Navy* (1979), *Maritime Strategy and the Nuclear Age* (2nd edn., 1984), *Modern Sea Power* (1987), and, with Bryan Ranft, *The Sea in Soviet Strategy* (2nd edn., 1989). He has recently completed, with D. J. Pay, *East–West Relations in the 1990s: The Naval Dimension* (1990). Most recently he edited a volume in the Brassey's Seapower series, *Coastal Forces*.

John Winton served in the Royal Navy in the Korean War, at Suez in 1956, and then for seven years in the Submarine Service. He retired as a Lieutenant-Commander to become a full-time professional writer in 1963. He has published 40 books which include *The Forgotten Fleet, Hurrah for the Life of a Sailor!, The Victoria Cross at Sea, The Death of the Scharnhorst*, and *The Naval Heritage of Portsmouth*.

PREFACE

The Royal Navy can claim a singular position in history. It is unique in many of its elements, and it has had effects that would have been difficult to predict by conventional logic. Its influence on the world's political and strategic arrangement, and even on its behaviour, was for over a century and a half profound. This influence was often directed by its political masters, sometimes with great precision; at other times it appeared to work as an outgrowth of the power and confidence of the service itself. However it operated, it seemed that the result was often disproportionate to the effort exerted.

Paul Kennedy in his authoritative work *The Rise and Fall of British Naval Mastery* explains this effect by making it part of a three-sided 'framework for national and world power': the navy and the trade it protected were one side, the empire (formal and informal) another, and the industrial base the third. Modern, not to say revisionist, historians would add a fourth—financial and banking—arm, and maybe a fifth, diplomatic one. No doubt when all these worked in concert they were of immense strength. The more extreme examples of the navy's influence would not have been possible without these other elements. Nevertheless it was an extraordinary instrument and it did extraordinary things.

It was also a peculiarly British instrument. That is to say, it evolved from no theoretical concept or plan but through a blend of fear, ambition, curiosity, and trial and error; it went through times of experiment and conservatism, boom and recession. It was immensely pragmatic, but also, generally, practical and energetic. Its large-scale organization was better than it looked, while in detail, particularly at sea, its operation reached exemplary levels of excellence. Of course at any given time the navy reflected the spirit of the age; the parallel between the technical brilliance and resourcefulness of the late eighteenth-century navy afloat and the industrial revolution ashore is one obvious example, and on the negative side the confusion and maladministration of the early Stuart navy has its counterpart in the low morale and poor economic state of the country as a whole at that time.

A final singularity was the navy's position in the national consciousness. For two and a half centuries it commanded a degree of respect and affection that was probably unrivalled by any fighting service anywhere; but more than that, it pervaded the national life. It is perhaps unfair to refer to a novel like *Persuasion*, because Jane Austen numbered many naval officers among her acquaintance; but a comb through not only British literature but aspects of British culture from forestry to folk-song, centuries before and after 1800, reveals so much material at so many levels that the case is made. On one matter, however, Jane is more informative than most: the navy was one of the instruments of social mobility that even then distinguished British society from some of its European counterparts. Indeed there is evidence that when this mobility was blocked—for example, when the navy became a fashionable service towards the end of the nineteenth century—its health suffered.

All these elements in the historical phenomenon that is the Royal Navy are covered by the 14 authors in this volume. Dr Rose traces the fluctuating strength and fortunes of the Royal ships, which so often followed—or did they lead?—those of the general state of England in early and medieval times. Professor Loades shows with particular strength the way in which the Navy Royal's administration developed: a development at least as important as the technical innovations and private and/or patriotic enterprises of the Tudor period. Dr Davies brings the navy out of its early-Stuart trough and into a fertile time, under Commonwealth and Restoration, of rationalization: of tactics, organization and training, all honed and tempered by the Dutch wars.

The eighteenth and early nineteenth centuries were the Royal Navy's operational zenith and here this history has been served by two eminent scholars from the United States, where so much recent work on the subject has gone on. Professors Hattendorf and Baugh deal respectively with the operational/strategic and the administrative aspects. Those words are, as the text will show, to be interpreted in the broadest sense.

The complexities of the nineteenth century required several chapters. Dr Lambert's strategic and operational treatment shows how British power and influence rested firmly on a pre-eminent battle fleet, however it may have been manifested at its extremities by other craft. David Brown traces the unprecedented technical development of the century and finds in it more wise direction and common sense than have some earlier writers. John Winton addresses the social history of the navy at the time, and does not flinch from the sour aspects while acknowledging the improvements that were made. Finally, the General Editor must confess to having presented Admiral Morris with a terribly diverse and broad set of subjects in exploration, hydrography, counter-slavery, and disaster relief; but Admiral Morris has risen to the challenge in a way that makes, to this observer, an immensely satisfying chapter.

The Royal Navy entered the twentieth century as a fashionable, well-publicized, booming service with, soon afterwards, a specific and identifiable threat to counter, and all that carried its own dangers. Commander Goldrick gives his view of the period up to and including the First World War with an authority based on extensive scholarship acquired while still pursuing an outstanding career in the Royal Australian Navy.

After the First World War the navy went through difficult times—to which it should not have been unaccustomed through experience of peace, of course, though some of the problems, notably the emergence of international arms limitation and competition for resources from the Royal Air Force, were new. Professor Till describes this complex, uneasy period; and Eric Grove shows how in the Second World War, far from being dispirited by its inter-war privations, the navy rose to a multitude of challenges—not without mistakes and reverses, some of its own making and some of the politicians', but with ultimate success.

Finally, Norman Friedman and I, he in terms of the 'Technical Revolution' and I in the politico-military and organizational fields, have tried to bring the story up to date. Contemporary history is a hard taskmistress, and one's own experience, particularly at the centre of affairs, may not be the best background for objectivity. But we have done our best to describe a period that was full of incident in policy, operational and material development. Where it will lead is not a matter for a prologue: the reader is referred to the postscript, where there may need to be some plain speaking.

All the authors in this study have played their parts nobly, responded to every suggestion from the Consultant Editor and from me, and even delivered largely to time and specified length (administration must always creep in). My thanks are due to them and particularly to Professor Ranft, who has been the perfect guru and without whose wise counsel the book would have been a poorer and much less scholarly thing. That it is a visually exciting thing is due mainly to the efforts of Sandra Assersohn and the Oxford University Press, aided by the often imaginative suggestions of the authors themselves.

One final point may be made before the reader moves on—impatiently one hopes—to the main text. At intervals throughout the book will be found diagrams of 'Action Stations': sketches of the manner in which a ship of the period went into battle, with a broad description of the stationing of her crew in order to move, to fight, and to stay afloat. These are put in because they are interesting in themselves and because they demonstrate the diversity of tasks of the sailor in action over the centuries.

But they also help to make the point that this book is about the Royal Navy. Given that the navy was such an instrument of power in world affairs, it of course addresses matters of strategy, policy, economy, trade, and all the broader elements of sea power, and seeks to place the navy in that context. But it is first and foremost a study of the navy and its people, a fighting service that has endured and, so long as we live in a power-directed world, must endure.

J. R. Hill

Bishop's Waltham
1995

THE OXFORD ILLUSTRATED
HISTORY OF THE

Royal Navy

THE WALL OF ENGLAND TO 1500

SUSAN ROSE

*I*n the fifteenth century the anonymous author of the *Libelle of Englyshe Poly-cye* declared in a memorable phrase that the sea was the 'rounde wall' of England 'as thoughe England were likened to a cite | and the wall environ were the see'. That would not have been the perception of the inhabitants of England for much of the period before the Conquest. Rather, the sea was the source of danger, even terror, if we are to believe the words of the writers of the Anglo-Saxon Chronicle. How, therefore, can we characterize the naval history of England (meaning by that the history of operations at sea against the enemies of the realm) in the period after the establishment of the Anglo-Saxon kingdoms to the accession of the Tudors? It can, perhaps, be seen as the gradual progression from a situation in which no effective means of defence against a sea-borne invader existed to one in which the sea could be portrayed, in Shakespeare's words, as England's 'moat defensive', 'which serves it in the office of a wall'.

Under the year 793 we are told by the writer of the Anglo-Saxon Chronicle that 'terrible portents appeared over Northumbria and miserably frightened the inhabitants . . . fiery dragons were seen flying in the air'. On 8 January 'the harrying of the heathens miserably destroyed God's church in Lindisfarne by rapine and slaughter'. This is generally regarded as the first recorded incursion of the Danes or the Norsemen who were to assault English shores for much of the succeeding 250 years. Forgetting, perhaps, that their ancestors had come over the seas to England in much the same way, the major problem of defence for any English ruler seemed to be to protect coastal areas from seaborne raiders and to prevent these raiders from penetrating further inland. This would seem to be a situation in which we might expect to see sudden explosive growth in naval power and in the appreciation of how to use such a method of defence. However, this presupposes that the technical means existed to engage an enemy at sea. Looking at what we know of the ships of the ninth and tenth centuries and what we know of events in the same period, it seems very doubtful that this was the case.

The popular image of the Viking ship is not far from the truth. They were indeed beautiful and very seaworthy vessels with the high curving prow, the low freeboard, and the sleek hull we can see in the Gokstad ship. But how would such a ship have been fought? It had no ram or beak like an ancient trireme or later Mediterranean galley; its crew was also its source of power as oarsmen, though

SAXONS AND VIKINGS

I

the ship could be sailed in favourable winds. There is no evidence that a separate body of fighting men was carried and there is equally no record of their being armed with anything but hand weapons. For one oared ship to come alongside another and board her was a manœuvre fraught with difficulty even in sheltered waters. To achieve this in the open sea, particularly in the stormy waters around the British Isles, would seem to be very nearly impossible.

Certainly all the encounters between the invaders and the English which are described in any detail are in estuaries or harbours. In 851 nine ships were captured at Sandwich and the others in that particular 'host' (as the Anglo-Saxon Chronicle always describes an invading party), were driven away; the following year many Danes were slain and drowned on the Thanet shore. In 884 another raiding group was caught and defeated in the River Stour in East Anglia. Only in 885 does the chronicler state that Alfred 'went out to sea with ships and fought against four ships' companies of Danes', capturing and slaying the crews of two while the others surrendered. He does not, however, provide any details of how this was achieved.

In 897 Alfred took the action which has earned him the somewhat ill-deserved reputation of being the founder of the navy. In this year the situation along the south coast was very bad, with continual Danish raids from their bases in the north and east of the country. To meet this threat Alfred ordered the construction of large ships of a new design, 'as it seemed to himself that they could be most serviceable'. They had at least 60 oars and were both 'swifter, steadier and with more freeboard' than those of the enemy. Again, they were most successful in shallow and sheltered waters, blockading river mouths and landing parties to seize Danish ships when they were beached and the crews ashore.

Once this force had been established there is evidence that the succeeding kings put the financing and manning of a national naval squadron on a more formal footing. The obligation which rested on all men to defend their homeland, to do duty in the *fyrd* (the *levée en masse*), had long been acknowledged. By the eleventh century this duty could also involve an obligation to serve at sea but, as had happened in the case of the land army, there was by this time an element of selection involved. The obligation to serve in the 'select *fyrd*' (as this body has been termed by some historians), was normally discharged by each community being responsible for the supply of one properly equipped fighting man for each five hides of land; the size of a hide was not completely standard but was notionally enough land to support one family. The laws of Cnut and Æthelred talk of the obligation on a community to find a ship's crew. It has been suggested that this considerable burden was placed on units of 300 hides. The enigmatic phrase *ship-scot* occurs in one or two sources, leading to the supposition that this was a payment in lieu of finding the crewmen themselves, perhaps lying on inland areas. Some towns had separate obligations in this matter; Maldon in Essex, and ports in Kent and Sussex that were later part of the Cinque Ports confederacy, were all expected to furnish the king with a fully equipped vessel in the hour of need. Was this scheme successful? The underlying technical difficulties, mentioned above,

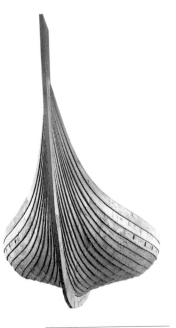

The Gokstad ship, excavated from a burial mound near Oslo in 1880. She is built of oak, 79 ft. long and 16.8 ft. in the beam.

had not been overcome; the strategic situation had undeniably worsened. Alfred had faced assaults from 'hosts' which, while made up of seasoned warriors, were not led by a rival monarch. Æthelred was opposed by the king of Denmark bent on a programme of conquest. Swein came in his 'wave-stallions', to use the Chronicle's evocative phrase, and raided at will: in 1003 in Wiltshire, the following year in East Anglia, and in 1006 all along the south coast from a base in the Isle of Wight. By 1016 the country seemed defenceless in the face of Danish attacks, with the invaders bringing their craft up the Thames to London Bridge and digging a channel to get them round this obstacle when it was held against them.

By the end of the same year, after Swein's death, England was part of the northern realm of Cnut—Swein's son, and ruler of England, Denmark, and Norway. Under him and his sons the Chronicle hints that a tax or *geld* for the upkeep of the fleet had become accepted; in 1040 it apparently raised £22,099, an enormous sum at the time, leading to the suspicion that the money was intended for other than strictly naval purposes. When Edward the Confessor became king in his turn in 1043, the Chronicle continues to mention royal ships but these seem to be now no more than the swiftest means of transport available, and by 1050 they had been paid off along with the household troops. It cannot be argued that this change had much effect on the ability of the English to defend themselves against an enemy coming, as was almost invariably the case, by sea. A deliberate battle at sea was still not a realistic possibility; apart from the problems caused by the deficiencies (for this purpose) in the design and armament of ships, there was also the great difficulty in getting news of the enemy. Was he intending to come? How many ships did he have? Where would he land? Had he landed? The answers to these questions were hard to come by.

If all went well an invader could be defeated within days of his getting ashore; this was the fate of the force of Harold Hardraada on 24 September 1066 at Stamford Bridge. If it all went wrong disaster threatened, a disaster like that at Hastings when the ruling class of Anglo-Saxon England was virtually wiped out. Harold Godwinson's ships, probably raised as the *shipfyrd* already described, had cruised

THE NORMAN CONQUEST

The Norman invasion fleet from the Bayeux tapestry, probably made by English nuns at Canterbury within ten years of the events it depicts. William's own ship has a signal lantern on the masthead.

off the Isle of Wight all summer, but had never had a realistic chance of preventing a determined enemy getting ashore. The English knew of the activity in Norman ports including the preparation of an invasion fleet. On this occasion, with so large a force being assembled, it would have been impossible to keep the news of the raising of the Norman fleet from their adversaries. The fact that William, duke of Normandy, was hiring ships in Flanders to get his troops across the Channel would have been known to English merchants trading in that part of the world. After a summer of contrary winds, the Norman ships finally assembled at St Valery at the mouth of the Somme about 14 September; on the 8th Harold's ships had finally abandoned their futile vigil off the Isle of Wight. William made an unopposed landing about 29 September at Pevensey when Harold was still in Yorkshire after his earlier victory against Hardraada. There was no decisive naval intervention at this crucial juncture.

FRENCH CONNECTIONS

Did the Conquest change the situation with regard to the defence of England by sea as it changed so many other things? A brief answer must be yes. The Channel was no longer a frontier between potentially hostile neighbours but an internal barrier between the two unequal portions of one family's possessions. Internal conflict between rival claimants to the throne did exist but the fear of coastal raiders who were also potential invaders diminished. What kings of England needed at this date was a secure means of transport across the Channel, not royal fighting ships. The only areas where different imperatives ruled were the marches between England and Scotland and England and Wales. The chronicles do sometimes hint that ships were used in support of operations in these places; in 1091 Florence of Worcester ascribes the failure of an attack on Scotland by William Rufus to the fact that the army suffered terribly from hunger after the supply ships had been dismissed. His operations in Anglesey similarly depended on controlling the Menai Straits. The major need, however—for transport across the Channel— was met, at least in part, by the obligations laid on the Cinque Ports of Kent and Sussex. Charters for these towns exist from 1155–6, but these probably confirm liberties existing from an earlier date, perhaps even from before the Conquest. Their duty, to provide the king with a total of 57 ships with crews of 21 men for a fortnight each, was highly useful as a ferry service but of little use for a Channel campaign.

The English loss of Normandy under King John again recast the strategic situation. The king of France, who had previously had very little direct control over lands bordering the Channel, gradually extended his power in this area. The valuable English colony in Gascony was only accessible from England after a long and often stormy voyage round the Brittany peninsula and into the Bay of Biscay. War, or the threat of war, became almost the normal state of affairs in these waters where English and French rivalry became intense. For the first time since the Conquest it becomes possible to find direct evidence of considered naval activity by the Crown.

The king had at least three possible sources of ships should he wish to contemplate any kind of naval action. He could acquire ships of his own; he could use the Cinque Ports' vessels in accordance with their obligations; and he could have recourse to the duty on all to defend their homeland. This, in effect, meant that the king had the power in time of need to requisition as many of the ships belonging to English owners lying in English ports as he judged necessary. This last power was by far the most valuable; no king could afford to have anything but a fairly small number of royal ships. The Cinque Ports' obligations were hedged around with restrictions reducing their usefulness as other than cross-Channel transports. None of these sources of possible naval power was new; there are traces in the Pipe Rolls of payments to a possible royal ship master, Alan Trenchemer, in the reigns of Henry II and Richard I. A fleet of considerable size was also raised to go on the Third Crusade. John, however, established something much more like a permanent administrative framework for naval activity by the Crown.

William de Wrotham, the Archdeacon of Taunton, had been involved with others in the raising of a special tax on foreign merchandise in 1204 in seaport towns. This local knowledge was put to good use in the following years when Wrotham appears frequently in royal accounts as the clerk in charge of the force of royal galleys which John had acquired to use against his bitter enemy Philip Augustus of France. By 1212 some sort of base for these vessels had been set up at Portsmouth. The exact number in royal ownership is uncertain but was undoubtedly as many as ten, including the king's flagship, the *Dieulabeni*, and his horse transport the *Portjoy*. Wrotham was also responsible for the large number of merchant ships pressed into the king's service and the Cinque Port squadron

NAVAL ACTIVITY IN THE TIME OF KING JOHN: THE ZWYN

A small drawing from the margin of a MS, which appears to show one way of dealing with the problem of disembarking horses from a ship.

(made more useful by an arrangement whereby fewer ships served for a longer period). John's building up of naval forces was justified the following year; the French raided into Flanders whose nobles asked for English help. A fleet under the earl of Salisbury set sail and reached the Zwyn two days later. The French had beached or anchored their vessels in the muddy, shallow port of Damme while they sacked the town. The English ships stood off but the fighting men took to the boats, and when they had taken all the booty they could manage, set fire to all those ships at anchor. The chroniclers claimed that 400 ships were captured or destroyed.

John was not able to use his navy in the civil strife with his barons, but in 1217 the regents for his son, Henry III, were able to turn it against French forces attempting to reinforce the Dauphin whose army was still on English soil. The French commander was Eustace the Monk, hated by the English as a traitor; news that he had sailed reached Hubert de Burgh, the English commander at Dover. The French fleet may even have been in sight from the castle since the chronicle describes it as sailing swiftly to the English coast before a violent wind. The English slipped out of the harbour with a force of only 16 large and about 20 small ships; to their great good fortune there was a shift in the wind and they were able to come up on the sterns of their enemies, grapple and board them. Eustace the Monk was found concealed in the bilges of his ship, dragged on deck and executed on the spot. This action did take place on the open sea even if near the coast. The account by Matthew Paris in his chronicle makes particular mention of wind shifts, of a sudden deliberate change of course, and of the tactic of cutting the halyards and stays to cause the collapse of the sails of at least one French ship.

DEVELOPMENTS IN SHIP DESIGN

It is clear that the technology of ship design had begun to advance from that of the Danish longship to something rather more manœuvrable. Two broad types had developed. The galley had a Mediterranean ancestry and was, as might be expected, a long, low vessel propelled principally by oars, though sails of the triangular lateen type were used as well. The alternative roundship, developed in northern Europe, was a single-masted sailing ship with a deep, beamy hull; we are largely reliant for early representations of this kind of vessel on the seals of seaport towns on which the ship is distorted to fit a circular frame. This gives the hull the look of a walnut shell, an impression which is probably erroneous since all the evidence suggests that these vessels were successful and seaworthy. They were steered at first by a steering oar at the stern; by the mid-thirteenth century this was on the way to being replaced by a sternpost rudder. Their single sail could be reefed; reefing points are clearly visible in many pictures; bowlines leading from the clews of the sail to the bow also allowed them to sail a little nearer to the wind. From the military point of view the most important change was the introduction of castles—raised platforms or half-decks at the bow and stern. From these the fighting men could hurl missiles at the enemy or fling the grapnels when about to board; a top-castle was also eventually fixed to the mast, well stocked with iron bolts or stones to throw down on to the crowded deck of an enemy with great effect. From the payments made to crews of royal ships we can also see that these ships, if going to sea for warlike purposes, were manned by both sailors and fighting men. The ship was in charge of a master; the soldiers had their own captain. There was virtually no structural difference between royal ships and pressed merchant vessels; it has been suggested that the castles were temporary structures fitted only if a battle loomed, but there is little evidence to support this view. Royal ships were perhaps only distinguished by being elaborately painted in red and gold with heraldic devices on their sails and prows.

Seal of the Barons of Dover, one of the Cinque Ports; this shows a typical early medieval round ship. The castles and the steering oar are clearly visible.

Even though the royal galley force disappears from the records at about 1250, in the late thirteenth and fourteenth centuries the tentative lines of development in naval affairs already sketched out seem to come together in a more definite manner. Particularly during the Scottish and Welsh campaigns of Edward I, the role of the navy in support of land campaigns was made abundantly clear. In the Welsh campaigns ships were used to supply and support the army; the squadrons used were made up of contingents from the Cinque Ports, local ports, the south coast, including Southampton, and Bayonne. The involvement of Bayonne in Gascony is a notable feature of English naval forces throughout the fourteenth century; this may well have something to do with their expertise in the fields of ship design and building; a particular form of swift handy craft known as the balinger was probably developed there, originally for the whale fishery.

These ships were used most dramatically in the conquest of Anglesey in 1282; Luke de Tony seized the island and built a bridge of boats across the Menai Straits. The soldiers then fell into an ambush by the Welsh in Snowdonia, but this did not alter the effectiveness of the naval element in the campaign. In Scotland the need for ships was equally evident, transporting men and supplies, including such things, in 1303, as pontoons for a prefabricated bridge for the crossing of the Forth. A form of blockade of the Scottish coast was also attempted to prevent them getting aid from their French allies. Edward II's difficulties and ultimate humiliating defeat in Scotland may, perhaps, be partly attributed to his failure to maintain control of the sea supply routes to his armies. On the west coast, Angus Og, the Lord of the Isles, supported Robert Bruce; Scottish forces also seized the Isle of Man for brief intervals giving them further opportunities to harry English supply lines. On the east coast the fall of Berwick to the Scots in 1318 was largely the result of their successful naval blockade.

During this period the Crown spent quite large sums on the navy, meaning by that term all ships, no matter in whose ownership, which were being used for defence. The king also set up in 1294 a shipbuilding programme intended to produce 20 galleys of 120 oars apiece; the obligation to build the ships was laid on certain seacoast towns. Those promised by Dunwich, Ipswich, Lyme, Newcastle, York, Southampton, and London (2) were built, but not to a uniform design, the number of oars varying from 54 (Lyme) to about 100 (the London galleys). These were, significantly, constructed under the direction of a master shipwright from Bayonne. This work was undertaken at the expense of the towns concerned (anything from £75 for the small Lyme galley to £326 for the largest London one).

This drawing from a Chronicle shows the reefing points fitted to the sails of early medieval roundships.

It illustrates how naval campaigns placed a heavy burden on all those who got at least part of their living from the sea; the requisitioning of merchant ships could disrupt trade for months; piracy, an ever-present danger, was worse in times of war when the merely greedy could hide under a cover of patriotism. Nor did campaigns always come to anything even after a prolonged period waiting for favourable winds or the assembly of a sufficiently large fleet. In 1297, for example, Edward I determined on an expedition to Flanders. The ships, pressed from all the south- and east-coast ports, were ordered to assemble at Winchelsea on 27 April. Nearly 300 were involved, but the expedition never set sail against the enemy, the only result being a bitter fight between the men of the Cinque Ports and the men of Yarmouth in the Zwyn in which 37 Yarmouth ships and 171 men were lost.

OUTBREAK OF THE HUNDRED YEARS WAR

The growing tension between England and France which resulted in the outbreak of what became the Hundred Years War brought these potential problems between the Crown and the maritime and merchant community into closer focus. Considerable naval forces were needed to transport armies to France and supply them once there. Action was necessary against pirates in the Channel and North Sea and protection was vitally needed by coastal communities open to swift and destructive raids by the French and their allies. Such forces always included a substantial number of arrested merchant ships with all the disruption to trade that that implied; royal ships, increasingly used to provide a core for sea-keeping squadrons, were expensive and added to the burden of taxation. As fortunes varied in the French wars we can see the Crown responding in various ways to the pressures exerted by these urgent needs.

At first the most important problem seemed to be that of coastal defence; the French had two galley bases equipped to build and repair this type of ship with the help of Genoese shipwrights, acknowledged masters of the craft. One, the *Clos des Galées* at Rouen, had been established as early as 1294, the other was set up by

Philip VI at La Rochelle in 1337. From both, these swift shallow-draught vessels could make a rapid Channel crossing in favourable weather and attack port towns, looting and burning before making as rapid a get-away. Portsmouth, Southampton, the Isle of Thanet, Dover, Folkestone, Rye, Sandwich, Plymouth, and even Bristol suffered in this way in 1337–40. A system of land-based coastal defence did exist in English maritime counties, originally set up by Edward I to deal with the fear of a French invasion in 1294. This involved the use of a coastal militia under the control of commissioners called the 'keepers of the maritime lands', with the ringing of church bells to warn the people of danger. The king also made strenuous efforts to gain allies for the English in the Low Countries with the discreet use of subsidies to friendly rulers. At sea English forces seemed to have few successes against the raiders except for one counter-raid by the Cinque Port men against Boulogne in 1339.

In 1340, however, by mid-June, the king had assembled a fleet in the Orwell; the core of this was his own vessels including the flagship, the *Cog Thomas*, under the command of Richard Fille. Most of the remaining 160 or so ships had been arrested in the usual way. The fleet sailed for the Flemish coast on 22 June; by the evening of the 23rd they were off Blankenberg and scouts had seen the whole French fleet, including a group of Genoese galleys, at anchor in the estuary of the Zwyn very near the port of Sluys. The English waited till the morning when the tide would be in their favour; they then sailed towards the enemy lines. At this point there is some conflict between the accounts of the battle; the Zwyn was a difficult sea-way much encumbered with sandbanks; the French probably had the tide and the wind against them at least in the initial phases of the battle. Some chroniclers state that the French commander, wishing perhaps to make things as easy as possible for his men in the hand-to-hand fighting against an enemy attempting to grapple and board their opponents which was to be expected in these circumstances, had ordered his vessels to be chained together in three large groups. The Genoese seem to have withdrawn as soon as this order was given; the result was a complete victory for the English forces. Edward III himself wrote the official report of the victory on board the *Cog Thomas* on 28 June. Despite a tough struggle the English had taken 180 French ships and French corpses were being washed up all along the neighbouring coasts; the royal ship *Christopher*, captured earlier by the French off Middelburg, had been retaken.

The king had every reason to be pleased with this outcome, later reinforced by the issue of a gold noble showing himself on the obverse on board a cogship bedecked with the quartered arms of England and France. It did not, however, end the danger of French coastal raids or win for the English anything that can be called the 'command of the Channel'. Such an outcome was not possible within the technical limitations of the shipping and the communications of the day; it did, however, associate national success with victory at sea, perhaps a welcome innovation. The importance of naval resources was further demonstrated in the Crécy

THE BATTLE OF SLUYS

SUPPORT OF THE CRÉCY
CAMPAIGN AND LES
ESPAGNOLS SUR MER

campaign of 1346 and the siege of Calais in 1347. In 1346 we can get a clear idea of the complexities involved in assembling an expeditionary force against France at this date. At first orders were sent out that arrested ships should be ready at Portsmouth by 14 February 1346; the date was put off until mid-Lent, then until a fortnight after Easter. The fleet finally sailed on 3 July; all had been ready by the end of May, but it was delayed by contrary winds. All this time soldiers and seamen had to be fed at royal expense while merchants chafed as a whole summer season wasted away with trading voyages impossible. The collection of victuals and stores for the army and navy could involve sheriffs from distant counties; ships also had to be adapted especially if needed as horse-transports. Hurdles to make stalls, tethering rings, ropes, and canvas were all needed, as well as fodder for the animals and special wide gangways to get them on board. For the siege of Calais in 1346–7 supplies went largely from Sandwich and a force was also needed for the successful blockade of the port. By this date we are fairly well informed of the detail of the naval aspect of these expeditions from the Exchequer Accounts. From 1344 an enrolled account appears under the name of an official identified as the Clerk of the King's Ships. The first holder of this post, William Clewer, had started as the clerk (or purser in modern terms) of the royal ship *Cog Thomas* before taking on responsibility for all royal ships, and at least some payments to arrested shipping. A total of 34 royal ships are named in the accounts of 1344–52, of which eight were in action in the major engagement against the Spanish in the Channel off Winchelsea in 1350.

This engagement, known to contemporaries as *Les Espagnols sur Mer*, is perhaps crucial in understanding the developments in naval warfare at this time. This battle was fought in mid-Channel; the account in Froissart's Chronicle may be more reliable than usual since he seems to have got some information from a knight who was present—Robert de Namur. The French had negotiated an alliance with Castile in 1349, largely, it is suggested, because of the vital help their well-found navy could give the French; the Castilian fleet had first of all sailed to Flanders for the purpose of trade. It was on their way home from this expedition that Edward III and his ships in harbour at Winchelsea caught sight of them in the Channel and came out to intercept them. The two fleets, intent as ever on a boarding action, encountered each other at speed so that the king's own vessel, manoeuvring to get alongside the leading Spanish ship, struck her with such force that her top-castle was sheared from the mast. The seams of the royal ship herself were sprung by the force of the impact and she, in fact, foundered just as her knights successfully seized the Spanish ship they had grappled. The English, including the king, escaped to safety on the captured ship. Other equally desperate boarding actions raged between the Spanish and the English, with the former to some extent having the advantage because of the greater size and height of their vessels. Sir Robert de Namur and his crew on board the ship carrying the royal household only escaped abduction and capture in their turn, grappled firmly to the side of a larger Spanish ship, when one Hanekin managed to leap on board their opponent and cut the halyard and shrouds so that the sail collapsed on deck. By the end of the day the English had taken 14 Spanish ships, but the rest got away. This was not a land battle fought out on the unsteady decks of a ship but one where skill in ship-handling, and the strength and design of the ship itself, were as vital to success as the courage of the fighting men.

The general decline of English fortunes in France in the 1370s also affected naval matters. As early as 1371 the Commons had petitioned the king complaining of the problems caused by the arrest of shipping by the Crown; the merchants are 'so impoverished that they can no longer keep up their business'; crews and masters have been taken for royal service leaving merchant ships with no men. Even so, the master of the king's *Dieulagarde* found he could not man his ship in 1374 since the crew were all in hiding. The increasingly reluctant support of the maritime community for royal endeavours may have had something to do with the losses suffered off La Rochelle in 1372. The earl of Pembroke was dispatched with rein-forcements to Poitou but had with him 20 generally small ships (less than 100 tons capacity) of which only three were ships with castles. The Castilians caught him off the port and employed the tactic of firing the English ships with blazing arrows. Pembroke himself was captured, a relief force from England got nowhere because of contrary winds, and as a result the French retook Poitou and Sain-tonge. A shipbuilding campaign was mounted by the Crown demanding help from seaport towns, this time aiming to provide barges or the moderate-sized balingers, which were efficient both under sail and as oared ships. About 14 were completed, including the *Paul of London*, but the advantage in some respects continued to lie with the French commander, Jean de Vienne, who, with his own great 'nefs' and the galleys of his Castilian allies could raid the English coast, terrorize the local population, burn their homes, and severely disrupt trade, even if he could not prevent some English expeditions successfully reaching a French port. The fact that royal ships were seen as the king's personal possessions to be disposed of at his death contributed in 1377 to the melancholy situation as far as the royal ships were concerned. By this time there were only six based in the Thames at Ratcliff, of which one, the *New Saint Marie*, was a mere hulk; the others were either given to a royal servant or sold for the best price available. It is, perhaps, not surprising that the same summer the French and Castilian galleys came unopposed up the Thames and burnt Gravesend, including the warehouses on the river-front, causing panic in London itself. Urged on by the Commons, the Crown instituted another barge-building campaign in the port towns on the understanding that the vessels would revert to the ownership of the towns concerned after one sea-keeping season, but with little positive result. If the figures had been available to any contemporary he might well have wondered at this juncture whether there was anything to show for the 23 per cent of royal revenue between 1368 and 1381 it has been estimated was absorbed by naval warfare—a total of some £246,000.

By the 1380s the war with France seemed to have drifted into a state of stalemate with neither side in a position either to prosecute it energetically or to make

DEFEAT, STALEMATE, AND DECLINE

peace; for all this the threat of invasion by France was seen as real, especially in 1385 when Jean de Vienne and his fleet sailed to Leith to support a Scottish move across the border. The next summer French actions seemed even more threatening; Charles VI massed an invasion fleet at Sluys but by October no move had been made and the danger receded. The English government in response had collected arrested ships at Dover and Yarmouth but seemed to have no coherent policy for naval defence. Their only real innovation was the institution of payment to the owners of arrested merchant vessels at the rate of half a mark (3s. 4d.), later reduced to 2s., per ton for a three-month period. This was in addition to the crew's wages which had always been recognized as a charge on the Exchequer. One might speculate that this was the only way to ensure some co-operation from a merchant community heartily sick of this way of raising a navy. The attempt to profit from the help of an ally skilled in the operation of galleys which had been so successful for the French brought little concrete result. Portuguese galleys were in English waters between 1384 and 1390, but the crews mutinied at Southampton in 1387 and their Channel cruise of that year led to no notable naval action. Certainly little attempt was made to reconstitute the group of royal ships dispersed on Edward III's death. By the end of Richard II's reign there were only four; the *Trinity del Tour, Gracedieu, Nicholas,* and *George.* These were used more as the private transport of the royal house than as fighting ships. The *Trinity,* for example, a large ship of about 300 tons, was used on voyages to Bordeaux for wine for the royal household and then, in company with the *George,* took Richard II on his last fatal journey to Ireland.

In the first half of his reign Henry IV seemed in this respect to share the attitude

The south-east coast of England: bases and battles in medieval history to 1500.

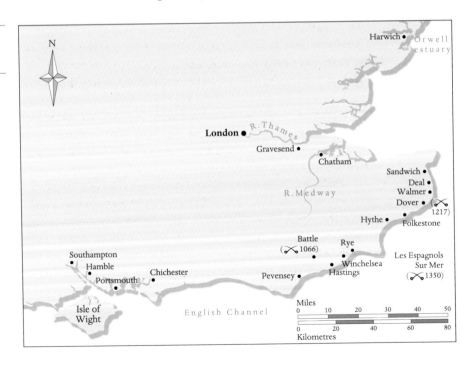

of his predecessor. The Clerk of the King's Ships, John Chamberlain, who had also briefly served Richard in the same post, devoted most of his energies to the decoration of the royal ships for the ceremonial voyage of the king's new bride Joan from Brittany to England. The *Trinity* was beautified with gilded escutcheons of the royal arms and the Garter and St George; banners of cloth of gold were prepared to fly from her masthead, while red and gold images of St George, St Antony, St Katherine, and St Margaret adorned the stern and two gilded eagles the prow. By 1409, however, the royal ships had been reduced in number to the *Trinity del Tour* and a ceremonial river barge. However, we cannot deduce from this that there was no need for naval activity in the Channel or elsewhere, or that the truces in operation with the French prevented any violence between the enemy nations.

The situation, in fact, seems to have been that a state of covert or semi-official war existed in the Channel with 'pirates' or, perhaps, more exactly 'privateers', preying on the shipping of the enemy nation. At times Henry IV seems to have condoned the activities of commerce raiders mainly based in south-western ports; the most notable of these, John Hawley of Dartmouth (both father and son of the same name), Richard Spicer, and Harry Pay almost had the status of folk-heroes. It has been suggested that they were far too valuable to the Crown as naval leaders to suffer any great intervention in some of their dubiously legal activities. Pay's fame had spread to the far side of the Channel; the chronicler of the adventures of an exiled Castilian nobleman, Don Pero Nino, who was raiding the English coast in 1406, was sure that 'Arripay' was the leader of English resistance during a raid on Poole. The activities of these men and others could cause difficulties for the king, especially in his relations with the Flemings, but it is perhaps an exaggeration to state that piracy in the Channel was out of control at this time; rather it was almost an instrument of policy for both sides and indicative of the fact that the conflict with France, while officially in abeyance, could at any time break out afresh.

    The accession of Henry V reinvigorated naval policy in much the same way that it abruptly changed general attitudes to the war with France. Henry was fortunate in that the strategic necessity for a much stronger naval presence, principally in the waters between England and France, coincided with developments in the building and arming of ships. The available documentary evidence is also much more copious and includes some shipbuilding accounts and full inventories of the equipment of the royal ships. The hazy ship names of the fourteenth century—when it is extremely difficult to find out whether there were any really substantive differences between, for example, cogs, hulks, and nefs—are replaced with much more clearly definable types. Pictorial representations are more accurate and, finally, the remains of the greatest of all the royal ships of the fifteenth century, the *Gracedieu*, still lie in the Hamble river at Bursledon, just visible at low water at equinoctial spring tides, and have been excavated.

    By the second decade of the fifteenth century, royal ships (and the vast major-

A view of the remains of the *Gracedieu* in the Hamble river at Bursledon. Only the hull lies there as all her equipment had already been removed before she was burnt to the waterline in 1439 after being struck by lightning.

ity of those arrested for royal purposes as well) were of three broad types: barges, balingers, and *naves* or ships. Very highly desirable for military purposes were carracks; those in the king's possession were all prizes captured from the Genoese—the loyal, or at least the well-paid, allies of the French. No ships of this type were built in England and there was, in fact, some difficulty in repairing them. Barges and balingers were quite similar, the only real difference being the barge's greater capacity for cargo and her comparative slowness. These vessels were clinker-built, with sternpost rudders. The largest balingers were of about 120 tons; the smallest of 24–30 tons. Balingers could be rowed in circumstances when sailing was difficult, but both barges and balingers normally relied on sail power. At the beginning of the century all were single-masted, but by the 1420s newly built examples (e.g. the royal *Anne*) had a main mast and a mizzen; the *Petit Jesus* was refitted in 1435 with three masts. Balingers were the eyes and ears of a fleet, often used as scouts or for rapid voyages with ambassadors or news for the court. In other than royal hands they had a bad reputation as the craft favoured by pirates. We have no reliable figures for their dimensions, but in many ways they seem like an amalgam of the virtues of longships and roundships.

The class known simply as 'ships' was as varied as one might imagine. Leaving aside the special group known as the 'king's great ships' which will be considered below, the term seems to be used of the merchantmen of the day, the cogships of the earlier century, beamy single-masted vessels of from 70 to 330 tons, on the verge of being made obsolete by the new designs influenced by Mediterranean ideas. Carracks epitomized this influence; those in Henry V's fleet, as has been said, were all built in Genoa. All were by contemporary standards very large, from 500 to 600 tons, looming above the general run of English vessels in a boarding action. In the early fifteenth century, they seem to have been rigged as two-masted ships with a square sail on the main and a lateen mizzen, but by mid-century would have had the rig shown by WA in his drawing (see p. 21). Their hulls were carvel-built (with edge-to-edge planking), something with which north European shipwrights were completely unfamiliar. As capacious cargo vessels well able to defend themselves in a hard-fought action they had no rivals at the time.

Henry V's great ships were in many ways an attempt to find a local answer to the threat to English shipping posed by carracks in the hands of France's allies. They were built between 1413 and 1420: the *Trinity Royal* (540 tons) at Greenwich,

Known as the Mataro model, this was probably made as an 'ex-voto', to be hung in a church as a token of a prayer answered; it seems to date from the early fifteenth century and was probably made in Portugal or the Atlantic coast of Spain. It allows us a clear view of the hull and the rigging of a medieval ship from an area which had much influence on English ship-builders.

the *Jesus* (1,000 tons) at Winchelsea, and the *Holyghost de la Tour* (740–60 tons) and *Gracedieu* (1,400 tons) at Southampton. All were clinker-built, the *Gracedieu* using a special three-layer style of planking which was probably unique. The *Gracedieu* had three masts and the others two. More than any other vessels before the sixteenth century, these were specialist warships, not adapted merchantmen. Their armament included cannon (something also found on other royal ships), though it is not at all clear how they were used. They were probably mounted on the deck so we may speculate that they were set up as something akin to stern-chasers, though given the unreliability of contemporary guns and their tendency to explode on firing they may have been a greater danger to their own crew than to the enemy. The vessels also carried quite large quantities of cross- and long-bows, spears, and the essential iron darts for hurling from the top-castle. The ships were also equipped with basic navigational instruments, a compass contained in a binnacle, a running-glass, and a lead and line. The master was in charge of sailing the ship but not of any soldiers on board; he was assisted by a 'constable' and a 'clerk' or purser; these officers received 6*d.* per day plus, occasionally, a further payment called a *regardum* of 6*d.* per week. The generality of the crew were paid 3*d.* per day, and a few boys 1½*d.* Their victuals were bread, biscuit, beef, salt meat, salt fish, and some wine; not very different from that offered to later seamen. It is hard to be categorical about the size of crew carried in any particular ship, since this varied according to the type of voyage envisaged; all one can say is that by modern standards these ships were almost grotesquely overcrowded. In 1417, for example, the *Mary de la Tour*, a former Genoese carrack captured by the English, went to sea with 100 crewmen, 62 men-at-arms, and 134 archers embarked in a ship of around 500 tons.

At its peak Henry V's group of royal ships totalled 36 vessels. From 1413 to 1420, the Clerk of the King's Ships, still based in the Thames below the Tower, was William Catton. From 1416 he was increasingly assisted by William Soper at Southampton, where the *Holyghost*, *Gracedieu*, and her attendant balingers the *Valentine* and *Falcon* were built. During the years of highest naval activity, 1415–20, these men received in total some £20,000 from the Crown, an average of about £5,000 a year. From 1420 Soper became Clerk himself on an increased salary (£40 per year instead of 1*s.* per day). He transferred all work on the royal ships to Southampton and Hamble; the great ships were anchored there with a bulwark garrisoned by a company of soldiers to protect them. In Southampton itself was some sort of dock area together with a stone storehouse for equipment and a forge. All must have seemed set for a further development of England's naval resources now based around a group of truly formidable royal ships.

At this point, however, the strategic imperatives shifted if for the best of reasons: what looked like victory in France. Naval actions had had their part to play in that victory. In a series of battles in the Channel English ships had given a good account of themselves. Henry had sailed on the expedition which was to end triumphantly in the battle of Agincourt in August 1415; some 9,000 men were accommodated in perhaps as many as 1,500 ships. If this seems an excessive number it should be remembered that some were very small and that horses, artillery, and all other supplies also had to be transported across the Channel. The king himself made the voyage in the *Trinity Royal*, which was accompanied by at least seven other royal ships. After the victory at Agincourt and Henry's return to England from Calais, an English garrison still remained at Harfleur, the original landing-place of the expedition. Its position became increasingly precarious in the early months of 1416 as the French and their allies, the Spanish and the Genoese, laid siege to it by both land and sea. At least eight Genoese carracks cruised off the town making any attempt to supply it from England very difficult; a further force of galleys under Gioanni de' Grimaldi were in the Channel and the French even felt bold enough to raid the Isle of Wight and Portland. On 15 August 1416, when conditions in Harfleur were very bad, an English fleet commanded by the Duke of Bedford, perhaps as large as 300 vessels, was sighted off the town. The correspondent in Bruges of the Italian chronicler Antonio Morosini reported to him that the ensuing battle was very cruel, with heavy casualties on both sides, but in the end Bedford was victorious, capturing three carracks; another was driven aground and wrecked. The English were more numerous than their opponents; the Spanish in fact had fled before the fighting began because they saw they were outnumbered, but the defeat of carracks by much smaller vessels was a notable success. A larger high-sided vessel still had an advantage in a boarding action but one that could be overcome. The spirited account, by the Spanish chronicler already mentioned, of an otherwise minor encounter with English ships in the Channel in 1406 makes clear the importance also attaching to seamanship by this time. He describes how his hero's galley escaped almost certain capture through the skill and daring of a balinger master who unexpectedly changed course so that he was able to ram and disable one of her attackers.

English balingers faced a more formidable task later the same summer when a group of six came up with a Genoese carrack off Calais. The balingers chased her all day and got boarding ladders up on to her deck, but even though many of her crew were wounded the ladders were thrown back on to the much lower decks of the balingers and she made good her escape with eight dead and 50 wounded out of a crew of 62.

The year 1417 was probably the most successful for English naval expeditions and their commanders. Three separate squadrons cruised in the Channel; Lord Huntingdon captured a further four carracks, Lord Castelhon three ships including two Spanish ones, and Thomas Carrew the ships known in English hands as the *Holyghost of Spain* and the *Agase*. While these expeditions were at sea the *Gracedieu* was building at Southampton. The *Holyghost de la Tour* and the *Trinity Royal* had themselves taken part; the need for more great ships to challenge and overcome the carracks must have seemed obvious. By the time the *Gracedieu* was ready for sea, however, the situation had changed. By the Treaty of Troyes signed in May 1420, the English king was the acknowledged heir of the French king and the Channel would soon be, as it had been before the reign of John, not a frontier but an internal barrier within one realm. Nevertheless the *Gracedieu* did lead an

England, France, and Flanders:
objectives and battles,
1200–1500.

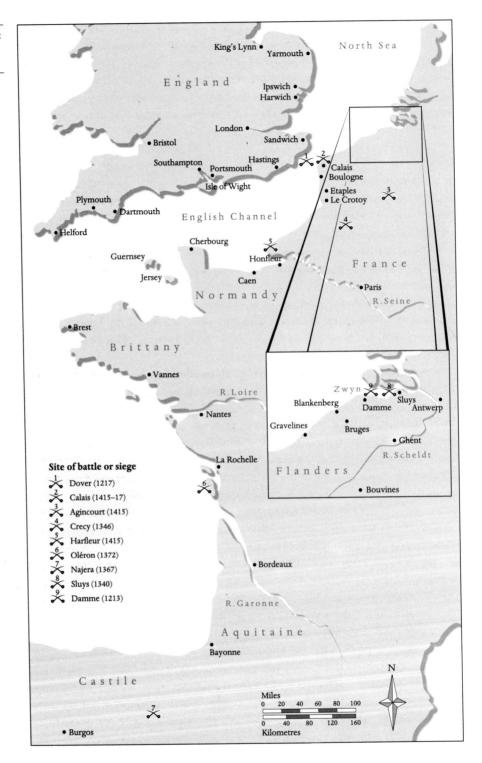

Site of battle or siege

1. Dover (1217)
2. Calais (1415–17)
3. Agincourt (1415)
4. Crecy (1346)
5. Harfleur (1415)
6. Oléron (1372)
7. Najera (1367)
8. Sluys (1340)
9. Damme (1213)

expedition in the Channel later that summer. Humphrey of Gloucester visited her in Southampton as she was about to set sail and reported back to his brother the king that she was 'the fairest that ever men saw, I trow in good faith'. At this time, there was no hint of her eventual melancholy decline, first of all anchored in the Hamble river, then laid up on the mudflats with her sister ships, and finally a burnt-out wreck having been struck by lightning and burnt to the waterline in 1439.

If Henry himself had not succumbed to dysentery in 1421, the royal ships of the English crown and the administration necessary to keep them at sea in good repair might have survived but, on the king's death, they were very vulnerable. There was a plausible, if short-sighted, argument that they were no longer needed. They were also the personal possessions of the late king and, therefore, as in Edward III's time, liable to be sold to pay his debts. Soper thus found himself in charge not of a busy dockyard preparing ships for war and other royal business but of the sale of the ships and their equipment for the best prices to be had. By 1442, when Soper finally gave up his post, there were no serviceable ships in royal ownership and his expenditure amounted to £4 16s. 4d. for his final two-and-a-half years in office. Ten years later the office of Clerk of the King's Ships ceased to exist and was not to be revived until 1480.

It is, perhaps, ironic that at this low point for the royal ships and for the naval defence of England as a whole there was a much better understanding of what was at stake. The anonymous author of the *Libelle of Englyshe Polycye*, already quoted from, fully understood the value of the 'grette dromons', as he called the king's great ships; why had the king built them? Because he intended to be, 'lorde rounde aboute environ of the see'. With these ships to hand no 'carrikys orrible, grete and stoute' could harm the English. As far as he was concerned the aims of English naval policy should be to protect trade, ensure the safety of the realm, and command the 'narowe see', that is, the Dover Straits. The fact that this author, writing between 1436 and 1438, felt that the royal government was singularly failing to pursue these aims did not make them any less desirable. Sir John Fortescue, writing in about 1470, put it just as plainly. In his view it was evident that 'though we have not always war upon the sea, yet it shall be necessary that the King have always some fleet upon the sea, for the repressing of rovers, saving of our merchandise, our fishers and the dwellers on our coasts.' The king particularly needed 'great and mighty vessels' because only these could 'board with carracks . . . and break a mighty fleet gathered of purpose'.

In the period from the sale of Henry V's ships to the later years of Edward IV's reign this is precisely what the English Crown did not have. The king used either arrested shipping or the method of commissioning an individual captain by indenture to raise a naval force to 'keep the seas'. There is no record of the use of the Cinque Ports ships after 1444–5; many of the ports were so badly silted up by this time that they could not supply suitable vessels. In 1442, the Commons, perhaps in desperation, put forward their own detailed scheme to the king to have a squadron of eight great ships each attended by a barge and a balinger and four pinnaces on the seas continually from Candlemas to Martinmas. They envisaged a force of 2,260 men at a total cost of £4,568 6s. 8d. for six months. Such a scheme had little chance of implementation considering the current state of the royal finances and all the other calls being made on them. The earl of Warwick, as captain of Calais, seemed more able to grasp the need for a squadron of fighting ships than the Crown, and maintained a fleet at his personal expense. These vessels engaged Spanish and Genoese ships in 1458 and 1459 with varying fortunes and formed the core of fleets keeping the seas for the crown in the early 1460s.

THE BEGINNINGS OF
REVIVAL

By the 1470s, when perhaps the worst strains of the civil wars in England were over, the number of royal ships slowly began to increase once more, Edward IV having from five to seven; all, however, were purchased from merchants and in no sense the specialized war ships developed by Henry V. When war broke out against Scotland in 1481, the need for naval support of the land forces which had been evident in the campaigns of the 1290s was again appreciated. Thomas Rogers, an experienced royal ship master, not a court official like Catton or a wealthy merchant like Soper, became the first Clerk of the King's Ships since 1452. He was responsible for some 15 ships including at least two of Spanish and one of Portuguese origin.

His work for the Crown continued after the turmoil of 1483; naval activity at first, however, was still on a very modest scale. No real centre for the royal ships existed like that built up at Southampton by Soper. Royal ships were repaired when necessary in the port where they happened to be lying. From early 1487, however, definite steps were taken to increase England's naval strength. The *Regent* was laid down at Reding, a creek on the river Rother in Kent. This, and her sister ship, the *Sovereign*, were the largest ships built for the Crown since the days of the *Gracedieu*. Both, certainly, would have qualified as Fortescue's 'great and mighty vessels'. Both were rigged and armed in a way which showed how much shipbuilding and handling skills had developed since Henry V's reign. The *Regent* had 225 serpentines (cannon) on board, some of iron and some of brass. In the *Sovereign*, the Clerk's accounts make clear that her guns (141) were grouped on the fore- and after-castles on at least two levels and at the waist of the ship. We are not dealing here with the gun decks of later ships with the ability to fire a broadside through gunports, but artillery was clearly of much greater importance than in earlier vessels. As regards their rigging, both ships had foremasts with fore topmasts, main, and main topmasts, main and bonaventure mizzens, and also a sprit sail on the bowsprit. There is no hint that these features were in any way unusual; this was the level of equipment expected in the best ships of the day.

What was more unusual was the dock that Henry VII also built at Portsmouth in 1495–6 at a cost of £193 0s. 6¾d.; this was not the kind of temporary resting-place on the 'wose' (mud), dug out and hedged around with brushwood made for Henry V's ships at Southampton or Bursledon, but a dry-dock with gates equipped with an 'engine to draw water out of the said dock'. Since the king had

The best drawing of a carrack still surviving. It seems to have been drawn by an artist (WA) who took some care over accurately recording the details of the hull and the rigging.

also paid £2,068 11s. 1d. to William Cope for fortifications at Portsmouth in 1494, we can, perhaps, see this as the real beginning of Portsmouth's role as a naval base.

To return, however, to the point of view with which we started: was England now defensible by sea in a way impossible in earlier times? Could she feel secure behind the moat of the Channel? It is certainly true that naval defence was now possible in a way that it had not been before. Ships could increasingly be fought at sea; developments in ship design, building and handling, and in gunnery meant that it was only a matter of time before the broadside and other tactics of the

The ship in the background here is a careful drawing of a late fifteenth-century armed vessel. Guns can be seen mounted on the decks.

sixteenth century became possible. Yet it was also clear that naval policy could not be made on the spur of the moment, or when a real emergency threatened. As Fortescue had stated some thirty years before, it was too late to build a fleet when an enemy was 'upon the sea'. If England was to become a naval power the Crown must show a sustained interest in this arm of the realm and be prepared to invest the money needed. The reliance on pressed merchantmen which had served the Crown so well for so long was beginning to come to an end. The disruption to trade was one factor and another was the need for true war ships carrying more weaponry than was practicable for merchant vessels. If the European situation seemed stable and the seas around England's shores devoid of enemies, however, no monarch was likely to take the initiative. One striking feature of the previous 200 years as far as naval matters were concerned was, perhaps, the speed with which the Crown could divest itself of its naval armament without, so far as we can tell, a moment's regret. At the end of the fifteenth century, the sea, therefore, could be the Wall of England, but only if the will to make it so existed and flourished in the minds of the king and his advisers. Without such determination it could be, as it had been in the past, the route by which raiders and invaders spread terror among the people.

2

FROM THE KING'S SHIPS TO THE ROYAL NAVY 1500–1642

DAVID LOADES

When Robert Brygandyne became Clerk of the King's Ships in May 1495 he assumed a limited responsibility for no more than three or four vessels. By the time that he retired in 1523 the navy had a peacetime strength of about 30 ships, varying in size from pinnaces to the *Henry Grace à Dieu* of 1,500 tons. By then the king's needs had long outgrown the administrative capacity of a single officer, but no proper structure had been created. Brygandyne's efforts were supplemented by those of a Clerk Controller, and by the piecemeal efforts of numerous other men who held no official appointments, under the general oversight of the ubiquitous Thomas Wolsey. In a sense the standing navy was created during this period, but it is impossible to identify any specific decision, or any action of conscious policy which brought that about. Edward IV had been no more equipped to 'keep the seas' with his own resources than he had been to police his realm with a standing army. Had Richard III possessed a standing navy Henry Tudor might never have landed, but he did not. Whether Henry VII learned from his predecessor's mistake is not clear. There was certainly no dramatic change of policy, but very soon after his accession, when there were many other calls upon his resources, the new king laid down two large and expensive warships. As was described in the previous chapter, these ships were not revolutionary in their design, but they were the largest to be built in England since the days of Henry V, and they were intended for fighting.

As a direct result of their commissioning, the new works were constructed at Portsmouth. This was partly necessitated by the fact that the Great Ships (roughly 400 tons and over) drew too much water to cross the Hamble bar and could not therefore be anchored or careened in that sheltered refuge, but it may also have been prompted by the king's desire to have something which would now be called a naval base. If such was his thinking, it was not pursued with any energy. Two ships, the *Margaret* and the *Bonaventure*, were added by capture; one, the *Carvel of Eu*, by purchase; and two, the *Sweepstake* and the *Mary Fortune*, were built. None of these was a Great Ship, and the last two, which were small barks, cost only a little over £200 (about £200,000 in 1995 equivalents), inclusive of rigging. Brygandyne's surviving account, which runs from 1495 to 1497, shows him spending a little over £2,000 on wages, maintenance, and repairs. Of this, £1,637 was spent on the *Regent* and the *Sovereign*. He was not responsible for the operational costs, or 'sea charges', and did not account for the hiring or equipping of additional vessels, so the total naval expenditure during these years was probably of the order of £8,000–£10,000, or £4,000–£5,000 a year. Henry was actually more interested in guns than he was in ships. He imported French gunfounders, and encouraged the establishment of two forges in Ashdown Forest. By 1497 he had nearly 50 gunners in his service, and in 1508 the first (unsuccessful) attempt was made to cast iron guns in Sussex. All of which makes it possible that it was under his encouragement, rather than that of his son, that the gun port was invented.

According to G. F. Howard, the idea of firing guns through ports cut in the side of a ship, as distinct from its castles, was evolved between 1505 and 1509, and the first ship to be designed in this potentially revolutionary manner was the *Mary Rose*. Henry VIII is normally given the credit for building both the *Mary Rose* and her sister ship the *Peter Pomegranate*, but it seems likely that they were laid down before his accession. If that was the case, then he inherited from his father seven operational ships, including the ageing *Regent* and *Sovereign*, two Great Ships in building, a modest naval establishment at Portsmouth, and an active Clerk of the Ships. Henry VIII had his own ideas about ships, about armament, and probably about naval strategy. Unfortunately, we do not know exactly what they were, as he never explained them. Consequently, we have to reconstruct his thinking from the progress of events, assuming in the absence of any evidence to the contrary, that the king was the driving force. War with France was his first ambition as king, and an increase in naval activity can be clearly seen through the fragmentary records of 1509–11. The *Mary Rose* and the *Peter Pomegranate* were launched, two new barks were rigged, and two small galleys called 'row-barges' were added to the fleet. By the time war broke out in 1512 Henry had at least 15 ships of his own, including the *Lion* and the *Jennett of Purwyn*, captured in the previous year from the Scottish rover Andrew Barton. During that year at least three further ships were built and six more purchased, including the *Katherine Fortileza* of 700 tons, at that stage the largest in service. In April 1512 Sir Edward Howard was appointed to succeed the Earl of Oxford as Lord Admiral, thereby converting the office from a figurehead to an active and effective command. Howard 'kept' the Narrow Seas throughout the summer, with a fleet which varied from 18 to 25 sail, of which about half were the king's own ships, attacking the French coast at will, and capturing large numbers of small traders.

HENRY VIII'S INHERITANCE

The reason for this success was not that French ships or guns were inferior, but simply that the English organization was more efficient, and Howard retained the initiative, even when the French finally got a fleet of 22 sail assembled at Brest in August. Moreover, the French admiral, René de Clermont, was unaccustomed to the idea of using his fleet for anything other than defensive purposes, and did not know quite what to do. On 10 August Howard solved that problem for him by moving in to the attack. The English had the initiative, and an element of surprise, but the French held much the stronger position within the harbour, and the weather was not in Howard's favour. The engagement was consequently inconclusive, and notable mainly for the toe-to-toe slugging match between the *Regent*

The *Regent* and the *Cordelière*,
locked in mutually
destructive combat, 1512.

and the *Cordelière*, which resulted in the destruction of both ships when the latter's
magazine exploded. There were other naval operations during 1512, in which the
remainder of the king's ships were involved; against the Scots, and escorting an
abortive military expedition to Guienne. But Howard's campaign was the main
effort, and the Lord Admiral accounted for almost £14,000 when the campaigning
season ended in October. In 1513 the French were better prepared to face an
aggressive English strategy. The veteran galley commander Prégent de Bidoux

had been in Brest with six of his vessels throughout the winter, and by early March 11 warships had already assembled there, with another 15 or 16 at various ports in Normandy. A new admiral, Guyon le Roy, had also replaced the unenterprising Clermont. However, he made no attempt to capitalize upon strategic advantage, simply waiting for the English to attack as his predecessor had done. Howard was thus presented with a tactical advantage, but he was unable to exploit it because on this occasion his ships were inadequately provisioned, and because he had no answer to Prégent's galleys in the enclosed waters of the harbour entrance. Attempting to resolve the latter problem by personal example, the Lord Admiral was killed, and the English fleet, although in no sense defeated, retreated in some confusion.

Thomas Howard, appointed to replace his brother as Admiral on 4 May, arrived at Portsmouth a few days later, but by the time that his fleet was again ready for the sea its main task was the traditional one of escorting the army which was to capture Tournai and Thérouanne. By November, when John Heron rendered his account for the war, £63,879 10s. had been expended upon the navy, of which Sir Thomas Wyndham, the Vice-Admiral and Treasurer of the Fleet, had dispensed £22,560 since 15 March. The year 1514 started much as 1512 had done. By the end of March an English fleet of 45 sail had secured unchallenged command of the Channel, but hostilities were suspended in June before any significant French response had emerged, and peace was concluded soon after. However, despite this anticlimax, and the absence of high profile victories, two-and-a-half years of war had transformed the English navy. Henry now had over 30 ships of his own, led by the massive *Henry Grace à Dieu*, and armed with an unprecedented quantity of heavy guns. Regular winter patrols had also been established, and a proportion of the smaller ships kept in active service all through the year. New docks were constructed at Limehouse, and at Erith in Kent in 1512, and storehouses to receive rope, canvas, and other equipment at Erith and Deptford. In February 1512, a second administrative office was created, that of Clerk Controller, and a gentleman usher of the Chamber, John Hopton, was appointed to fill it. Most important of all, the decision was taken, presumably by the king himself, to maintain the navy at this augmented level after the war was over. Henry saw Great Ships as an essential part of his *maiestas*, and particularly appropriate to the king of an island people. These successes, and the defeat of the Scots at Flodden on 9 September, made the year a good one for English arms, but the fleet accomplished nothing as an independent arm. Several ships had been deployed against the Scots, but had seen no action.

Ordinary expenditure over the next few years probably amounted to some £2,000–3,000 a year, dispensed mainly by the two Clerks. In 1517 shipkeepers' wages alone accounted for £700. Although some ships disappear from the records during the years of peace, and may have been either scrapped or sold, others were added. The *Great Galley*, Henry's huge and innovative galleasse, was launched by Queen Catherine in October 1515, and three smaller ships were purchased between 1517 and 1519. In 1519 the first royal yacht, the *Katherine Pleasaunce*, was

built at a cost of £323. New facilities were also added. A basin was excavated at
Deptford in 1517 to provide safe anchorage for some of the Great Ships, and storehouses were added at Woolwich, Portsmouth, and Southampton. A *de facto* division of labour between the Clerk and the Clerk Controller located the former at
Portsmouth and the latter at Deptford.

By 1520 an embryonic admiralty had come into existence, but it could not yet be
described as a department of state. There was no budget, and no unity of control
except in the hands of the king himself—or, more realistically, Cardinal Wolsey.
From 1522 to 1525 the navy was again on a war footing, but the action was sporadic
and the records very incomplete. At one point 16 ships were noted as being in
commission at Portsmouth; on another occasion 21, and we can catch numerous
glimpses of movement between Portsmouth and the anchorages of Deptford and
Woolwich. The fleet probably expanded somewhat, but not as dramatically as
between 1512 and 1514. At the same time war placed increased pressure upon the
administration, and some potentially important developments took place. When
Robert Brygandyne retired in 1523 he was replaced by Thomas Jermyn, whose
patent of appointment specifically confined his operations to Portsmouth. More
importantly, the Clerk Controller, John Hopton, died in 1524, and his work was
divided between two officers. Thomas Spert, an experienced master mariner, was
appointed to the Controllership proper, while a separate post of keeper of the
Erith and Deptford storehouses was created for William Gonson. Gonson was a
man of many parts: variously described as 'gentleman usher of the king's chamber', 'Teller of the Exchequer', and 'merchant of London', he was to become the
key man in the developments of the second half of Henry VIII's reign. His energy
and ability commended him, first to Wolsey and then to Thomas Cromwell. By
1527 he was regularly receiving money on warrant from the Exchequer and other
revenue offices, and distributing it to those who were actually paying the bills.
Both Spert and Jermyn received a proportion of their money through Gonson,
although they also continued to receive some direct. Gonson clearly occupied a
position of exceptional trust. In 1528 we find him organizing escorts for Imperial
merchantmen, and by 1532 he was sending out his clerks and messengers with a
general commission 'to seek the king's ships of war with money and commissions
for fulfilling all things according to the king's pleasure', as though he held a
general oversight of naval administration.

During the long peace from 1525 to 1542, facilities were gradually extended, and
management practice developed. By 1533 routine expenditure amounted to some
£5,000 a year, of which Gonson was handling about £4,000. A second dry-dock had
been constructed at Portsmouth in 1522–3 at a cost of £553, and in 1527 the original
dock was refurbished and extended. Thomas Spert seems to have compensated
for his lack of administrative responsibility by becoming the chief technical expert,
and there are several references to his inspecting ships being considered for
purchase, or in need of repair. Victuallers were appointed on a contract basis for
the various ports used by the king's ships. Brewhouses and bakeries were built,

and rented out to private businesses when not required. The first royal shipwright was appointed in 1538, and designated 'king's merchants' began to be commissioned at the same time to purchase hemp, canvas, and cordage, particularly in the Baltic. Most of the shipbuilding programme was contracted out, and it is difficult to discover what was actually spent as the contractors were almost invariably paid direct. Nine ships of various sizes were commissioned between 1522 and 1524, and four more, mostly large, between 1534 and 1539. The ten-year gap may be an indication that the king's attention was distracted, and indeed in 1533, Eustace Chapuys, the Imperial ambassador, reported with a hint of satisfaction that Henry had been informed that none of his ships could be fit for service in less than 11 months. Perhaps the regular maintenance programme which had certainly been in place before Wolsey's fall had lapsed in the interval between chief ministers. If that was the case, it had been resumed by August 1533, and by April of the following year Cromwell was reminding himself of the need to keep the naval administrators up to the mark. In June 1536 he noted with some satisfaction that since his joining the king's council in 1532, six ships had been rebuilt, and provision made for two more. The *Henry Grace à Dieu* was completely refitted in 1535, and in April 1538 Henry was gleefully showing off three large and newly refurbished vessels to the French ambassador.

The activity of the navy during these years was undramatic, but continuous. Escorts were provided for the Merchant Adventurers, shipping their cloth to Antwerp, for the North Sea herring fleet, and for the deep sea fishermen ventur-

One of the *Mary Rose*'s heavy guns. A bronze culverin recovered in 1979, resting on a replica gun carriage.

The *Mary Rose*. Henry VIII's first great man-of-war; commissioned in 1511 and rebuilt in 1536. Sunk in the Solent in 1545. There is hearsay evidence of negligence resulting from a drunken brawl on board. The story is told in William Harrison's 'Chronology' (1581). The illustration is taken from the Anthony Roll (*c.*1545).

ing to Iceland. Sir William FitzWilliam, who became Lord Admiral on the duke of Richmond's death in 1536, led a large fleet to Calais in January 1540 to bring the king's new bride, Anne of Cleves, back to England with suitable honour. Regular forays were conducted against pirates, and casual aggression deterred by sporadic patrols in the Channel, the North Sea, and the Irish Sea. Above all, every time there was increased diplomatic tension the king's Great Ships were re-rigged and armed, although a harmless cruise was usually the only outcome. In 1539, convinced that the emperor and the king of France were about to combine against him, Henry assembled about 100 armed ships at Portsmouth, of which some 40 were his own—almost the full complement of the navy at that time. The following year, ostensibly for defensive purposes, but really to put pressure on James V, about 20 ships were sent to control the channel between Argyll and the north east of Ireland, and two years later, protected by the renewal of hostilities between France and the Empire, Henry launched a full-scale attack against Scotland.

By 1540 Henry had about 45 ships of his own, all but ten of them over 100 tons. Between 1543 and 1545 three further large ships (over 400 tons) were built, and three more purchased. Several prizes were taken, including two large French galleys in 1546, and several merchantmen of over 100 tons were built with the subsidy of 5s. a ton which the king had intermittently paid throughout his reign when he wished to encourage his subjects to add to the pool of available fighting ships. The warships of the 1540s were also much more heavily armed than those of Henry's first war in 1512. The *Great Bark* of 500 tons carried 87 guns of different calibre, of which about a dozen could be classed as heavy, and the *Lion* of 160 tons carried 33, of which eight were heavy. The serpentines which the *Regent* had carried in such vast numbers were obsolete, and could now be classed as pea-shooters. At long last tactics had also developed to take account of the new technology, and Lord Lisle's fleet orders of 1545 not only showed an understanding of the latest Spanish thinking, but also a keen awareness of the need for discipline in manœuvring in order to make the maximum use of synchronized gunfire. It seems likely that the first broadsides were fired in anger during the encounters between the English and French fleets in the summer of 1545. Having disposed of the Scots at Solway Moss in November 1542, Henry then entered into a new Imperial alliance, as much to recapture the zest of his youth, it would seem, as for any more tangible reason. In so doing he committed himself to invade France with 42,000 men in the summer of 1544. The campaign which resulted was a mess, but it succeeded in capturing Boulogne, to the king's disproportionate delight. However, before the end of the year the emperor had signed a separate peace at Crespy, leaving the English to defend their new conquests as best they could.

To Francis I this presented an unequalled opportunity for revenge. Without Imperial support, Henry made no attempt to launch a new invasion, but by June 1545 Lord Lisle had 160 ships and 12,000 men at sea, prepared to intercept whatever attempt the French might make. The wind, however, favoured the French, and early in July Admiral D'Annebaut managed to get past Lisle with 130 ships, intent on capturing Portsmouth and the Isle of Wight. He failed, largely because Lisle was able to recover his position, and as the fleets encountered off Portsmouth, the wind died away completely. The encounter which followed is more famous for the loss of the *Mary Rose* than for any action, because apart from a little sporadic gunfire there was no fighting. Because the French, in spite of their large mobilization and landing troops and burning villages on the Isle of Wight, failed to achieve any of their objectives, this campaign must be counted as an English victory, but it was far more notable as a logistic triumph than as a military one. There was no more serious campaigning before the treaty of Cercamp in the summer of 1546 brought the French war to an end, and in January 1547 Henry VIII died possessed of some 55 ships, almost all in serviceable condition, a fortified naval base, and three or four substantial dockyards. He also by then had an administrative structure capable of coping with such equipment and premises.

In February or March 1545, a memorandum was drawn up proposing a new institution which was already being called (although not in the memorandum) 'the king's majesty's council of his marine'. This was to consist of the three existing admiralty officers, the Clerk of the Ships, the Clerk Controller, and the Keeper of the Storehouses, plus four others, to be called the Lieutenant (or Vice-Admiral), the Treasurer, the Surveyor and Rigger, and the Master of Naval Ordnance. The object was clearly to establish a formal structure and chain of command, with proper accountability in place of the somewhat *ad hoc* manner in which the individual officers had hitherto operated. Whether the idea had originated in the bureaucratic mind of Thomas Cromwell, in some idea of the Lord Admiral, or in the actual position occupied by William Gonson, we do not know. However, it looks as though Gonson's death in 1544 provided the impetus for the reorganiza-

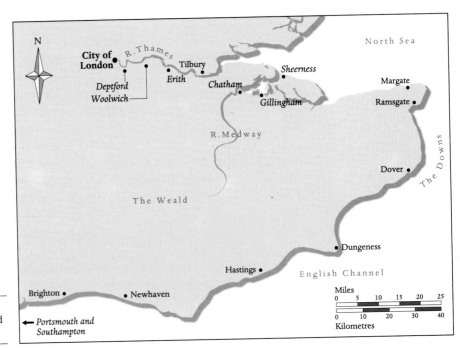

North Sea

N

City of
London
Deptford
Woolwich
R. Thames
Erith
Tilbury
Chatham
Sheerness
Gillingham
Margate
Ramsgate

R. Medway

The Downs

Dover

The Weald

Dungeness

Hastings

English Channel

Brighton
Newhaven

Miles
0   5   10   15   20   25

0   10   20   30   40
Kilometres

Portsmouth and
Southampton

Thames and Medway
dockyards in the Tudor period
(shown in italic type).

Facing: This vivid
representation of a boarding
action in the late fourteenth
century gives a clear idea of
medieval naval warfare. The
event shown is the battle of
Guernsey, a skirmish off the
Channel Islands.

tion, and the model was possibly the Ordnance Office, which had a staff of six senior officers. The first appointments seem to have been made in the summer of 1545, but the situation remained fluid for some months, and it was not until 1546 that the first patents of appointment were issued, and we can be sure that the Council for Marine Causes was in being. All the officers were well paid, and were expected to be full-time professionals. The Vice-Admiral, for example, who presided over the Council, received in fees and allowances some £275 a year; the Treasurer £220; and the Clerk of the Ships £133 6s. 8d. The men appointed were also suitably experienced seamen and administrators; Sir Thomas Clere as Lieutenant; Robert Legge, Treasurer; Benjamin Gonson (son of William), Surveyor and Rigger; William Broke, Controller; Sir William Woodhouse, Master of the Ordnance; and Richard Howlett, Clerk of the Ships. The significance of this reorganization can hardly be overemphasized. For the first time the Admiralty could function as an autonomous department, with 'line management' and properly defined functions for its various members.

WAR IN THE NORTH

Scotland had not been included in the treaty of Cercamp, and no sooner had Edward Seymour established himself as duke of Somerset and Lord Protector for the young Edward VI than he began to make plans to renew the war in the north. The strongly Catholic stand being taken by the Francophile party ensured that the reformers who were beginning to make headway in Scotland would incline to England. One such group succeeded in murdering Cardinal David Beaton, and barricaded themselves into St Andrews castle. However, their reliance upon English protection turned out to be misplaced. On 27 July, a French fleet

de laude et confort que le roy
dangleterre deuoit faire a
elle. Dont messire loys
despaigne messire charles
germaulx messire othon
ornes estoient establiz
sur la mer. A lencontre de
greuese auecque quatre
mille geneuoys moult

bien en point et mil hommes
darmes et trente deux gros
vaisseaulx
Cy parle listoire de la
bataille de greuese qui fut
entre messire robert dar
tois et messire loys des
paigne. Le iiijxx vije. chap̄

Ainsi que messire
robert du tois le
conte de penne
brot le conte de
sallebrin le conte de suf
fort le conte de kenfort le

huon de stanfort le seigt.
despensier le seigneur de
bourfier et plusieurs aultres
cheualiers dangleterre et
leurs gens auecques la
contesse motfort nagoiet

Plus ch en tous

Above: Henry VIII at the Field of the Cloth of Gold (1520). This was a magnificent example of the competitive display of Renaissance princes, in which Great Ships, men, clothes, and weapons were all used to demonstrate and enhance the monarch's *maiestas*.

Right: A picture of the raising of a siege in Africa which provides us with an image of the heraldic decorations of fourteenth-century ships.

commanded by the veteran Leo Strozzi bombarded them into surrender. English warships were at sea, but too few and in the wrong place. At the beginning of September, Somerset marched north from Berwick with a well-equipped army almost 20,000 strong, shadowed by a fleet of 65 sail under the command of Edward Fiennes, Lord Clinton. Thirty-four of Clinton's ships were warships, of which about 30 belonged to the king, and the remainder were victuallers, which took supplies north and carried the plunder south at the end of the campaign. The expedition was a complete success. Scotland's field army was destroyed at the battle of Pinkie Cleugh, near Musselburgh; the town of Leith and numerous villages and forts were destroyed; and the fleet captured the island of Inchcolm in the Firth of Forth, seizing and destroying a number of ships in the process. The French had been wrong-footed by the speed and efficiency of the English campaign, but Francis I had died in April and the new king, Henry II, was determined to recover Boulogne. He would lose no opportunity to exploit the resentment which a humiliating defeat had left in Scotland. In the summer of 1549, when Somerset was in desperate trouble with domestic rebellion, and when his attempts to follow up his victory in Scotland had totally collapsed, Henry took his opportunity and declared war.

By the time that he did so, Scotland had fallen into his hand like a ripe plum. The garrisons with which Somerset had attempted to preserve his ascendancy had served only as an irritant. In June 1548, in spite of the presence of Clinton in the north-east with some 70 ships, 16 French galleys and about 30 transports evaded his loose blockade, landing some 6,000 French troops at Leith. Under this protection (or threat) the treaty of Haddington was then concluded, whereby the young Queen Mary was to be betrothed to the Dauphin and a personal union of the Crowns was to follow in due course. A few weeks later Mary left Scotland by the western route, and in spite of a half-hearted attempt at interception, arrived safely in Brittany. By the time that Somerset fell, the victim of a palace revolution, in October 1549, the record of the navy since Henry's death had been rather less than glorious. This may have been just bad luck, given the uncertainties of locating any enemy at sea in sixteenth-century conditions, or it may have been the fault of the sea commanders. It was certainly no reflection on the Council for Marine Causes, nor upon the Privy Council which provided the funds. During these two-and-a-half years the navy was kept on a full war basis, at a cost of some £40,000 a year, exclusive of victualling. Apart from the sale of a dozen small galleys, or 'rowbarges', which had been Henry VIII's particular toys, but which no one else valued, the strength of the fleet remained undiminished. No less than £18,824 was spent at the main dockyard of Deptford between March and December 1547, and a ship list of January 1548 shows no fewer than 52 vessels, displacing over 11,000 tons. The value of this careful husbandry was shown during the autumn and winter of 1549–50, when repeated French attacks upon Boulogne were defeated by retaining command of the sea.

When the war was ended in March 1550, a reduction in the level of activity inevitably followed. But the peace was uneasy, and in spite of its financial difficulties the council took the view that it would be a false economy to cut back heavily on naval expenditure. The revenue commission which reported in 1552 recommended, among other things, a substantial reduction in the Admiralty budget, but no such reduction took place. In the three years from 1548 to 1551, about half of which was wartime, Benjamin Gonson received and dispensed about £66,000, while expenditure for the last two years of the reign ran at about £20,000 a year. Henry's rowbarges were replaced with six new pinnaces between 1547 and 1549, and four new ships of between 140 and 300 tons were added between May 1550 and August 1552. Also, in June 1550, Edward Baeshe, an experienced victualler, was appointed to the newly created office of Surveyor General of the Victuals. As such he became a member of the Marine Council, and thus brought the last important

A selection of personal items representative of those which might have been carried in compartments of decorative pouches, possibly belonging to officers of the *Mary Rose*. Items are: pouch, showing front flap decorated with silk embroidery and silver initials; boxwood comb; wooden seal; pocket sundial; die; thimble ring; decorative clasp; tokens; wooden whistle; and a rosary.

A selection of rigging items recovered from the *Mary Rose*. Includes: (*left, front*) bronze sheave; (*left, centre*) parrel spacer and trucks; (*middle, back*) single sheave rigging block; (*middle, centre*) shoe block, with sheaves at right angles; and (*right, front*) deadeye, for securing the lower end of shrouds.

aspect of naval provision under the council's surveillance. Evidence of specific activity between 1550 and 1553 is scrappy, but this certainly does not reflect a neglected or decaying force.

The most important maritime enterprise of this period had nothing to do with the navy, but can hardly be passed over without a mention. As a result of a voyage undertaken by Sir Hugh Willoughby and Richard Chancellor in 1553 the Muscovy Company was founded in 1555, and the major reorientation of England's commercial priorities had taken a small beginning. This had no immediate effect upon the navy, but when the same spirit of enterprise took John Hawkins to West Africa and the Caribbean in the 1560s, the queen's ships and the queen's money were soon engaged.

The early part of Mary's reign must have put the admiralty under considerable strain. There was no doubt about the loyalty of the fleet, and the spontaneous decision by the crews of six ships stationed in the Orwell helped to clinch the queen's bloodless victory over her unpopular rival, Jane Grey. However, the Lord Admiral, Lord Clinton, had been rather too close to the Grey party for anyone's comfort, and he was replaced by Lord William Howard in March 1554. Sir Thomas Wyndham, the Master of Naval Ordnance, had died during the summer of 1553 and had not been replaced; but worst of all the energetic and competent Surveyor, William Winter, was arrested on 20 February for his part in Sir Thomas Wyatt's rebellion, and disappeared into the Tower. Nevertheless, in 1554 there were 29 men-of-war in commission, patrols were mounted in the Channel throughout the winter, and a fleet sent to Spain to escort the queen's intended consort, Philip, to England in the early summer. The first full year of the reign, from October 1553 to September 1554, saw £26,126 pass through Gonson's hands; a figure which, even if it includes victualling, represents no diminution from the latter part of Edward VI's reign. The commonly held view that Mary neglected the navy until 1555 is based partly upon her failure to fill the two vacant offices on the council, partly on the fact that a considerable number of old ships were sold in 1554, and partly upon the loss by fire of the ancient but still awe-inspiring *Henry Grace à Dieu*. This loss was attributed to 'neckclygens and lake of over-syth', although on whose part is not clear. Between Michaelmas 1554 and Christmas 1555 no fewer than 14 vessels were disposed of, but with the exception of the 300-ton *Primrose* they were all old, small, and fit only for scrap. Their disposal should consequently be seen as good husbandry rather than the reverse. During those same 15 months the 'ordinary' charges of the navy (that is maintenance, materials, and shore wages) amounted to £11,000; 'sea charges' to £4,000; and victuals to nearly £12,000; an annual expenditure of about £21,000.

In 1555 the building programme was resumed after a gap of about four years. There was nothing unusual about such an interval, and no special explanation is needed for the council's decision to resume. The *Philip and Mary* of 500 tons and the *Mary Rose* of 600 tons were laid down, probably in September 1555, and launched towards the end of 1556. Deptford was still by far the most important

MARY I'S NAVY

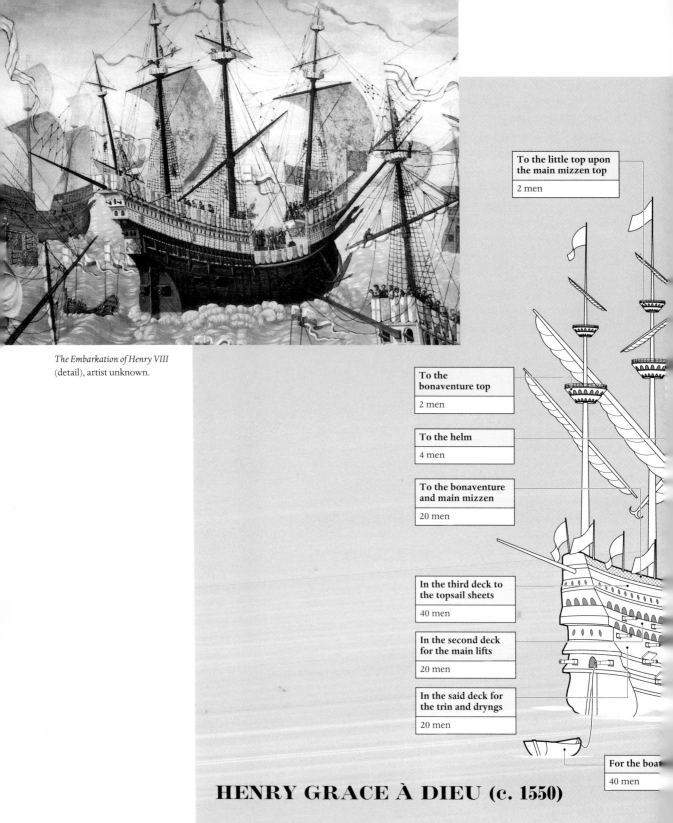

The Embarkation of Henry VIII (detail), artist unknown.

To the little top upon the main mizzen top

2 men

To the bonaventure top

2 men

To the helm

4 men

To the bonaventure and main mizzen

20 men

In the third deck to the topsail sheets

40 men

In the second deck for the main lifts

20 men

In the said deck for the trin and dryngs

20 men

For the boat

40 men

HENRY GRACE À DIEU (c. 1550)

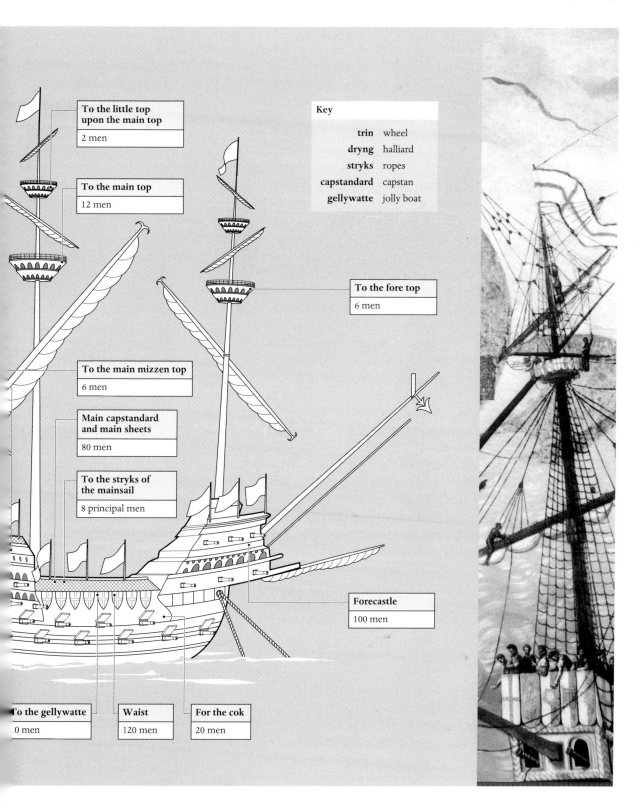

To the little top upon the main top

2 men

To the main top

12 men

Key

trin	wheel
dryng	halliard
stryks	ropes
capstandard	capstan
gellywatte	jolly boat

To the fore top

6 men

To the main mizzen top

6 men

Main capstandard and main sheets

80 men

To the stryks of the mainsail

8 principal men

Forecastle

100 men

To the gellywatte

0 men

Waist

120 men

For the cok

20 men

dockyard. Virtually nothing was spent at Portsmouth, emphasizing the fact that its use was largely confined to wartime. By the end of 1555 Gonson's account was £5,000 in deficit, and there are signs that the Privy Council was becoming dissatisfied with the efficiency of naval management. On 11 January 1556 the Lord Admiral was instructed to take a 'secret muster' of all the soldiers and mariners then serving in the queen's ships, without informing the Council for Marine Causes. The results of the muster and the causes of the mistrust are alike unknown. Surviving evidence does not suggest any lack of effectiveness. About the middle of July the queen's ships caught up with a substantial pirate fleet, manned by French corsairs and English exiles, which had been causing heavy losses to English and Imperial ships, and captured or destroyed the greater part of it. During the year Gonson received a little over £28,000, and by the end of December had reduced his deficit to £1,155. In spite of this the Privy Council's concern persisted, and in January 1557, with Philip pressing for English participation in his war against France, major reforms were decreed. The Lord Treasurer, the marquis of Winchester, was placed in overall control 'with the advice of the Lord Admiral', and Gonson and Baeshe were to account annually and separately, presumably to him. At the same time Winchester stipulated that an 'ordinary' of £14,000 a year should be paid to Gonson in two instalments, thus creating a proper budget for the first time. It is unlikely that these reforms were implemented before the outbreak of war in June, but they represent the culmination of almost half a century of progressive and constructive development in naval administration, and demonstrate that Mary's Council, no less than those of her father and brother, showed a proper concern for the well-being of England's first line of defence.

During the war, which lasted from June 1557 to March 1559, the navy performed adequately, without greatly distinguishing itself. A fleet of 21 ships was at sea when hostilities broke out, and they provided the escort which 'wafted' an expeditionary force of about 8,000 men to take part in the St Quentin campaign. At the same time an attempt by Sir John Clere with nine ships to stage a landing in the Orkneys ended in a disaster which cost the lives of Clere himself and three of his captains. The navy cannot, however, be held responsible for the greatest set-back, which was the loss of Calais in January 1558. Normal patrols were actually at sea when the imminence of the danger was belatedly understood, but could not be contacted swiftly enough. Sir William Woodhouse was instructed on 31 December to get five ships to sea 'with all possible diligence', but by the time he tried to enter Calais harbour on 3 January it was already too late. An attempt was made to counter-attack, but Woodhouse's ships were scattered and damaged by a severe storm on the night of 9–10 January, and by the time that had been remedied the French position was too strong to be challenged. The English Council displayed little enthusiasm for a war which most of them had explicitly advised against, and Philip's envoys denounced their indolence and incompetence, but by June 1558 140 ships were ready in Portsmouth to take part in a combined attack on Brittany. At the same time letters of marque were issued to 23 privateers to take their own rewards at the expense of France. Lord Clinton, who had by this time recovered the office of Lord Admiral, landed 7,000 men near Brest at the end of July, but determined resistance and inclement weather broke up the attack. Only a small detached squadron under Vice-Admiral John Malen achieved anything of note, when he was able to assist the Count of Egmont to overcome a French force threatening Dunkirk by bombarding them from the sea. Between 30 December 1556 and the same date in 1558, Benjamin Gonson dispensed almost £145,000, so the cost of the war by the time that the peace of Cateau Cambresis was signed in April 1559 probably amounted to about £170,000 for the navy alone.

Mary had died in November 1558, but that had made no significant difference to the Admiralty, and in March 1559, the Council for Marine Causes undertook a stock-taking, which was both a review of its achievement to date and a policy statement for the future. Thirty-nine ships were listed, of which 27 were in commission. Gillingham, rather than Deptford, was suggested as the most suitable anchorage, needing, it was claimed, fewer shipkeepers. A blueprint for the future of the navy was also provided, to include 30 ships, of which 24 should be over 200 tons, divided into five categories, foreshadowing the 'ratings' of the following century. A number of ships should always be either at sea or on fourteen days' standby, and the storehouses kept replenished with victuals for 1,000 men. The two documents which make up this review constitute unequivocal evidence of a carefully structured policy, but there is no sense of innovation. Broadly speaking, Elizabeth adhered to this policy over the following 20 years. She remained deaf to pleas for more money, and reduced the ordinary twice by 1564, first to £10,000 and then to £6,000. However, since she at the same time removed the charges of ordinary victualling to another account, this did not indicate any reduction in the repair or maintenance budget. Until 1565 the navy remained on what might be described as a 'quasi war' footing, first because of the intervention in Scotland and then because of the Le Havre campaign. In 1562–3 Gonson received £9,630 on his ordinary account, and £13,375 'extraordinary', indicating a shift away from the 1557 accounting structure which was to become ever more pronounced as the reign advanced, although no new policy statement was made. During these years the building programme also continued at a steady pace. The *Elizabeth Jonas* (already laid down in 1558), the *Victory*, the *Triumph*, and the *Aid* were launched, the *Jesus of Lubeck* and the *Great Bark* extensively renewed. By the last quarter of 1563 29 ships were 'aflote' at Gillingham, plus half a dozen at Portsmouth and a similar number elsewhere, three Master Shipwrights and scores of carpenters, caulkers, labourers, and clerks were 'in regular wages' at the various dockyards, and the admiralty had acquired that substantial permanent structure which it was never subsequently to lose.

During the long, and sometimes uneasy, peace which followed, only minor modifications occurred. When William Winter was restored to favour he was appointed to the Mastership of the Ordnance, as well as to the Surveyorship which he had held before, and on the death of Sir William Woodhouse in 1565 the office

ELIZABETH I'S INHERITANCE

39

Elizabeth I; the Armada portrait, attributed to George Gower.

of Vice-Admiral was discontinued. At that point Winter probably took over the chairmanship of the committee, and continued to hold it until his death in 1589. Between 1565 and 1589 the Council for Marine Causes therefore consisted of six officers, holding seven separate offices. Of the senior officers, Benjamin Gonson continued to serve until his death in 1577, when he was succeeded by Sir John Hawkins, and Edward Baeshe until 1587. Lord Clinton continued as Lord Admiral until his death in 1585, becoming earl of Lincoln in 1572. In 1564 the Victuals accounts were finally separated from those of the Treasurer, and if there was a policy for general funding, it is extremely hard to reconstruct. Baeshe began operating on a contract basis in 1565, when he entered into an agreement to discharge his duties in return for annual pro rata payments of 5d. a day for a man at sea, and 4½d. a day for a man in harbour or ashore. For this each man was entitled to receive a gallon of beer, 1lb. of biscuit or a 20 oz. loaf of bread, and a main ration which might be 2lb. of fresh beef, ½lb. of salt beef, ½lb. of bacon, ¼lb. of stockfish or 4 herring, with butter and cheese, according to circumstances and the day of the week. Obviously his profit had to depend upon cost-cutting, and it was an arrangement which invited fraud and malfeasance. Baeshe seems to have been an honest man, and complaints against him were few, but that was no thanks to the system. On several occasions he got into financial difficulties, and the rates were raised to 5½d. and 6d. in 1573. His complaints of poverty were probably exaggerated, but any attempt to improve the quality of the victuals could only be at his own personal expense.

Astronomical compendium by Humphrey Cole (c.1569).

Hawkins's contract of 1579 worked in a similar fashion. He undertook to maintain 25 ships to a certain standard, but not to repair them or provide shipkeepers, in return for £1,200 a year. At the same time the queen's master shipwrights, Peter Pett and Matthew Baker, undertook to ground and grave the ships, and to carry out certain types of repair for £1,000 a year. The contractors were not required to account, and clearly expected to make a profit. Theoretically this ensured that about two-thirds of the work then classed as 'ordinary' would be carried out for £2,200 a year—a substantial saving. However, the first result was furious quarrels between Pett and Baker on the one hand, and Hawkins on the other, which must have undermined his authority as an officer. Secondly, because the contractors were only responsible for a proportion of the work, they tended to shift expensive jobs on to other officers or craftsmen who did not have contracts, thus preserving their profit margins at everyone else's expense, and eliminating the savings which the system was intended to achieve. The situation had become so bad by 1584 that a commission of inquiry was instituted, and both contracts were called in. In the event Hawkins was exonerated, and offered a fresh contract whereby he undertook virtually all the work of the old ordinary for £4,000 a year. Pett and Baker immediately renewed their assault, accusing him of gross negligence and malpractices of every kind. He was probably not guilty of corruption in its cruder forms, but undoubtedly continued to shift expensive work. This did not seriously undermine the efficiency of the Admiralty or its dockyards, but it did make the system largely pointless from the government's point of view, and it was abandoned in 1587, when hostilities with Spain had already commenced.

The most significant development of these years was the queen's ambivalent relationship with adventurers like Hawkins and Drake, who repeatedly and deliberately infringed the commercial monopolies claimed by Spain and Portugal in Africa and the New World. Until the end of Mary's reign royal ships had simply been hired out, when they were available, for a stipulated rent and against a bond of indemnity to cover loss or damage. In 1561, however, Elizabeth loaned to Sir William Chester the *Minion*, the *Primrose*, the *Brigandyne*, and the *Fleur de Luce* for a voyage to West Africa, the four ships and £500 worth of provisions constituting an investment which entitled her to one-third of the profits of the voyage. In 1564 this tactic was repeated, when the *Jesus of Lubeck* was loaned to Sir John Hawkins as part of the Queen's investment in a slaving voyage which was bound to cause offence on both sides of the Atlantic. In 1568, when Hawkins was caught *in*

ADVENTURERS
AND PIRATES

Sir John Hawkins, Treasurer of the Navy, 1577–1595.

flagrante delicto at San Juan d'Ulloa, not only was he again using the *Jesus*, but he was bearing the queen's commission and flying her standard. It is not surprising that Philip became increasingly sceptical of Elizabeth's disingenuous protestations and professions of good will. The queen's investment in Drake's voyage of 1577 was a private one, but when he eventually returned from what became a three-year voyage of circumnavigation, bearing plunder worth £150,000, she knighted him on the deck of his flagship in September 1580.

As relations with Spain deteriorated, it became increasingly easy to represent piracy as patriotism, or even as a part of a protestant crusade, especially when it took place in the Caribbean or the South Atlantic. In European waters, irrespective of pretext, it assumed the proportions of an epidemic. The English were not the only offenders, but they were almost certainly the worst, and the Spanish ambassador was not the only one to make incessant complaints. One group of Rouen merchants, plundered in 1576, pursued their case through the Admiralty courts on both sides of the Channel for nine years before finally resorting to letters of reprisal. The Council not only sent out royal ships to hunt down offenders, it also licensed syndicates of increasingly desperate English merchants to provide their own remedies. Part of the problem was the opportunist nature of so many of the crimes. For every full-time pirate there were dozens of part-timers—normal traders who, once in a while, could not resist the temptation of an easy prize. As most merchant ships had to be armed for self-defence, it was very easy to turn aggressor when the occasion served. The other problem was the collusive attitude of many officials. Justices of the Peace, and even Vice-Admirals, had one foot in the royal service and the other on the pirates' deck. Plunder could be landed and sold, weapons acquired, and legitimate complaints diverted or silenced. Gentry families such as the Killigrews and the Bulkeleys were deep in this unsavoury trade, and Elizabeth's government lacked the determination to deal with them as they would have done with common criminals. The navy was actually more successful at catching pirates afloat than the Council was in dealing with their aiders and abettors. Between 1581 and 1583, Studland Bay in Dorset became a notorious haunt, some 40 stolen vessels being unloaded and stripped there, but in the latter year the queen's ships eventually moved in. Over 30 prisoners were interrogated between 6 and 8 August, and a fortnight later nine were hanged at Wapping.

DEVELOPMENTS
IN SHIPBUILDING

Great Ships were of little use for this kind of operation, and it was the barks and pinnaces of the navy which were the hardest worked during the years of peace. Nevertheless, large ships continued to be built. After the launching of the *White Bear* in 1564 there was a lull until the *Foresight* in 1570, followed by the *Dreadnaught* in 1573 and the *Revenge* in 1577. Three smaller ships were also built in 1573 and one in 1577, so that the momentum was well maintained. It was also during these years that some innovations in design began to be adopted. The *White Bear*, like the *Triumph* or the *Philip and Mary*, was high-sided and broad in the beam, not unlike the earlier carracks, but the *Revenge* was longer, narrower, and carried much less superstructure. It used to be claimed that Sir John Hawkins invented these 'race-

Profile for the design of a galleon (*c.*1586). From Matthew Baker's 'Fragments of Ancient English Shipwrighting'.

built' galleons, but it now appears more likely that he adapted the design from a much earlier series of experiments carried out in the reign of Henry VIII. As early as 1515 Henry had caused to be built, and may have designed, a large galleass which he named the *Mary Imperial*, but which was normally known as the *Great Galley*. She carried 60 guns and was propelled by 120 oars. In spite of the king's enthusiasm, she was probably not a great success, and after about 20 years in service was rebuilt as a sailing ship and renamed the *Great Bark*. As such she appears in the Anthony Roll of 1545, her low profile in marked contrast to the towering carracks such as the *Mary Rose* and the *Henry Grace à Dieu*. At least two other ships, the *Tiger* and the *Bull*, both built in 1546, showed the same profile. They could be moved by sweeps, but were sailing ships rather than galleasses. The performance of these ships may well have commended them particularly to Hawkins, whose experience was mainly that of a privateer. Privateers were not interested in defence, but only in attack, so that speed and manoeuvrability became prime desiderata. Similar considerations also applied to the guns which were carried. Smaller ships, or a weaker fleet, would naturally resort to their guns to hold a more powerful enemy at bay. Consequently 'stand-off' tactics, using relatively long-range guns, were also developed by the privateers, and it was by recruiting privateers, such as Robert Reneger in the 1540s, Peter Killigrew in the 1550s, and Hawkins himself in the 1560s, that such tactics found their way into the repertoire of the English navy.

In spite of having a long and distinguished tradition of naval warfare in the Mediterranean, Spain was slow to develop an effective Atlantic fleet. In 1580, however, Philip secured the succession to the throne of Portugal, and with it a fleet of ocean-going warships. He also acquired an enemy in the person of Dom Antonio, the Prior of Crato, an illegitimate son of the Portuguese royal house. Dom Antonio attracted unofficial support from both England and France, and in 1583 a fleet operating in his name challenged Spain's recently augmented navy at

43

Terciera in the Azores. Thanks partly to skill and partly to luck, the Spanish commander, the Marquis of Santa Cruz, was completely victorious. Thereafter, Santa Cruz believed that he could beat any power which might challenge him upon the seas. These were successful years for Spanish arms. In the Netherlands the Duke of Parma inflicted one defeat after another upon the rebellious provinces, and the murder of William the Silent in the same year left the Netherlanders desperately casting about for allies. The result was the treaty of Nonsuch in 1585, which bound England to support the Dutch with a small army, and thereby guaranteed that the long-running cold war with Spain would swiftly convert into open hostilities. Philip did not, however, react immediately. There were serious shortages of grain in Spain in the spring of 1585, and the king authorized large-scale imports, giving full safe conduct to all the ships which came, mainly from England and the Baltic. Then, on 29 May, he ordered all such ships to be seized, their guns confiscated, and their crews imprisoned.

The English reacted with predictable and self-righteous fury. Claims for hugely inflated losses were lodged with the Admiralty court, and no time was given for those claims to be pursued by peaceful means. On 9 July, letters of reprisal were issued for about £125,000, thereby launching a campaign which was virtually indistinguishable from open hostilities. Even more swift was the commission issued to Sir Francis Drake on 1 July to undertake a punitive expedition to the new world. This expedition was set up with exemplary speed and efficiency, along the same lines as Hawkins's voyage of 1567. A joint stock capital of £60,400 (about £40 million in 1995 equivalents) was raised, to which the queen subscribed £10,000, and a fleet of 22 sail assembled, of which the *Bonaventure* (600 tons) and the *Aid* (250) belonged to the navy. Drake sailed from Plymouth on 14 September, and returned, with his fleet severely depleted by weather and disease, on 28 July 1586. The impact of this protracted raid was out of all proportion to its real achievement. The truth is that it was a soft triumph, because the Spanish officials were at odds among themselves, and the defensive preparations were everywhere seriously defective. Nevertheless, the psychological effect on Spain was immense. Europe buzzed with the report of how an audacious Englishman had bloodied the king of Spain's nose, and Philip was mortified as he had not been by two decades of provocative pinpricks.

Probably the most important consequence was the king's decision that he would have to deal a crippling blow to England before he could be sure of victory in the Netherlands. A plan for just such an armada already lay on his desk, submitted some time earlier by the Marquis of Santa Cruz. This envisaged an invasion direct from Spain, requiring 55,000 men, over 500 ships, and costing about four million ducats. Philip had not responded, for the simple reason that he knew such an undertaking to be beyond him, but in April 1586 the Duke of Parma, his commander in the Netherlands, came up with a much more realistic alternative. He would himself invade England with 30,000 of the men already under his command. All that would be needed from Spain would be a warfleet capable of defeating the English navy, and holding the Narrow Seas long enough for Parma's army to make the crossing. This plan would require some 150 ships, and Philip was under no illusions about the strain which such an enterprise would place upon his logistic resources. However, anger, and a mounting sense of religious purpose, overcame his doubts, and the relevant orders were given. The only man capable of leading such an expedition was Santa Cruz, and he still wanted to take a force direct from Spain. It was therefore agreed that some 19,000 men would sail with the armada itself, which would thus require transports as well as fighting ships. August 1587 was fixed as the target date.

By the autumn of 1586, the English knew that a major attack was pending, and naval preparations were expanded and accelerated. Eleven new ships were completed in 1586, and two more in 1587—the largest building programme since the 1540s. Several of these were capital ships, including the *Vanguard* and the *Ark Royal*, and expenditure for a year in which there was very little action at sea stood at £44,000. The queen herself continued to hope that the crisis could be ended by negotiation, but this time her ministers knew better, and were able to persuade her to keep up the momentum of her defensive preparations. In March 1587 Drake was commissioned to command a fleet of 23 sail, to which six large warships were contributed by the queen, and the remainder mainly by the city of London, to 'impeach' in whatever way he could the hostile preparations which were being made. The result was a microcosm of the best and worst features of the Elizabethan navy. Drake sailed on 2 April; on the 12th, perhaps knowing it was already too late, the queen cancelled his commission. Ostensibly sailing to an unknown destination, Drake knew perfectly well that his target was Cadiz, and that his purpose was military. This could hardly have been admitted to the London captains, who made no secret of their priority to make a profit out of the voyage. On 19 April Cadiz was taken completely by surprise; 24 Spanish ships, together with large supplies of arms and equipment, were either destroyed or captured, and 14 royal galleys which attempted to defend the harbour were knocked out in a few hours by the weight of the English guns. It was a complete but limited victory. No attempt was made to take the city, and no plunder, in the ordinary sense, was taken. Having achieved his objective, Drake was then compelled to remain at sea in order to satisfy the privateers. He took Sagres in southern Portugal, and briefly blockaded the coast, causing panic but no great material loss. An attempt to intercept the Spanish silver fleet failed, and it was eventually more by luck than judgement that he took the Indies carrack *San Phelipe*, with a cargo worth at least £150,000. Honour now being satisfied on all sides, he returned to Plymouth on 7 July, just in time to avoid the Marquis of Santa Cruz, who had set out to look for him with a powerful fleet. By the time the damage had been made good and Santa Cruz's fleet returned from their futile chase, it was too late to launch the armada in 1587.

By that time, the enterprise was in serious danger of devouring itself. There were not enough ships for the men who had been mustered, and the men were

THE CADIZ RAID AND THE ARMADA, 1587–1588

45

The *Ark Royal* (commissioned 1588). A typical English galleon of the late sixteenth century.

consuming victuals faster than they could be provided. The set-back at Cadiz had also demonstrated the limited usefulness of galleys, and the urgent need for more heavy guns, which Spain did not have the capacity to produce. In the middle of February 1588, Santa Cruz died, leaving his gigantic operation in a state of apparently terminal chaos. The man who redeemed this situation was Alonso Perez de Guzman, duke of Medina Sidonia. With costs running at £100,000 a month, he had to get the armada to sea by the summer, or bankrupt the state. Remarkably, he succeeded, but the imposing fleet which he led down the Tagus on 30 May was riddled with more or less concealed deficiencies. Enough guns had been found, but many were old or unsuitable, and they were of every conceivable size. Most of the sailors were conscripted landsmen who had never been to sea before, and the soldiers were raw recruits. Only about 30 of the 130 ships were properly equipped warships, and some of those had been over-hastily prepared. Once at sea the fleet would have to be self-sufficient in terms of food, water, powder, and shot, and since no one knew exactly how long the expedition would take, supply was largely a matter of guesswork. To make matters worse, between Lisbon and Coruña the armada was scattered by a gale, necessitating emergency repairs, and a further desperate search for fresh food. Medina Sidonia had no illusions about

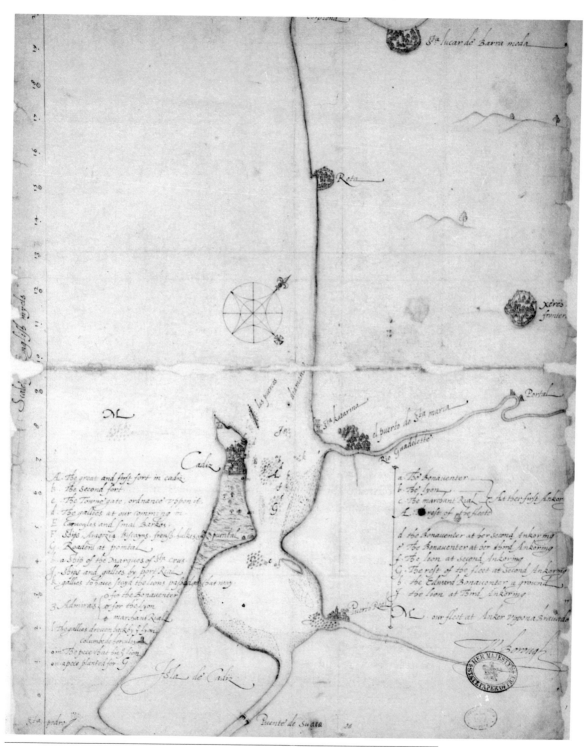

Cadiz harbour at the time of Drake's raid, showing English and Spanish positions; from a map drawn by
William Borough, one of Drake's officers (1587).

the difficulty of his task. A superb organizer, he was an inexperienced commander, and had no option but to stick rigidly to his instructions. On the other hand his captains were skilled and experienced seamen, and both discipline and morale in the fleet were exceptionally good, thanks largely to the prevailing sense of divine mission. The king was elated by his own crusading spirit, but it was not satisfactory to leave the details of his rendezvous with Parma in the hands of God. The simple fact was that the Spaniards did not control a deep water port of adequate size in the Netherlands, which meant that Parma would have to be absolutely ready at the very moment when the armada appeared. Given the uncertainties of weather and sixteenth-century communications, the odds against that happening were astronomical. Moreover, Parma's troop barges would have to make their way out through the shallow coastal waters before they could come under the protection of the armada's guns, and lying in wait was a squadron of heavily armed shallow-draft Dutch warships under the command of Justin of Nassau.

The English fleet lay partly in the Downs, and partly at Plymouth. About 150 reasonably well-found ships had been assembled, of which some 40 were the queen's own, at a cost of about £200,000. The Lord Admiral, Howard of Effingham, was not particularly experienced, but he was well respected, and he was supported by some of the most skilful captains of the period—Drake, Hawkins, and Frobisher. Some of the English ships may have been better sailers than the Spaniards, but that was not particularly relevant. The two critical advantages which Howard enjoyed were first, superior guns and gunners, who could maintain a much higher rate of effective fire at a greater range, and secondly, home waters. Not only did his captains know where they were without benefit of charts,

The advancing armada harried by the English fleet (July–Aug. 1588).

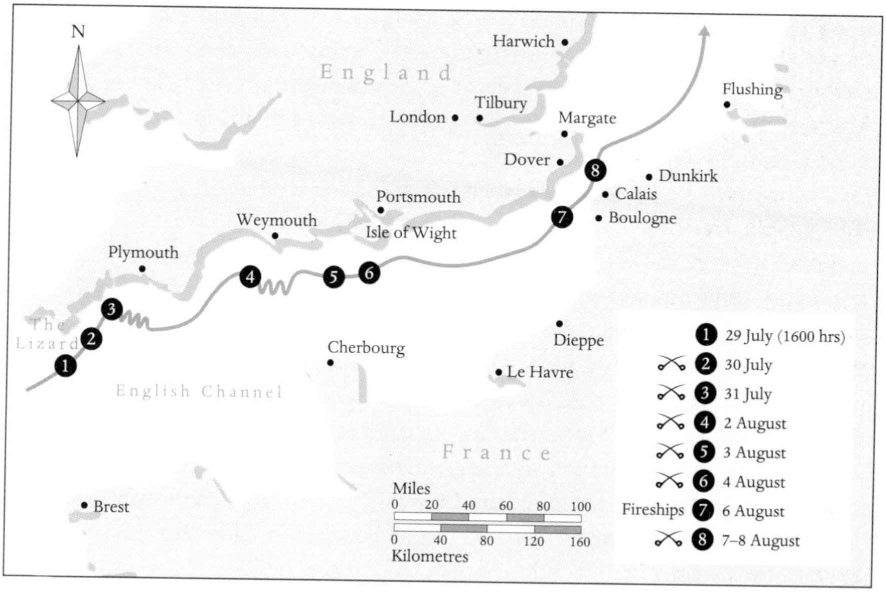

The Armada Campaign, 1588.

Map legend:
- 1 — 29 July (1600 hrs)
- 2 — 30 July
- 3 — 31 July
- 4 — 2 August
- 5 — 3 August
- 6 — 4 August
- Fireships 7 — 6 August
- 8 — 7–8 August

they could also take on fresh supplies if the original provisions had been miscalculated. English discipline was nothing like as good as Spanish, but that was relatively unimportant given the very high level of motivation which had been achieved. The armada was sighted off the Lizard on 29 July, and a running battle ensued over the next six or seven days as it moved slowly up the channel. The honours at this stage were pretty even. The English had alarmed and impressed the Spaniards with their line-ahead formation and raking gunfire, but they had done little real damage and had not disrupted the armada's formation. The critical point was reached on 6 August, when Medina Sidonia reached Calais roads, to discover that Parma would not be ready for another week. The king's plan had broken at its weakest point. The armada had no option but to anchor in an unprotected position, exposed both to the weather and to English attack. On the night of 7 August the English sent in the fireships. It was an old and expected tactic, and Medina Sidonia was prepared, but when the attack came, the discipline of his fleet, so admirable up to this point, broke down. The following morning the armada was scattered along the Flanders Banks, at the mercy of an onshore wind, and the English moved in for the kill. 'Kill' is an exaggeration, because few ships were sunk and the attackers soon ran out of powder and ammunition. Nevertheless, much damage and heavy loss of life was inflicted, and when the wind changed, Medina Sidonia drew his battered ships off to the north, intent only on getting home. Many of the galleons which were subsequently lost on the Scottish and Irish coasts had been rendered unseaworthy by English guns.

England breathed a sigh of relief, but her seamen were left with little to celebrate. AFTERMATH, 1589–1603
Plague had broken out on the ships before Howard called off the chase, and within a few days men were dying by hundreds in the streets of the seaport towns as they

49

struggled ashore. The Admiralty, which had coped so splendidly with the problems of mobilizing the fleet, and rather less splendidly with the problems of keeping it at sea, failed entirely when it came to demobilization. Nor were the English ships unscathed, and it seemed in the winter of 1588–9 that England would be in no better case than Spain to mount a campaign in the following year. Nevertheless, such was the resilience of the English system that by the beginning of April over 90 ships and 17,000 men had been assembled at Plymouth for a counterstroke. On this occasion, however, the weakness already apparent in the Cadiz expedition undermined the whole operation. No more than six ships, and about a quarter of the money had come from the queen, so although the commanders, Sir Francis Drake and Sir Henry Norris, bore the queen's commission, they treated her instructions with scant respect. The expedition's financial backers, mainly in London, decided that the time had come to capitalize on Dom Antonio, and misled by his fulsome promise of financial reward, set out to capture Lisbon and place him upon the throne of Portugal. Elizabeth's orders were explicit. Drake and Norris were 'first to distress the ships of war in Guipuzcoa, Biscay, Galicia and other places . . .', in other words to destroy the remains of the armada. Instead, they sailed straight to Lisbon, found that Dom Antonio had no following to speak of, and came home without having achieved anything. After 1590, the war returned to equilibrium. Repeated English attempts to take the *flota* failed, and one such effort, in 1591, resulted in the heroic, but pointless, sacrifice of the *Revenge*. Privateers, encouraged by occasional bonanzas such as the taking of the *Madre de Dios* in 1592, continued to operate, but with decreasing effectiveness as Spanish commerce, apart from the *flota*, virtually disappeared. In 1595, Drake and Hawkins set out, like a pair of ageing gun-slingers, to repeat the Caribbean exploit of ten years earlier. They found the Spaniards ready for them, and both died on the voyage. Nevertheless the expedition to Cadiz in 1596 was one of the most efficient acts of war carried out by any Tudor government. Led by the Earl of Essex and Lord Thomas Howard, and containing a useful Dutch squadron, the English fleet surprised and captured the city, which they ransomed for £180,000, destroyed 24 large ships and innumerable smaller ones, captured 1,200 pieces of ordnance, and inflicted damage estimated at 20 million ducats.

By that time Spanish naval power had long since recovered from the set-back of the armada, but it accomplished remarkably little against England. There was a small-scale raid on Cornwall in 1594, and an unsuccessful landing in Ireland in 1601, but nothing to set against the frequent and damaging raids upon the Spanish coast. Luck certainly played its part in this. The second armada of 1597 was well conceived and competently carried out. The English knew nothing about it, and only the gales which dispersed the fleet and frustrated its reassembly prevented Philip from scoring a belated success in the last year of his life. By the end of the century, the English Admiralty was probably past the peak of its efficiency, but a small event in 1599 demonstrated its capability. An alarm was raised by Spanish naval preparations against the Netherlands, and the Privy Council ordered an emergency mobilization. Eighteen warships were rigged, victualled, and fitted out in 12 days. 'The Queen', wrote Sir William Monson, 'was never more dreaded abroad for anything she ever did.' When peace was signed in 1604, King James had some 45 ships of his own, and almost 200 merchantmen of over 100 tons to call on. He also had commanders who were capable of formulating a full 'blue water' strategy of blockade, even if they were not fully capable of carrying it out. Unfortunately, he also inherited an Admiralty far gone in corruption, which was to reach the proportions of national scandal within four years.

The root of this problem lay in Elizabeth's pathological reluctance to spend money, already to be seen in the contracts of 1564 and 1579. The fees of Admiralty officers had not been raised since the offices were created in the 1540s. As in other public offices, inadequacy of official reward led officers to exploit their positions for whatever profit might be had. The navy cost about £54,000 a year from 1590 to 1603, which was remarkably little for wartime, but the price was paid in botched operations like the expedition to Lisbon, and in the steady deterioration of Admiralty morale. James was not greatly interested in the navy, and no new ship was laid down until the *Prince Royal* in 1609. Charles Howard, by this time earl of Nottingham, remained Lord Admiral until 1618, but once the war was over he paid little attention to the routine which then supervened. This indifference, itself unchecked, permitted Sir Robert Mansell, who became Treasurer in April 1604, and Sir John Trevor, Surveyor since 1598, to turn the navy into a private milch cow. Mansell and Trevor carefully placed their own servants in many subordinate positions, and were joined by the principal shipwright, Phineas Pett, in what was virtually a conspiracy to defraud the Crown. Exorbitant travelling expenses were claimed; every officer who went to sea paid himself at an admiral's rate; timber, pitch, and other supplies were bought privately at standard rates and then sold on to the Admiralty at a huge profit. In 1605 the conspirators fitted out one of the king's ships, called the *Resistance*, at the king's expense, and then leased it to the Admiralty as though it had been their own! 'The whole body is so corrupted', wrote Sir John Coke, 'as there is no sound part almost from the head to the foot.' The extent and notoriety of these abuses prompted a political intervention from Henry Howard, earl of Northampton, himself no very upright statesman, who was anxious to discredit his cousin the Lord Admiral. In April 1608, a commission of inquiry was established, headed by Northampton as Lord Privy Seal. The testimonies collected soon built up a damning case, which was presented in the form of a report under six headings. While this was going on, a second and related inquiry into malpractices in the building of the *Prince Royal* was also initiated, resulting in equally damning conclusions of scandalous mismanagement and shoddy workmanship. In May 1609, the king sat in judgement on the *Prince Royal* inquiry himself, and in spite of the overwhelming evidence, would find no fault with either Mansell or Pett. In the light of this it is not surprising that the findings of the main commission, and all its recommendations for remedial action, were left to gather dust.

In spite of this deplorable state of affairs, the Admiralty did not cease to function.

PROBLEMS AT THE BEGINNING OF THE STUART PERIOD, 1604

The Chatham Chest, inaugurated near the end of the reign of Elizabeth I, remained the main source of charitable provision for wounded and injured seamen for the next two centuries.

One new ship of modest dimensions, the *Phoenix*, was laid down in 1612, and another in 1615, while four Elizabethan capital ships were rebuilt during the same years. The quality of all this work was heavily criticized by interested parties, but it does not follow that all the criticism was justified. More seriously, such modest naval activity as took place required a mixture of cajolery and force to man the ships. Frauds over victualling and pay were so entrenched that seamen vowed they would face hanging rather than serve in the king's ships. However, rescue of a sort was at hand. By 1618 the king was at last convinced that rigorous economy was necessary, and Lionel Cranfield's efforts in the household had shown how it could be achieved. A fresh commission of inquiry was announced, and the fast-rising favourite, George Villiers, earl of Buckingham, declared an interest. Buckingham, perhaps genuinely alarmed by the navy's miserable failure to respond to the king's call for a campaign against pirates in 1617, saw himself as the saviour of the navy. The commission's task was eased by the fact that the regrettable Mansell was now Lieutenant rather than Treasurer, but its success was really guaranteed by Buckingham's support. A team of 12 experienced royal servants was named, headed by Cranfield, and between June and September 1618 the Admiralty was meticulously taken apart, and all its shortcomings exposed. In contrast to what had occurred ten years earlier, when the report was considered in November, the Privy Council acted decisively. The existing officers were summoned, and asked whether they were prepared to implement the commission's recommendations. As these recommendations amounted to a total indictment of their stewardship, it is not surprising that they refused. They were then suspended from office and the Admiralty was put into commission. The investigators themselves became the Admiralty commissioners, with responsibility for the implementation of their own proposals, under the guidance of the Lord Admiral. The aged earl of Nottingham was exonerated of any responsibility for what had occurred, but on 28 January 1619 the earl of Buckingham became Lord Admiral.

Over the next five years this arrangement functioned with admirable efficiency. Ten new ships were built, and the annual charges of the navy, which had risen to over £50,000 a year between 1610 and 1618, were pegged back to £30,000. Officers were given a pay-rise in 1618, and the monthly rate for the seamen was raised from 10s. to 14s. in 1624. In 1618 there had been 23 serviceable ships and 10 unserviceable ones, and in 1624 there were 35 serviceable ships. Several old vessels had been disposed of by then, and the building operations confined to Deptford, where two ships could be worked on simultaneously. The new constructions were, with one exception, all substantial capital ships, although none equalled Phineas Pett's controversial construction, the *Prince Royal*, which was of 1,200 tons and the first three-decker built for the navy.

Operationally these years were rather less impressive. The only major expedition, against the Algerian corsairs in 1620–1, was a miserable and expensive failure, possibly because Sir Robert Mansell was no more honest and efficient as a sea commander than he had been as an administrator. By 1628, when the commission

The *Sovereign of the Seas.*

expired and the navy returned to normal administration, James was dead and England was again at war with Spain. In 1625, perhaps desiring to emulate one of the finest achievements of the previous generation, an expedition had been sent against Cadiz, and in spite of the reforms of the previous five years, the contrast could hardly have been more painful. The commander, Viscount Wimbledon, was a soldier whose fleet orders show signs of tactical skill and imagination, but he was inexperienced, and his fleet was disgracefully under-prepared. The ships were seaworthy, but the officers were inadequate, and the victuals and clothing alike deficient. The more energetic of the pressed seamen deserted before the expedition sailed, and the remainder died like flies of disease and malnutrition. The voyage certainly cost at least £250,000 and possibly twice as much, achieving absolutely nothing except to expose the weakness of Buckingham's well-intentioned regime. Undeterred, the Lord Admiral tried again two years later, sending a relief expedition to La Rochelle. The result was the same, military failure and heavy loss of life resulting from putrid victuals and inadequate clothing. The Lord Admiral's death put an end to further grandiose plans, and the man who had appeared to be the saviour of the navy in 1618 ended by presiding over some of its most conspicuous and avoidable failures. In concentrating on ships, dockyards, and financial controls, he had failed to realize that an operational navy depends upon skilled manpower, which needed to be managed with the same care as that bestowed upon the guns or rigging. It was not so much corruption as sheer incompetence which ruined the navy in the late 1620s.

The following decade was one of peace, which lightened the burden upon the maritime community but did not much improve the Admiralty record for man-management. The shipbuilding programme had faltered after 1627, but was picked up again in 1632, and a further 12 ships were built between then and 1637, including the great *Sovereign of the Seas* in the latter year, at a cost of over £40,000.

The English fleet in front of La Rochelle, 1627.

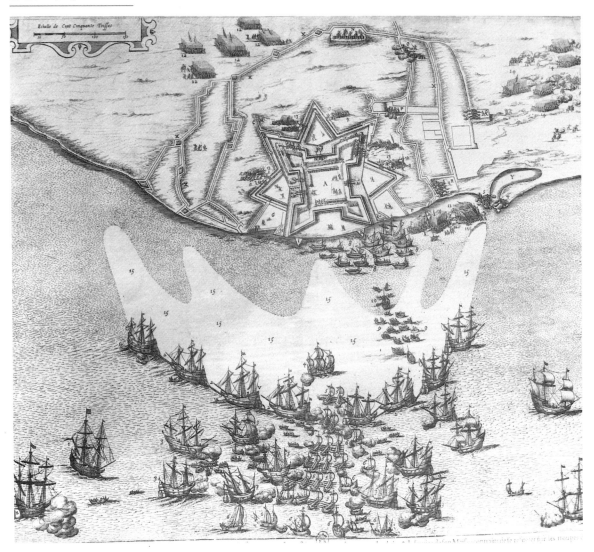

The collection of Ship Money after 1634, and the fact that it was actually spent upon the navy, eased the financial difficulties which had contributed to all the problems since 1618, and helped to ensure the maintenance of the *matériel*, but could not in itself guarantee efficiency. This is a period about which a great deal is known, because of the survival of the records, but little is of great significance. The corruption which had so dogged the Admiralty before 1618 had largely disappeared. The dockyards functioned efficiently and the ships were kept in seaworthy condition. But in spite of further pay rises, naval service continued to be bitterly unpopular, and the 1630s saw no naval operations of significance. Part of the problem was that the professional experience, which had been such a strength of naval administration throughout the Tudor period, had never been recovered after the bleak years which followed the end of the long Spanish war. The whole sixteenth century had been a time of development and achievement, not always even, but very pronounced and taking place in each Tudor reign. The early seventeenth century saw a partial disintegration, followed by a partial recovery, but the early Stuarts never cared for their navy the way the Tudors did. It was not until the 1650s, under the Commonwealth, that significant developments began again, and the navy resumed a career of success. The navy was almost solid for parliament in 1642, which is perhaps the most succinct verdict upon the naval administration of James and Charles I.

3

A PERMANENT NATIONAL MARITIME FIGHTING FORCE 1642–1689

J. D. DAVIES

*T*n August 1642, a navy consisting of 35 vessels declared for parliament at the outbreak of the English civil war. In December 1688, a navy of 151 vessels surrendered to the authority of William of Orange. These raw figures alone hint at the importance of the intervening period for the expansion of the service, but they conceal many other, equally significant, developments. Faced with often entirely new domestic and foreign conflicts, and responding to both changes of regime and to parliamentary and other pressures, the navy witnessed major developments to the nature as well as the number of its vessels; to the tactics which those vessels employed, as well as the sea areas in which they were deployed; to its higher administration; and to the means of recruiting and retaining both officers and men. Central to these changes were a number of individuals whose influence over ways of thinking within and about the service is less easy to measure than their concrete achievements in battle or on paper. Robert Blake, for instance, contributed a legacy of victory unknown since the Armada, but thereby impelled his successors to live up to that legacy at all costs. King Charles II and his brother James made the term 'royal navy' more real than symbolic by taking a deep and genuine interest in all aspects of the service. Peculiarly, however, this is the one age in the history of the navy when the single name which could be regarded as being most synonymous with it is that of a civil servant. Samuel Pepys may have been even more effective at telling posterity how important he was to the navy than he was in reality, but from 1660 onwards much of what went on in the service was touched by the hand of the East Anglian clerk. It was both typical of Pepys, and indicative of this period's significance in transitional and formative terms, that when he reluctantly resigned the secretaryship of the Admiralty in February 1689, he claimed the official letterbooks as his personal property and took them with him, prompting a prolonged and acrimonious correspondence from his successors. Pepys's navy was still a relatively small, personal force in which everyone knew everyone else, rather than a great, impersonal, bureaucratic machine—but only just.

Facing: The battle of Leghorn, 1653. The humiliating destruction of a small English squadron ensured that, in future, either sufficiently large forces were deployed in the Mediterranean or (as happened in the second and third Dutch wars) English merchants there were effectively left to fend for themselves for the duration.

SHIP DEVELOPMENT

In the 1640s parliament had relied on the smaller ships of its navy to bottle up royalist harbours, to supply its own beleaguered outposts (such as Plymouth), and to support occasional amphibious operations, such as the attack on Milford Haven in 1644. The execution of Charles I in January 1649 forced a new and larger set of naval priorities on the government of the fledgling commonwealth. A royalist squadron had been at large since a naval revolt in the previous year, and the state was faced by a daunting array of potential enemies—the United Provinces of the Netherlands, Protestant but no friend to England after decades of commercial and other rivalries, and France and Spain, the great Catholic military powers of Europe, both of which were sympathetic to the exiled house of Stuart. Expansion of the fleet therefore began immediately, with greater priority being given to larger vessels which could contend in battle with the Dutch in particular. By the end of 1651 over 40 new ships had been added, and the first Anglo-Dutch war in 1652–4 and the subsequent war with Spain gave further impetus to naval construction. Moreover, the new tactic of fighting in line, developed in the Dutch war and considered later in this chapter, demanded the employment of larger and stronger ships which could fight broadside-to-broadside, inevitably taking heavy punishment by so doing; the hired merchantman, the mainstay of so many earlier fleets, would not be up to the task. As a result, between 1652 and 1660 one first-rate (the *Naseby*), three seconds, thirteen thirds, and fifteen fourths were added to the fleet, together with many lesser craft.

When the Stuarts were restored in 1660, their personal preferences for the

Ship model of the first-rate *St Michael*, built at Portsmouth in 1669, which served at the battles of Solebay, Schooneveld, and the Texel in the third Dutch war. Built as a second-rate but later re-rated, she carried between 90 and 98 guns.

prestige inherent in 'Great Ships' and the continuing expectation of conflict with the Dutch ensured that naval construction resumed as soon as funds permitted. A substantial programme had already been commenced in 1664, but the second Dutch war which began in that year forced an acceleration, both to expand fleet numbers and to replace war losses. Even in the period of retrenchment after the war ended in 1667, five new first-rates were built before the next war broke out in 1672. The 1670s saw a significant change in the political aspect of naval shipbuilding. Previous warship construction since the Restoration had been undertaken on the king's initiative, and was directed largely against the perceived threat from the Dutch. However, the 1660s had seen a rapid and alarming growth in the French navy, which completed six great first-rates in 1668–9 alone, and by 1675 it was apparent that England was falling numerically behind a force which had barely existed at all 15 years earlier. Therefore, although the proposal to build the 30 new ships necessary to regain parity with France and the Netherlands again originated with Charles II, it was only pushed through a suspicious House of Commons by placing heavy emphasis on the French threat (despite the king's strong francophile tendencies) and by permitting the House to debate the minutiae of size and armament in unprecedented detail. By 1685, one first-rate, nine seconds, and twenty thirds had emerged from the royal dockyards and private yards at Blackwall and Bristol. These, together with the survivors of the earlier building programmes of the 1650s and 1660s, formed the backbone of the fleet which went to war with France in 1689.

Two particularly significant types of ship were developed in this period. As the distribution between the rates of the '30 ships' suggests, the third-rate, generally mounting between 56 and 70 guns, came to be regarded as the most useful of the larger types of ship. Derived from a successful prototype, the *Speaker* (built by Christopher Pett at Woolwich in 1650, and which survived as the *Mary* until 1703), the two-decker third-rate could perform a greater variety of tasks than the firsts and seconds, which could not normally be employed outside the summer months. The thirds were powerful enough to lie in the line of battle, but could also be deployed in winter squadrons (as in the fleet mobilized to defend against William of Orange's invasion in 1688), as flagships of detached forces in the Mediterranean or Caribbean, or individually as escorts to important convoys. However, the *Speaker* had originally been regarded as an enlargement of another new category of warship, the frigate. Drawn from the design of the successful Dunkirk raiders, the frigates of the 1640s and 1650s were low, fast, and well armed; while the largest were again considered suitable to lie in the line, the frigates of the fourth and fifth rates were more usually employed on the protection of trade. Relatively few were built after the Restoration, partly because of the versatility of the new third-rates and partly because of the longevity of the Commonwealth's ships, but the need to increase the number of vessels in these rates was recognized soon after the war with France began in 1689. Fireships were employed *en masse* in the Dutch wars, with such incidents as the destruction of the first-rate *Royal James* by a Dutch fireship in 1672 seeming to prove their utility, and it became more

One of the fourth-rate frigates built in the 1640s and 1650s, perhaps the *Constant Warwick* of 1645, the original prototype vessel modelled on Dunkirk privateers. She survived until 1691 when she was taken by the French.

common to retain a few of these craft on the strength of the fleet in peacetime. Charles II's love of yachting introduced the concept of the royal yacht into the navy; the first, the *Mary* of 1660, was a gift from the Netherlands, and a whole fleet of new yachts, usually named after Charles's family and mistresses, followed in the 1660s and 1670s. There were a few less successful innovations, such as the *Margaret Galley*, built in Italy in 1671 and manned by foreign officers and native American rowers, but her failure led the navy to experiment with a more robust design, the galley frigate, which in turn inspired many of the frigate designs of the eighteenth century. Another oddity was the *Kingfisher*, effectively an early 'Q-ship' which could masquerade as a merchantman and which distinguished herself against the North African corsairs. Finally, the names of ships changed to suit the changing political climate. Republican names, such as *Liberty*, *Naseby*, *Dunbar*, *Marston Moor*, and *President*, came in after Charles I's execution, only to be changed again at the Restoration in favour of such more-or-less blatantly royalist ones as *Royal Charles*, *Royal Oak*, and *Happy Return*.

ADMINISTRATIVE CHANGE

The changing names of ships reflected more fundamental changes in the administration of the navy. Until 1649, the traditional framework of royal naval administration had been in place, albeit with a parliamentarian Lord High Admiral in the shape of Robert Rich, earl of Warwick. Doubts over Warwick's loyalty led parliament to remove him in February 1649, and his powers were divided. His seagoing responsibilities were given to three 'generals-at-sea', the army officers Robert Blake, Richard Deane, and Edward Popham, all of whom had limited seagoing experience but whose main task was to secure the fleet's political loyalty to the new regime. The executive powers of the Lord High Admiral devolved upon a succession of Admiralty committees, dominated initially by the efficient and outspoken Sir Henry Vane (although the composition changed to reflect the many changes of government in the 1650s), and under them a body of navy Commissioners carried out the functions of the old principal officers of the navy. Throughout the 1650s, the main problem facing the naval administration was an often chronic shortage of money. The existence of a greatly enlarged navy was demanded by the parlous international situation, but it proved to be increasingly difficult for the state to finance both this great fleet and the large standing army which it needed to suppress domestic discontent. By the end of the 1650s several ships' companies were owed three to four years' pay, and the victualling system was in similarly dire straits—hardly surprising when the government's income was roughly half its expenditure on its two armed forces. This apparent break-down of naval finance contributed greatly to the fleet's passive acceptance of the restoration of the monarchy in May 1660: fairly or unfairly, the naval administra-tion of the commonwealth was regarded at the time as an unmitigated disaster.

The new Stuart regime attempted to create an administrative framework which combined elements of both traditional and republican practice. The four principal officers of the navy—Treasurer, Controller, Surveyor, and Clerk of the Acts—were revived, but they were to be joined in a navy board by three commis-sioners, echoing commonwealth practice. The king's brother and heir, James, duke of York, was installed as Lord High Admiral. The navy board remained essentially the same throughout the period, except in 1686–8 when a 'special commission' took its place to expedite a major programme of refitting. However, the admiralty went through several metamorphoses. James's tenure lasted until 1673, when the Test Act, parliament's attempt to drive Catholics out of public service, put paid to his office. Charles II then took on the powers of the Admiralty himself, assisted by a commission of leading ministers which met occasionally in an advisory capacity and above all by the ubiquitous Samuel Pepys, who had served as Clerk of the Acts since 1660 and now became Secretary to the Admiralty. Further political dramas in 1679 led Charles to attempt to mollify his parliamen-tary critics by creating a new commission drawn largely from their ranks, and resigning to them all the powers of the Lord High Admiral. Despite serious factional rivalries among its members the commission survived in this form, struggling against yet another bout of severe Treasury-inspired retrenchment, until 1684, when Charles dismissed it. From then until the 'glorious revolution' of 1688 the powers of the admiralty rested with the king, first Charles, then (from February 1685) James, with Pepys installed as a more powerful and exalted Secre-tary of the Admiralty than before; but this experiment with what was effectively a French-style ministry of the marine was swept away by the revolution, when commissions along the lines of those of 1679–84 were revived and became the norm.

The divisions of responsibility between the various officials and departments which ran the navy were often unclear—as indeed they were between them and their subordinates in the dockyards or at sea, and between them and the other bodies (notably the privy council) which had a tangential role in naval adminis-tration. Much depended on individual personalities, and after 1660 the most important of these were always the royal brothers. Despite an erratic workrate, Charles II had a technical interest in such matters as shipbuilding and sailing qualities, a genuine personal concern for the careers of officers and men, and a natural political interest in where his ships were and what they were doing. These interests were shared by the more dour but also more hard-working James, who added to them a genuine bond of loyalty between himself and his subordinates (a consequence of his having commanded in person at the battles of Lowestoft in 1665 and Solebay in 1672). With their cousin Prince Rupert of the Rhine also actively involved, first as a fleet commander in the second and third Dutch wars and then as first commissioner of the admiralty in the 1670s, the royal patronage of the service was considerable and far-reaching in its effects. It raised the status of the navy in the nation, and certainly inspired loyalty among many of those who served in it, but it also had its drawbacks. If the policies of the royal brothers were subjected to critical scrutiny in parliament and the nation as a whole, as they were from the mid-1670s onwards when charges of aiming at an absolute monarchy were aimed at them, then the navy became an obvious target, and many suspicions of the extent of royal control over it were aired in the House of Commons.

Two administrators in particular did much to satisfy the Stuarts' ambitions for their navy. Sir William Coventry only became involved with the service in 1660 through his appointment as secretary to the duke of York, but it quickly became clear to him that most of his work in that office would be concerned with Admiralty affairs. Meticulous and efficient, he came under attack on several occasions in the 1660s but survived because of his talent for assembling a weight of paper evidence to disprove charges of corruption, and he eventually moved to the Treasury in 1667. Coventry co-ordinated much of the difficult task of restructuring the navy in 1660–1, weeding out extreme republicans and replacing them with royalists, and he was also heavily involved in the planning of the second Dutch war in 1664–5, when he served at sea alongside James. In several respects, he served as a role model and mentor for the young man who had become Clerk of the Acts in 1660 on no better grounds than his Huntingdon connections with Edward Montagu, earl of Sandwich, one of the commonwealth's last 'generals-at-sea' and one of the chief architects of the restoration. Samuel Pepys's early attempts to master the intricacies of naval administration were bedevilled by the alternative attractions of the theatre, heavy drinking, and accommodating women. Nevertheless, by the time of the second Dutch war he was already regarded as the main man of business in the navy, a reputation founded, like Coventry's, on his mastery of voluminous amounts of paper—a very small part of which was his famous diary, running from 1 January 1660 to 31 May 1669, which was to become by far the most widely read and best-loved work ever produced by a naval bureaucrat. His elevation to the Admiralty in 1673, and his return there in 1684, gave him the chance to put into effect many of his ideas for the reform of the navy and to attempt (not always successfully) to rebuke recalcitrant officers and subordinates for not living up to his ideal of a dedicated and professional service. Opponents and critics (such as the Admiralty commissioners of 1679–84, who had driven him

Samuel Pepys (1633–1703), Clerk of the Acts to the Navy Board, 1660–73, Secretary to the Admiralty, 1673–9 and 1684–9. Both his diary and his surviving papers at Magdalene College, Cambridge, and the Bodleian Library, Oxford, reveal much about the workings of the navy in this period.

from office) were rubbished by a barrage of carefully manipulated statistics and the inevitable mass of paper evidence, while Pepys carefully concealed any evidence which pointed to any lapses of dedication or professionalism on his part. An insatiable dabbler in science, music, or indeed virtually any intellectual pursuit, Pepys was truly, as his friend John Evelyn said of him at his death, 'a very worthy, industrious & curious person, none in England exceeding him in the knowledge of the navy'.

Pepys and Coventry served what was to a considerable extent a peripatetic naval administration. In effect, the Admiralty was where the duke of York was in the 1660s, where the king was in the 1670s, and where Pepys was in the late 1680s, a fact which caused some problems in 1689 when Pepys refused to move from what had been at one and the same time his private house and the office of the Admiralty. The navy board's office stood in Seething Lane in the City until 1673, when it was burned down, but its replacement was erected on the same site; the victualling office was further east, next to the Tower of London. The nearest dockyards to London—Deptford and Woolwich—were now too small to service the largest vessels, but were employed in shipbuilding and refits of lesser rates. Slightly further away was the largest yard, Chatham, where much of the fleet was laid up 'in ordinary' in peacetime years, and which was the best situated of all the dockyards for the conditions existing in the Dutch wars. The requirements of those wars had also led to the development of two small and temporary dockyards at Harwich and Sheerness. Furthest away of the main yards was Portsmouth, always a centre of shipbuilding and refitting, and of particular value as an operational base when an enemy threatened to come into the Channel, as in 1666 and 1690. Portsmouth's case often proved the point that the deficiencies of naval administration multiplied in proportion to distance from London. Although much of the administrative framework looked impressive by the 1660s, with printed pro formas and detailed instructions galore, the reality in the yards often failed to match these appearances. In particular, the government's frequent shortages of cash could cause chaos, with bills and workmen going unpaid (often until

Chatham dockyard in 1690. The largest industrial concern in Britain, with, in the foreground, several of the ships laid up 'in ordinary', i.e. those out of service and manned by skeleton crews only.

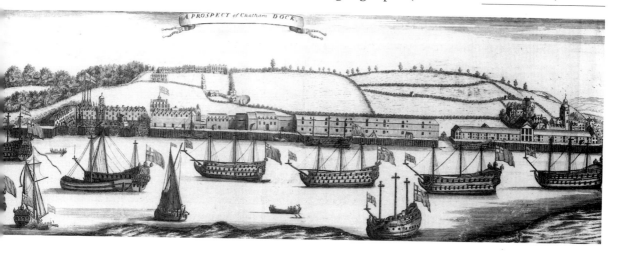

they were several years in arrears) and tradesmen refusing to supply the navy any longer. The more distant yards invariably lost out, a fact which put far-away naval outposts like the small but important servicing facilities at Plymouth and Kinsale, or the victuallers of the Mediterranean fleet, in an especially invidious position. Finally, the location of the yards determined to a considerable extent the mechanics of the most difficult task facing the naval administration of the time: the provision of manpower for the fleet.

MANNING THE FLEET

Apart from a 'hard core' of men who preferred service in the fleet, the late seventeenth-century navy relied on recruiting from the whole 'maritime community' of the nation—from merchant ships, colliers, fishing vessels, or barges. The peacetime service required up to 4,000 men and could usually meet this number from volunteers, but in wartime over 20,000 men were needed and other means, especially impressment, were employed. Technically, most of the navy's manpower was obtained from London, the Thames, and the east coast of England, but this was not the complete picture. Given the geography of the wars with the Dutch, seamen who wanted to serve in the navy would naturally gravitate towards the dockyards in the east; even those who did not had to stay in the same area, as most of the opportunities for merchant service were also concentrated there, and this made them vulnerable to the press. Other large 'pools' of seamen were concentrated in Bristol and Devon, but significant numbers could be obtained from outlying counties, Wales, Ireland, and Scotland.

It is impossible to generalize about seamen's experience of the navy in this period. Of course, many pressed men hated it—attempts to avoid the press ranged from migration inland to overt violence against the press officers, and once in the service desertion was rife. On the other hand, even some pressed men seem to have looked on a spell of naval service as just another brief episode in what for them were always episodic careers; many of those who volunteered had ambitions to gain advancement to petty rank or beyond, to obtain pensions for themselves and their families, or simply to serve a respected local officer. In that sense, too, the navy in this period was still very much a personal institution, with many officers taking large parts of their crews with them when they changed ship. On the other hand, a 'turnover' from one ship to another, an expedient often used by the administration to counteract manning problems, was a prospect to be greeted with alarm, partly because any interest gained with the previous ship's officers was lost, partly because a turnover meant a difficult pursuit of one's back pay. Indeed, pay was the greatest single determinant in naval recruitment. For much of the period, and quite justifiably, the navy had a reputation as a miserly and tardy paymaster, and in wartime merchant shipping (competing for the much smaller pool of available labour) paid greatly inflated wages which provided a substantial incentive to avoid or desert naval service. Expedients such as the use of tickets (effectively IOUs) only increased resentment, and the various measures which were tried to encourage recruitment, such as bounties for volunteers, had only limited success. Attempts were also made to streamline the recruitment system as

Above: The *Sovereign of the Seas* (commissioned 1637), and her controversial designer, Phineas Pett (1570–1647). A 100-gun ship which looked ahead to the navy of Charles II.

Left: James, Duke of York (1633–1701), Lord High Admiral of England, 1660–73, Commander-in-Chief at the battles of Lowestoft (1665) and Solebay (1672); King James II, 1685–8, when he was overthrown in the 'glorious revolution'. Godfrey Kneller's painting includes in the background James's flagship flying the admiralty, union, and red squadron flags.

a whole: schemes which would have involved local government officials drawing up registers of the seamen in their areas sank without trace. Faced with all these problems, successive naval administrations struggled to make up the numbers of seamen they needed to man their wartime fleets.

To a certain degree, the opposite problem affected recruitment of commissioned and warrant officers: there were often too many of them to fill the places available. In the 1650s, the continuous deployment of a large fleet in successive wars against the Dutch and Spanish made the navy an attractive source of employment. The commonwealth drew its commissioned officers largely from professional seamen, notably warrant officers and merchant ship masters, with an occasional leavening of transplanted army officers; a few, such as Robert Blake, had some experience on both sea and land before being catapulted into high command. At the restoration, Charles and James retained many of these 'tarpaulins' in their officer corps, but also sought to enhance that corps' political loyalty by promoting the younger sons of gentry and aristocratic families. These 'gentlemen officers' were often fiercely criticized for their supposed ignorance, arrogance, and reliance on court connections to achieve unjustifiably rapid promotion, and cases like that of Sir William Berkeley, a vice-admiral at 26 after barely five years at sea, seemed to give credence to such accusations. Even so, many of the 'gentlemen' proved to be able and popular captains, and their overall level of proficiency was raised as time went on—partly by various reforms imposed from above and partly by the pressures of competition for places. In peacetime there were far fewer commissioned posts available than there were supplicants for those posts, in itself an indication of the popularity of a service so directly and obviously favoured by royalty. Increasingly, 'gentlemen' officers won these battles for commissions: the charge of depending on great patrons had considerable weight in a service which relied heavily on the workings of patronage. Members of great and well-connected landed families such as the Berkeleys, Herberts, Russells, and Stricklands gained many of the plum commands in the 1670s and 1680s, when even one of the king's own sons (admittedly an illegitimate one), the duke of Grafton, was sent to sea. However, the door was still not quite shut to able 'tarpaulins', especially if they could gain influential patrons themselves or if they gained the king's attention by performing some signal act. Sir John Berry, Sir John Kempthorne, and Sir John Narbrough, among others, all gained high commands despite their humble origins, and they in turn promoted younger men like Narbrough's protégé and kinsman Cloudesley Shovell.

It would be disingenuous to suggest that all these officers competed for places solely to honour themselves and their family name by serving their country, though this was undoubtedly a strong motivating factor for many 'gentlemen' in particular. Officers' pay was comparatively unattractive, but naval service presented many other opportunities for enrichment. Wars were always to be prayed for: they meant not only more places to fill, but prize money as well. In peacetime, dull convoy duty could be made tolerable by a gratuity from the merchant interests under one's care. Above all, there was the 'good voyage'—the

Robert Blake (1598–1657). The greatest admiral of the commonwealth period was born in Bridgwater, engaged in maritime trade in the 1630s, and made his name as a soldier at the siege of Lyme Regis in 1644. At his death, shortly after his great victory at Santa Cruz in 1657, he was given a spectacular state funeral in Westminster Abbey.

Facing above: A royal visit to the fleet, 1672. Charles II paid several visits to his ships during the second and third Dutch wars; on this occasion his royal standard flies in his brother James's flagship, *Royal Prince*. Several of the royal yachts also appear in this painting by van de Velde.
Facing below: The Battle of the Texel, August 1673. Cornelis Tromp's *Gouden Leeuw* attacks Sir Edward Spragge's *Royal Prince* at the height of the battle; Spragge was killed shortly afterwards.

Sir William Berkeley (1639–66), one of the first and most rapidly promoted 'gentlemen captains' of the restoration period. He owed his quick rise largely to the influence of his brother Charles, one of King Charles II's early favourites, but died in the four days' fight of 1666 by making an over-impetuous lone attack on the Dutch.

carriage of valuable cargoes aboard warships, with a suitable commission for the captain concerned. Orders for Cadiz, Genoa, or other ports where bullion cargoes could be expected were highly sought after, and Pepys, for one, believed that the priorities of captains in the Mediterranean had been entirely perverted by the mad scramble for 'good voyages'. Occasional attempts to clamp down on the practice had little success: in the 1670s in particular, demand for the service from English merchants alarmed by the state of war in the Mediterranean was simply too great to be ignored, and in any case Charles and James sympathized with the officers' financial aspirations and rarely issued more than half-hearted admonitions. Taken together with the extensive range of abuses in which any captain might indulge (such as the 'false muster', which entailed entering non-existent seamen on the ship's books and claiming their pay), the 'good voyage' enabled many captains to enrich themselves considerably and to set themselves up as landed gentlemen on respectable estates in the counties around London.

The new situation in which the navy found itself after 1650, and more markedly after 1660—a much larger service, requiring many more officers and men in wartime but with many of its officers unemployed in peacetime—demanded a range of responses to encourage and retain good men. In the 1660s and 1670s a system of half pay was introduced, establishing for the first time the principle that naval officers should be paid when not actually in service at sea, and superannuation arrangements were also put in place. A system of giving unemployed commissioned officers supernumerary places as 'midshipmen extraordinary' was developed during the peacetime years. Above all, more formal arrangements for training young officers were introduced in an attempt to remedy some of the more obvious problems of promoting young 'gentlemen'. At first all were notionally to undergo some training in the rank of midshipman, but in 1677, a formal examination for the post of lieutenant was put in place, a development which has long been regarded as one of Pepys's greatest contributions to the navy. In future, officers aspiring to that rank would have had to have served at least three years at sea and have been at least 20 years old before sitting a gruelling examination before a board of senior officers. Another officer whose method of appointment was reformed in this period was the naval chaplain. Previously in the hands of individual captains, the selection of chaplains was entrusted to the bishop of London in 1677, but the effect of such formal regulation was debatable. From the 1650s to the 1680s, puritan chaplains under the commonwealth and royalist anglicans alike found many officers and men profane, irreligious, and superstitious; but then, officers and men could also recount many examples of drunken and inadequate chaplains, even after the reform of 1677. Indeed, the other great 'naval diary' of the age, that of the chaplain Henry Teonge, reveals an author more concerned with the menu for the evening meal than with the interpretation of the word of God.

Fewer reforms were concerned directly with the seamen. In 1652, the distinction between 'able' and 'ordinary' seamen was introduced in an attempt to encourage recruitment, but the newly raised wages of 24s. and 19s. a month respectively then remained the same for the rest of the period and well beyond. Payment of gratuities to seamen or their families in the event of mutilation or death was also formalized, though the traditional instrument of charitable provision, the Chatham chest, became increasingly hard pressed as a result of the number and frequency of short but bloody wars throughout the period. Encouragement went hand-in-hand with a more formal system of discipline. Following the disastrous performance of the fleet at the beginning of the first Anglo-Dutch war, a set of formal Articles of War was promulgated at the end of 1652, and, in modified and expanded form, these were codified in statute law in 1661. They covered a wide range of offences ranging from sleeping on watch to murder by way of theft, desertion, sodomy, and striking an officer, and specified death as a mandatory or optional punishment in many cases. In reality, the lash was often used as a substitute and by no means all offences ever got as far as a court martial, but for most of the period both officers and men served under an unprecedentedly formal disciplinary framework. Above all, the Articles of War were designed to achieve one end: to ensure that the fleet functioned effectively in time of war.

The causes of the first Anglo-Dutch war, which began in 1652, could be traced back through almost half a century of commercial rivalry. Conflicts had ranged from the North Sea herring fishery to the East Indies, where the execution of several Englishmen on the island of Amboyna in 1623 generated a depth of ill-feeling on the English side which was to be kept alive by regular reprintings over the following 50 years of accounts of this 'massacre'. These commercial tensions were intertwined with more ideological ones. Dutch advocacy of freedom of the seas, regarded as an essential buttress of their domination of the global carrying trade, conflicted directly with England's long-held but tenuous claim to sovereignty over the 'British seas', a somewhat amorphous area supposedly extending as far as Cape Finisterre. John Selden's *Mare Clausum* of 1635 gave weight to this claim, demanding that foreign vessels should strike their flags to British warships in those seas as a mark of respect. Underpinning English concerns was both jealousy of Dutch success and the fear that the Dutch were aiming at a world trading empire which would surely strangle England's commerce. These fears were exacerbated after 1649. An alliance between the two Protestant republics might have seemed logical, especially after the temporary downfall of the pro-Stuart house of Orange in 1650, and English deputations to The Hague even proposed the creation of a confederation. However, supposed slights and insults directed at the English delegates only reinforced an increasingly belligerent attitude at home, where prices were high and maritime trade was in difficulty, but where there was also a navy powerful enough to enforce Selden's claims. In October 1651 parliament passed the Navigation Act. Under its terms, only English vessels or those of the country of origin could import goods into Britain or her colonies; similarly, colonial exports could only be transported in English ships. Within six months, both sides were mobilizing large fleets for sea.

In May 1652 a Dutch fleet under Marten Harpertszoon Tromp, at sea to protect Dutch merchant shipping against English searches, encountered Robert Blake's force off Dover. A dispute over the 'salute to the flag' led to an exchange of fire, and two Dutch ships were lost. Several other clashes, taking the form of English attacks on Dutch convoys, took place over the next three months: these early encounters immediately revealed the immense strategic advantage which England possessed throughout the Dutch wars, namely its position astride the only routes which incoming and outgoing Dutch trade could take. The Dutch dependence on commerce forced them to 'run the gauntlet' and therefore pre-ordained the strategies of both sides, with the English trying to intercept Dutch trade and the Dutch trying to save it by eliminating the main English fleet. The first serious Dutch attempt at an all-out attack ended in defeat at the Kentish Knock in September. Discipline in the Dutch fleet was comparatively poor, partly as a result of rivalries between the different maritime provinces, and they were driven back to port with the loss of three vessels. However, this set-back was

The battle of Scheveningen, July 1653. On the right, the English vessel *Andrew* is in flames after being attacked by a Dutch fireship; just to her left and in the middle distance, the Dutch flagship *Brederode* (M. H. Tromp) engages the English flagship *Resolution* (Monck).

reversed with a vengeance off Dungeness in November, when Blake foolishly engaged Tromp's far larger force and was forced to withdraw to the Thames, leaving Tromp in effective control of the Channel for two months. The defeat at Dungeness led to drastic action, with an increase in seamen's pay, the publication of articles of war, and a reorganization of the fleet into three squadrons: the van, middle, and rear, each with three flag officers—the system which quickly became 'the white', 'the red', and 'the blue'. These reforms bore fruit in 1653, when the better discipline of the English fleet was again revealed in a three-day running fight off Portland. Tromp lost 12 warships, 43 merchantmen, and 1,500 men out of a fleet of 70 warships and 150 merchantmen. In June, the two fleets met again off the Gabbard shoal, and the Dutch were again driven off with heavy losses (20 ships lost as against none for Blake and his fellow 'generals-at-sea', Richard Deane and George Monck). A second decisive victory followed off Scheveningen in July, when Tromp himself was killed along with about 4,000 of his men. The Dutch were only saved from complete economic ruin by the fortuitous, unhindered arrival of several large convoys by the difficult route 'northabout' around Scotland, and by the elevation of Oliver Cromwell to Lord Protector in December 1653. Always a staunch critic of a war against fellow protestants, he rapidly concluded the Treaty of Westminster in April 1654—a treaty which was widely criticized for being too lenient with an enemy who was 'on his knees'—but in reality the Dutch had given way over both the Navigation Act and the salute to the flag.

The reasons for the English success in the war were not attributable simply to geography or better discipline. The ships of the commonwealth's new navy tended to be larger and more heavily armed than their Dutch counterparts, which had to be small enough to transit the shoal waters off Holland and Zeeland: the largest Dutch vessel in the war, Tromp's flagship *Brederode*, carried only 59 guns, a figure easily eclipsed by many vessels in Blake's fleet. To emphasize this advantage, the instructions which the English fleet employed in the battles of 1653 placed the new squadrons in a line formation, in which weight of fire from the broadside would be at a premium. Line tactics were too new to be consistently successful in the Dutch wars; apart from the possibility of any system breaking down in the confusion of battle, many captains naturally resented any restriction of their freedom of manœuvre (and of their freedom to seek out a prize for themselves), and a debate about the desirability of fighting in line continued intermittently into the 1660s and beyond. Apart from the fleet actions, the commonwealth's fast frigates had also proved their worth, taking about 1,500 Dutch prizes. Unfortunately, the scale of the victory and the abrupt ending of the war encouraged a sense of complacency: it had been an easy triumph, and it would be easy to finish the Dutch off 'next time'.

The restoration of the monarchy in 1660 provided the impetus for a revival of Anglo-Dutch warfare. To the continuing and long-standing commercial tensions were added new conflicts over trading outposts in Africa, with both nations seeking to exploit the Guinea coast, and a new version of the Navigation Act was passed in 1660 to counter the commercial dominance which the Dutch had quickly re-established. The new regime needed to prove itself the military equal of the old republic: many royalists, notably the duke of York himself, were eager for a war to make their names, while survivors of the first war (notably Monck, now duke of Albemarle after playing the pivotal role in securing the restoration) expressed a soaring confidence that they could attend to the unfinished business of 1654. Pressure from merchants and from factions within parliament and at court culminated in what can only be called hysterical war fever, which increased steadily throughout 1664 and was exacerbated by a series of colonial incidents—notably the capture of New Amsterdam, soon to be renamed New York, in August. Attacks on Dutch convoys over the winter of 1664–5 began the war in home waters, although an official declaration only came in March 1665. The respective strategic advantages and disadvantages of the English and Dutch, and the responses they made to them, remained essentially the same as in the first war. The English fleet was now larger, and still possessed the advantage in terms of the fighting power of individual ships. The Dutch were still hampered by the inter-provincial jealousy which forced them to have 21 flag officers for the 1665 campaign as against England's nine, although those nine bore witness to the political changes which had altered the complexion of the navy since the end of the first war. Old commonwealth flag officers like John Lawson and William Penn now served alongside old 'cavaliers' like Prince Rupert and Thomas Allin, and young 'gentlemen' like Sir William Berkeley and above all the duke of York himself, Lord High Admiral and admiral of the red aboard the *Royal Charles* (formerly the *Naseby*). Aboard the English fleet was another recent innovation:

the Admiral's Regiment, founded in 1664 to be a more permanent maritime force than the usual body of regular troops posted temporarily on warships, and which eventually developed into the Royal Marines.

The first battle, off Lowestoft in June 1665, seemed to suggest that the pattern of the first war was to be repeated. Although the English line formation quickly broke up, the superior weight of gunnery again proved its worth, with the Dutch flagship *Eendracht* blowing up and another 16 ships lost with her. However, the victory was not followed up. Confusion or conspiracy aboard the *Royal Charles* led to an order being given to shorten sail, allowing the Dutch fleet to get away in the night, and over the following weeks the opportunity was entirely squandered.

The First War, 1652–4

1. Dover, 29 May 1652
2. Plymouth, 26 August 1652
3. Kentish Knock, 8 October 1652
4. Dungeness, 10 December 1652
5. Portland, 28 February–2 March 1653
6. Gabbard, 12–13 June 1653
7. Scheveningen, 8–10 August 1653

The Second War, 1664–7

1. Lowestoft, 13 June 1665
2. Bergen, 2 August 1665
3. Four Days Fight, 11–14 June 1666
4. St James's Day Battle, 4–5 August 1666
5. 'Holmes's Bonfire', 9 August 1666
6. The Medway, 19–23 June 1667

The Third War, 1672–4

1. Solebay, 28 May 1672
2. Schooneveld, 28 May & 14 June 1673
3. Texel, 11 August 1673

Battles of the Anglo-Dutch Wars.

The earl of Sandwich, who had replaced James in the chief command, attempted to intercept the returning Dutch East Indies fleet as it sheltered in the neutral harbour of Bergen before making its dash for home, but the attack was a fiasco and contributed to Denmark's entry into the war on the Dutch side in 1666. No further major fleet operations took place in 1665; the English had to contend with the twin problems of the great plague and vicious recriminations among the flag officers over the perceived failure of the year's campaign. In an attempt to achieve greater success in 1666 Charles II effectively revived the concept of 'generals-at-sea' by appointing Rupert and Albemarle as joint admirals, but their first action together came perilously close to a disastrous outcome. France had declared war on England at the beginning of the year in fulfilment of treaty obligations, and (unfounded) rumours that a French fleet was approaching the Channel from the west led the joint admirals to divide their fleet. Rupert sailed west with 20 vessels, leaving Albemarle with only 56. The gamble that the Dutch were not ready for sea backfired; on 1 June Michel de Ruyter with 93 ships bore down on Albemarle's depleted force off the North Foreland. The English ships acquitted themselves well despite the odds, holding their own while retreating slowly towards the Thames. Rupert, who had been hastily recalled from Portsmouth, joined on the third day, and a final fourth day of fighting left both sides withdrawing in exhaustion. The English had come off worse, losing ten ships (including the great first-rate *Royal Prince*) to the Dutch four. A victory in an engagement on St James's day and a successful raid on the Vlie and Terschelling, in which 150 Dutch merchantmen were destroyed, could not redeem another unsatisfactory and indecisive campaign.

By the end of 1666 war weariness had set in—a feeling strongly enhanced by the disastrous fire of London in September—and the state's ever-parlous finances were increasingly stretched to meet the cost of the war. Peace negotiations started in March 1667, and the government believed that these and a favourable international situation would allow them to adopt a defensive (and cheaper) strategy until the war ended. The main fleet was laid up in the Medway, but a belated programme of strengthening that anchorage's defences got nowhere. The Dutch, meanwhile, were already planning one final blow to avenge the raid on the Vlie and to make peace from a position of strength. In June, de Ruyter attacked the ships in the Medway, exploiting the weakness of the defences and the divided, indecisive English command, and towing away the *Royal Charles* after burning three other first- and second-rates. As a result, the second Dutch war ended with the Dutch in undisputed command of the seas. The Treaty of Breda of July 1667 made several concessions to the Dutch over the navigation acts, but the salute to the flag remained and England retained New York.

In the four years after the treaty, international alignments were drawn and redrawn. The triple alliance of 1668 actually bound England and the Netherlands (with Sweden) together to counter the alarmingly successful advance of Louis XIV's France through the Spanish Netherlands. However, Charles II was soon negotiating with Louis for an Anglo-French alliance to crush the Dutch once and

The *Royal Charles*, built as the *Naseby* in 1655 and the duke of York's flagship in the 1665 campaign, being towed away by the Dutch after de Ruyter's audacious attack on the Medway in June 1667. Her sternpiece still hangs in the Rijksmuseum, Amsterdam.

for all, and the outcome was the Treaty of Dover (1670). Apart from going to war, Charles also promised to restore his kingdom to the church of Rome. Historians have long debated the motivation which underlay the treaty. There were no obvious and immediate colonial or commercial problems dividing England and the Netherlands, unlike in 1652 or 1664. An incident over the 'salute to the flag' in 1671 was blatantly manufactured (the tiny royal yacht *Merlin* deliberately sailed through a Dutch fleet). Ultimately, probably the most plausible explanation of Charles's anti-Dutch policy is a desire for revenge for the humiliations of the previous war. The republican regime of the Netherlands was to be destroyed and replaced by a diminished rump of a puppet state under his suitably grateful nephew, William of Orange. Whatever Charles's motives, their consequences—especially an alliance with the hated and feared Catholic might of France—would be very difficult to 'sell' to parliament and the nation as a whole. Charles and his ministers had to have a quick victory; but with a huge French army, the strongest military force in Europe, poised to cross the Rhine, and a combined Anglo-French fleet with many powerful new vessels ready to smash a smaller, ageing Dutch navy, the prospects for such a victory looked more than promising.

73

In fact, no aspect of the third Dutch war went according to plan. The French attack on land conquered most of the country but failed to take the vital provinces of Holland and Zeeland, which cut their dykes to save themselves. At sea, de Ruyter launched a pre-emptive attack on the combined fleet in Sole Bay on 28 May 1672. The allied manœuvring was hampered by a lack of communication between the English and French, and the Dutch exploited the situation skilfully. Sandwich's *Royal James* was burned to the waterline and the admiral himself drowned; James had to shift his flag twice when successive flagships were badly damaged. By the end of the day, when the two exhausted fleets broke off the action, the losses were roughly equal. Thereafter, no further major operations took place at sea in 1672, and William, now in command of Dutch land forces, continued to fight a successful rearguard action against the French. Prince Rupert took over command of the combined fleet in 1673, but his dislike of his French allies bedevilled the year's operations. A new strategic objective, the landing of an invading English army, still entailed the same prerequisite, namely the elimination of the Dutch fleet, but two indecisive battles off the Schooneveld in May and June failed completely to achieve this result. Once again, superiority in numbers and weight of broadside had been negated by tactical confusion and poor communication between flag officers. The final battle of the Dutch wars took place off the Texel on 11 August 1673. For whatever reason (sceptical contemporaries favoured conspiracy theories) the French van squadron never engaged properly, and the hardest fighting took place between the two rear squadrons, under the respective commands of Sir Edward Spragge and Cornelis Tromp. Spragge was killed while transferring his flag from the damaged *Royal Prince*, but the battle was broken off with no other notable loss on either side; on the other hand, the combined fleet had to withdraw to lick its wounds and to avoid bad weather, leaving the invasion project in tatters and a clear way home for the Dutch convoys. The French inactivity at the Texel brought to a head a violent popular reaction against what had always been an unpopular war. With parliament hostile, Charles II had little alternative to unilaterally withdrawing from the war by the Treaty of Westminster of February 1674.

In terms of all the main reasons why the English navy had gone to war, the third Dutch war was just as unsuccessful as the previous two. The Dutch state had survived, albeit seriously weakened by the ravages of the French invasion, and it retained a powerful grip on the carrying trade—although, ironically, England's abrupt withdrawal from the war benefited much of its merchant service more than had the war itself; with England now neutral and France and the Netherlands continuing their war until 1679, English merchantmen snapped up many Dutch markets and routes. These advantages were not readily perceived in 1674, however. Instead, Englishmen saw that few of the colonial disputes had been resolved to England's advantage. Even in the first war, the most successful of the three, the 'holy grail' of a self-financing naval war had proved elusive. Quite why this should have been so baffled contemporaries, given the English superiority in numbers of ships and guns, and their undoubtedly superior administrative system. Tactical uncertainty and downright incompetence played their parts, but so did sheer bad luck and the North Sea weather, while the near-bankruptcy of the English state ensured that it could never fight more than one or two naval campaigns comfortably. It is certainly true that the Dutch navy improved after the first war, building larger ships and improving its gunnery. Above all, skilled Dutch commanders made the most of inferior resources and usually displayed greater determination; but they had good reason to. In the final analysis, the Dutch simply had more to fight for—nothing short of the political and economic survival of their nation.

The Dutch wars were decided in the North Sea, but one of the characteristics of the period 1650–89 was the increase in the number of permanent or semi-permanent naval 'presences' outside home waters. In its early days the commonwealth was forced to deploy squadrons to Lisbon to hunt down Prince Rupert's royalist squadron and to the West Indies to deal with royalist islands there. The war with Spain on which Cromwell's government embarked in 1654 saw several major operations in distant waters. The 'western design', a scheme to capture Hispaniola by an elaborate amphibious operation, harked back to Elizabethan priorities; the island would provide the springboard for trade with the Americas and for further conquests of Spanish territory. In fact, the operation was a shambles. The army embarked in William Penn's fleet was racked by dysentery well before it attacked, and the 'western design's' only success, the capture of Jamaica, seemed at the time like a very poor consolation. Blake had greater success nearer to home. A sustained blockade of the Spanish coast failed to secure its main objective, the capture of the plate fleet bringing silver bullion from America (although two vessels from one of these fleets were taken in September 1656), but it did provide an impressive example of blockade tactics for future generations. In April 1657 Blake attacked the plate fleet at Santa Cruz on Tenerife. In a brilliant series of manœuvres in the teeth of both shipboard guns and land batteries, Blake destroyed all the vessels at anchor in the confined harbour and came off without significant loss—or significant material gain either, for the bullion had been hidden well inland.

OTHER NAVAL OPERATIONS, 1650–1689

The Mediterranean was a natural concern for all seventeenth-century governments. Many English mercantile interests, such as the great Levant company and the Newfoundland fishery, depended on the area, but they also faced a number of real or potential threats: the Dutch, French, or Spanish, depending on the international situation, and at all times the Barbary corsairs of Algiers, Tripoli, and Tunis, which regularly attacked western shipping. Following an inglorious English performance in the Mediterranean in the first Dutch war, when a squadron was trounced by the Dutch at Leghorn, Blake made a spectacular cruise through the sea in 1654–5, launching a successful attack on the heavily fortified harbour of Tunis. Thereafter, fleets of varying size were maintained against the corsairs on a virtually permanent basis by both the commonwealth and restoration governments. The greatest problem hampering these fleets was the absence of a natural operating base. Depending on which particular Barbary state was posing the greatest threat, suitably located refitting and victualling facilities could be leased from accommodating governments, such as those of Portugal, Tuscany, or the Knights of St John at Malta. An entirely new set of opportunities appeared to open up when England acquired Tangier in 1661 as part of the dowry brought by the Portuguese princess Catherine of Braganza to her marriage with Charles II. A substantial garrison was sent out and a fledgling merchant community was established in the hope of turning the town into an entrepôt for Mediterranean trade. Above all, a massive breakwater, the 'mole', was built at great expense to create a suitable harbour for the fleet. These grand plans came to little. Admirals found the mole unsatisfactory, and as a victualling base Tangier had the obvious disadvantage of possessing no hinterland (it spent long periods under siege by the Moors). Indeed, in the early 1680s the then commander of the Mediterranean fleet, Arthur Herbert, delivered the most damning verdict on Tangier by refitting his ships at Gibraltar, just across the straits. By so doing he revived an English interest in Gibraltar which had been suggested originally by Cromwell in the 1650s, but he also helped to seal the fate of Tangier, which was evacuated and demolished in 1683 in a Treasury-inspired economy measure.

The depredations of the Barbary corsairs in the Mediterranean, and of Dunkirk and Ostend privateers or Dutch warships in home waters, led to pressure from merchant interests for better naval protection, and this resulted in an increasingly elaborate system of convoy provision. By the 1660s and 1670s a sophisticated timetable had been established, with convoys allocated, for example, to the merchant fleets from Virginia or Barbados, to the Newfoundland and Yarmouth fishing fleets, and to merchantmen going to Mediterranean ports such as Smyrna, Genoa, and Zante. A 'western squadron' to escort merchant ships into the mouth of the Channel also became a permanent part of the peacetime naval establish-

England in the Mediterranean, 1649–89.

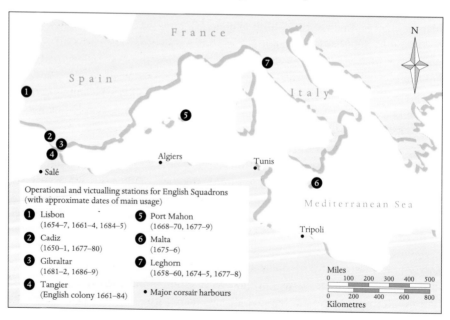

Tangier under English rule. In addition to the defences of the town, the 'mole' can be seen on the right. Several fourth-rates of the Mediterranean fleet lie at anchor on the left.

ment. In addition, English warships ventured further afield. There was a permanent presence in the Caribbean, though this rarely numbered more than two or
three ships, and other small vessels were based in New England. Ships occasionally operated on the African coast, and a squadron was sent to India in 1661 to
annex Bombay. The Dutch wars saw some bloody but small-scale actions on these
distant seas: St Helena, for example, was captured by the Dutch in January 1673
and retaken by an English squadron four months later. In the Indian ocean, the
conflict was fought out almost exclusively between the ships of the two nations'
East Indies companies. Occasionally, plans were mooted at Whitehall for sending
large squadrons to the West or East Indies, but lack of money and the uncertainty
over whether the situation might have changed completely before the ships got
there invariably led to the abandonment of such projects.

By 1688, King James II was in serious political difficulty. His attempts to improve
the lot of his Catholic co-religionists had only served to create a coalition hostile
to him in the 'political nation', and the birth of a male heir in June effectively eliminated the prospect of a Protestant succession in the form of James's daughter
Mary and her husband, William of Orange. By October, William had a fleet and
army ready to invade England, though the apparent reality of a Dutch invasion
(potentially reviving the old hatreds of the three wars) was concealed behind an
'invitation' from several leading English aristocrats, declarations of William's
interest in preserving protestantism, and cosmetic gestures such as giving the
exiled Arthur Herbert command of the invasion fleet. To counter the threat,
James and Pepys had assembled a powerful force at the Gunfleet under the
command of George Legge, Lord Dartmouth, who had been a young 'gentleman
captain' in the second and third Dutch wars. Dartmouth's strategy was hampered
by uncertainty over Dutch intentions (it was believed more likely that William
would sail to Yorkshire or the Thames rather than his eventual destination,
Devon), and by discontent among his own officers and men, many of whom
covertly supported and were in touch with William and their old admiral and
patron, Herbert. Above all, when William finally came out on an east wind the
same wind trapped Dartmouth behind the Gunfleet. The Dutch fleet sailed down

THE END OF THE
STUART MONARCHY

George Legge, Lord
Dartmouth (1648–91),
Commander-in-Chief of the
fleet sent to demolish Tangier
in 1683 and of that which failed
to intercept William's invasion
in 1688. Later arrested on
suspicion of Jacobitism,
Dartmouth died in the Tower.

the Channel with the English in belated and hopeless pursuit, and William made a successful landing in Torbay on 5 November.

Regardless of whether winds or conspiracies had been the major influences on the English fleet's ineffective movements, the outcome of the brief naval campaign of 1688 had far-reaching consequences. James was driven into exile, and Pepys had been too closely associated with him to survive under the new regime. William's accession as King William III in February 1689 effectively tied England to his anti-French policy, and therefore drew the navy into a French war which was to be far longer than any conflict it had experienced since the 1590s. However, for the first years of that war William's navy was inevitably that of Charles and James: most of the officers had risen under their patronage, and many of the ships were products of their building programmes. Indeed, there were still some relics of even earlier navies. Sir Richard Haddock, joint admiral of the fleet in 1690 in yet another revival of the concept of 'generals-at-sea', had been a captain as early as 1657, and was one of several veterans of Cromwell's navy who commanded ships for William of Orange. Several products of the first building programmes of the commonwealth were sent to sea against the French after 1689, and so was Charles I's *Sovereign of the Seas*, which had seen comparatively little service in the Dutch wars. Above all, many of the innovations and reforms of the 1650s, 1660s, and 1670s were integral to the navy after 1689. 'Gentlemen' and 'tarpaulins', both qualified by the lieutenants' examination, commanded ordinary and able seamen under the auspices of the Articles of War on board ships which fought line-ahead, in three squadrons of three divisions each. Similarly, many of the problems which had perplexed officers and administrators throughout the period, such as ineffective recruitment methods and bad pay, remained to be tackled by their successors in the 1690s and beyond. To these naval men in William's reign and later, the eras of Blake and Pepys came to seem like golden ages, when a victorious and efficient service had been created. Nostalgia certainly distorted this image, but it had a grain of truth in it: between 1642 and 1689 a small, impermanent force, heavily dependent on the ships of the merchant service, had been transformed into a larger, purpose-built, national navy.

4

THE STRUGGLE WITH FRANCE 1690–1815

JOHN B. HATTENDORF

*T*he employment and operations of the navy at sea cover a range from broad national policy, the general conduct of war, and strategy to matters of command, leadership, tactics, gunnery, navigation, and seamanship. These topics define, largely, the range of professional naval interest in the subject and, traditionally, they have constituted the main focus of naval history. Today, however, historians recognize them as only one definable segment in understanding naval power, which is clearly dependent on wider, shore-based support, including administration, finance, science, technology, supply and logistics, ship-design, construction, and social and political matters. Although these shore-based factors, the province of Chapter 5 of this book, are key elements that define and limit naval capabilities, the entire purpose of the navy rests on its functioning at sea, carrying out the purposes intended for it.

In traditional naval history, the period between 1689 and 1815 is the classic age of fighting sail, when the Royal Navy was the predominant naval force. In examining it, nineteenth-century historians focused mainly on the great battles between opposing fleets. Today, one looks at operations in a broader light, seeing a wider range of activities and results.

Combat was the ultimate test of a warship, and indeed combat was the first consideration in their design, but sea-keeping capability also mattered, and numerous other naval roles depended upon it. Such roles prompted the Royal Navy to develop a wide range of ship-types. Ships-of-the-line were the largest, most imposing, and most intricate to employ in combination, yet other ships, smaller and more humble, were essential to the exercise of naval power.

During this 126-year span, Britain's main rival was France, but this was neither an exclusive nor a continuous rivalry. While it is true that Britain and France fought one another in seven major wars during this period, there were periods of peace and even collaboration. Naval operations in these intervening times were different from the wartime operations and took on other characteristics. Most importantly, when France and Britain fought, they did so in the context of European alliances, rival imperial competition, and competing economic structures, all of which changed from time to time and were within the context of a slowly evolving regime of international, legal restraint on warfare. These factors formed the fundamental reasons for, placement of, and restraints on naval operations.

In peacetime periods, the navy played a role in diplomacy, deterrence, and peace-keeping operations as well as providing an imperial government with protection from piracy, combating illicit trade, preventing insurrection and domestic disturbances, maintaining communication and transport, and supporting the defence and acquisition of territory.

In wartime, the British navy's roles expanded to include supporting allies, attacking enemy trade, protecting Britain and her colonial outposts from invasion, carrying out combined operations with military forces, and convoying transports and merchant vessels in addition to facing the enemy battle fleet.

The repeated wars against France between 1689 and 1815 provide a recurring theme, across which one can trace a continuity in development and a wide range in naval operations.

HMS *Royal William*: A Royal Visit to the Fleet, 1705. Built in 1670 as *The Prince*, the 100-gun ship was rebuilt and renamed *Royal William* in 1692, serving as Cloudesley Shovell's flagship at Barfleur.

THE GLORIOUS
REVOLUTION

The accession of William III and Mary II marks the opening of this long period in naval history. As the leader of the European alliance opposing Louis XIV, William's election to the English throne, while remaining *stadhouder* in the Dutch Republic, created a revolution in English foreign policy as well as in constitutional issues. While the navy had been strengthened and reorganized under Charles II and James II, the change in sovereigns brought with it two key developments. First, it completely changed its strategic focus, turning it from a force in alliance with the highly efficient French navy to one opposing it. Secondly, it brought it into an alliance and direct co-operation with the Dutch navy.

The revolution began with a huge naval operation. On 1 November (New Style) 1688, 50 Dutch ships-of-the-line, 25 frigates, 25 fire-ships and 400 transports, carrying some 4,000 horse and 11,000 infantry, set sail from the mouth of the Maas under the overall command of an English admiral, Arthur Herbert. Originally intended to sail to the north and land at the Humber, wind and weather forced the expedition to sail down the Channel. In heavy fog and an east-south-east wind,

A royal yacht with a fleet getting under way in a breeze. Under naval protection, William III, George I, and George II each crossed the Channel several times to visit the Continent.

they sailed past the opposing naval forces under Lord Dartmouth riding in the Thames estuary, and some days later, on 16 November, Herbert landed William and his forces at Torbay.

This largely unopposed landing was a remarkable naval achievement. Despite several attempts to get down-Channel against the wind and tides, Dartmouth found himself unable to carry out his orders to oppose Herbert and the Dutch. Ten days later, Louis XIV declared war on the Dutch. Although not specifically mentioning William's movement to England, it was surely one of the series of 'extraordinary armaments' that the French saw as violating the 1678 Treaty of Nijmegen. Faced with a French threat to England, and confronted by James II's desire to use Dartmouth's fleet for spiriting the baby Prince of Wales into French hands, Dartmouth responded to a request from William and surrendered his naval forces to him on 12 December. With William's entry into London a few days later, the Admiralty and other offices of the navy came under his control. The Dutch fleet and the transports that brought William dispersed and returned to the Netherlands, while for a time William had two fleets operating separately, before Dartmouth relinquished command.

In March 1689 Dutch emissaries met in London to arrange the means of Anglo-Dutch naval co-operation. Reviving the arrangements of the 1678 treaty of defensive alliance and a series of subsequent modifications to it, they came to a new agreement by which England would provide five-eighths of a fleet and the Dutch three-eighths, all under the overall command of an English flag officer. In this way they planned to have an effective battlefleet which would equal the French line of battle. They also provided for support ships and a squadron of frigates for special duty. In addition to the overall numbers, the Dutch designated specific strategic stations, designating 50 ships for the Channel and Irish Sea, 30 for the Mediterranean, and a dozen for the North Sea narrows between Dover and Walcheren, suggesting, in addition, an expedition to America.

Meanwhile, in The Hague on 9 March, the States-General responded to the declaration of war which Louis XIV had made the previous November with one of their own. Parliament was reticent in replying to William's requests to assist the Republic, and, in the interval, James II sailed from Brest with a squadron of 30 French warships. Upon receipt of the news, William ordered Admiral Herbert to sail immediately and intercept the French. Arriving on the morning of 1 May in Bantry Bay, Herbert met a second French squadron, under the comte de Château-renault, and tried to engage them as they were landing troops. Approaching from leeward in restricted waters, Herbert forced an engagement, but was unable to find an advantage. After some hours of exchanging gunfire, and in danger of decisive defeat, Herbert withdrew to England, without being able to affect the French landing and without being seriously damaged. Both sides claimed victory and William rewarded Herbert by creating him viscount Torrington.

William III reacted immediately with a declaration of war against France on 5 May 1689, and finally agreed to the proposals for Anglo-Dutch naval co-operation that had been under discussion for nearly half a year, laying down the basis for naval operations in the next twenty-five years. Poorly prepared for a naval war in the western approaches to the Channel or in the Mediterranean, the Dutch navy was an essential complement to England's. The close co-operation was a part of the larger transformation, in both English foreign policy and English attitudes toward Continental politics, that came with the Glorious Revolution.

THE NINE YEARS WAR
(1689–1697)

William III moved quickly in ordering the fleet to sea, but naval co-operation in the alliance of the two maritime powers faced serious practical problems in operational as well as financial matters, and questions of priorities, seniority, and procedure, as well as language, in councils of war. As William III took diplomatic initiatives to form the Grand Alliance against France, he did so with a broad concept of naval strategy already in place. The Anglo-Dutch agreement had already laid out the placement of squadrons in the Mediterranean, Irish Sea, North Sea, and Channel, with the option of an expedition to America. At the same time, England had to concentrate on defensive tasks, transporting troops to recover Ireland. Delays and problems with support left the navy unable to respond quickly.

It was not until the following year that the two enemy fleets encountered one another again. At the moment when both the French and English fleets appeared in the Channel, William was in Ireland leading the troops at the Battle of the Boyne. Torrington, having at sea only half the number of French ships opposed to him, preferred not to seek a battle. In London, however, Queen Mary and the Council, in the light of political opinion, overruled his professional judgement and ordered him to stand and fight. It is the classic case of statesmen interfering in the professional judgement of a local commander on the scene. In the battle that followed on 30 June 1690 off Beachy Head, the French, under the comte de Tourville, defeated the Anglo-Dutch fleet. During the battle, the Dutch took the brunt of the French attack, while Torrington held back the English ships to preserve a strategic reserve against invasion. 'Whilst we had a fleet in being', Torrington argued, 'they would not make the attempt' to invade Britain.

The French named their victory the battle of Bévéziers. Called Beachy Head in English history, the defeat created panic at home and stress in relations with the Dutch, but William's victory at the Boyne offset some of the political and diplomatic problems which might otherwise have occurred. In the period that followed, the two Maritime Powers continued to work diplomatically in consolidating the Grand Alliance against France while expanding their naval forces. The French did not follow up their victory either by an immediate attempt at invasion or by maintaining a superiority at sea. Meanwhile, English naval forces remained on guard against invasion and concentrated on protecting their trade.

The emphasis on fleet operations and combat between opposing battlefleets brought with it a new formulation in naval tactics. In the summer of 1691, Admiral Russell issued his *Sailing and Fighting Instructions*. Based on the instructions the duke of York had issued in 1673, the new orders were significantly different. Incorporating Dutch as well as English experience, they formed the doctrinal basis of

British naval tactics up to 1783. Reflecting Torrington's views, they may well have expressed the consensus of the time, since they had been circulated anonymously in manuscript form before being printed. At the time they were issued, the new *Instructions* were thoughtful and efficient ways of promoting ideas such as massing fire power and maintaining effective control of the fleet. Responsibility for tactics always rested with the admiral in command. However, as tactical doctrine, the *Instructions* stressed the line of battle, laying down the basic principle that British ships should strive to approach an enemy from the windward position, engaging the full battle line, the enemy's van with the British van.

During the winter of 1691–2 the French laid plans to invade England, gathering ships, men, and supplies in northern French ports. Obtaining intelligence of their imminent departure, the allied fleet under Admiral Edward Russell intercepted the invasion fleet on 29 May 1692 off Barfleur on the Cotentin peninsula and engaged it for twelve hours. Russell pursued the French in light winds for five days. The main body of the French fleet sought shelter in the bay of La Hougue, on the east side of the Cotentin peninsula. Entering that bay on 3 June, Vice-Admiral Sir George Rooke attacked and burned 12 French ships at anchor, while James II watched from the cliffs above.

The battles of Barfleur–La Hougue marked a turning-point in the naval war. They stopped the French invasion and redeemed English pride after the defeat at Beachy Head. It was England's most spectacular sea victory for more than half a century. Following it, France altered her naval strategy from a *guerre d'escadre*, emphasizing fleet actions, to the *guerre de course*, concentrating on the attack on trade. In this, the French fleet remained a substantial threat, although it never sought a fleet action during the remainder of the war.

The English navy responded in a variety of ways to the changed emphasis in French strategy. The presence of the French fleet prevented England from moving with unrestrained freedom, and she had to maintain a force ready and capable of meeting the French fleet should it seek battle. The logical follow-on to the battle was an attempt to invade France, but when an attempt to organize an expedition with 14,000 troops on board to do this near Saint-Malo during the summer of 1692, it ended in fiasco. Temporarily eliminating the possibility of further combined operations, the English navy began to concentrate on other possibilities. Beyond the urgent need to protect trade from the depredations of both French privateers and warships, one of these was the early plan, mentioned even before the war, of sending an Anglo-Dutch squadron to the Mediterranean. Another was to focus on the French fleet base at Brest. A third was to bombard and burn various cities and towns along the French coast.

A new phase of the war began in 1693 when, after many delays, a naval escort set sail from England with a large fleet of merchant ships bound for Smyrna on the coast of Turkey. Thinking their charges beyond danger when they had reached the coast of Spain, the English left the merchants to sail independently. Unknown to them, the French Brest fleet had sailed south and was lying in wait off Cadiz. With no protection, the French were able to destroy or capture 92 vessels with

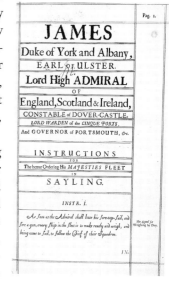

Title page of the Duke of York's *Sailing and Fighting Instructions*, 1673. The first great tactical book of the Royal Navy, these consisted of 26 sailing and 24 fighting instructions, collating various signals and instructions then in use. This set gave tactical guidance to the fleet for less than twenty years, but it established the basic form that remained in effect for a century. In particular, it stressed the importance of maintaining the line in battle, something which was much needed at the time, but gradually became a restraint on tactics.

cargoes valued at over £1 million while the remainder lay immobilized, unable to move out of Spanish ports while French warships cruised outside.

In retribution, the allies launched an attack on Brest with 7,000 men in the spring of 1694. Like the earlier attempt at Saint-Malo, this too ended in failure, marking the end of a series of plans that had persisted through this phase of the war. At the same time, the Smyrna convoy disaster provided the impetus for send-

The Battle of La Hougue, 23 May 1692, by Adrien van Diest (1677–1704). Following the Battle of Barfleur, 12 French warships took refuge in La Hougue Bay further down the east side of the Cotentin peninsula. On 2–3 June (New Style) 1692, Vice-Admiral Sir George Rooke sent in small boats and burned the ships lying close inshore under the forts at St Vaast and Île Tatihou.

ing a fleet to the Mediterranean. In the summer of 1694, using Cadiz as a staging-point, Admiral Russell arrived in the Mediterranean with a joint Anglo-Dutch force and remained until the autumn of 1695. While there, the balance of the war shifted to southern Europe as Savoy became a critical element in the Grand Alliance, and William was momentarily able to put into effect a grand strategy of surrounding France on all sides. Russell's activities laid the basis for a continuing British fleet presence in the Mediterranean, as he conducted operations in support of Anglo-Dutch diplomacy in the region, facilitated allied military efforts ashore, and protected trade.

At the same time, there was another development with long-term consequences for future naval wars. While the fleet was bombarding the French coast, Admiral Lord Berkeley became concerned that his operations might be interrupted in 1695, if the French came out of Brest. He suggested that a squadron cruise near Ushant to give warning should they appear. The following year, Admiral George Rooke suggested a more systematic approach, putting the fleet into the sheltered anchorages of Bertheaume and Camaret Bays to block up the French in their main port. During the wars against France that followed, this thinking formed the basis for operations off Brest and along the channel leading into Brest from Ushant (in French, the Île d'Ouessant).

Meanwhile, the allies' situation on the southern front deteriorated. The allied fleet failed to find a safe port for the winter in the Mediterranean, and was forced to return home. Its departure in 1695 left Savoy exposed to French power. Unwilling to accept the risks this involved, the duke of Savoy left the Grand Alliance and made a separate peace with France in May 1696. In October, the Emperor conceded defeat in Italy and began withdrawing his troops. These events foreshadowed the end of the war, and the opening of the Peace Conference at Rijswick in May 1697. Before the negotiations were concluded, the fleet continued its operations of annoying the French by spasmodic bombardments of coastal towns, protecting trade, and guarding against any invasion attempts. In early 1697 Admiral Sir George Rooke succeeded to command of the Fleet and, shortly thereafter, encountered and captured a fleet of neutral Swedish merchantmen in the Channel carrying war supplies to France. The incident created a minor diplomatic tempest, but led to establishing the principle of international law which more clearly limited the rights of neutral shipping in wartime and defined the rights of belligerents to attack trade.

Despite the initial military and naval set-backs and the serious loss of about 500 English merchant ships, the allies had won enough of a limited advantage over the French to end the war. The signing of the peace at Rijswick in September 1697 resolved only some of the issues that had brought on the war in 1689, but the peace that followed was complicated by an additional problem: the Spanish succession, an issue which threatened the fundamental structure of European relations as well as England's own political and economic basis. The imminent death of Spain's childless king, Carlos II, presented serious issues as to what coun-

THE PEACE OF RIJSWICK

try would come to control the economic sinews of Spain's empire as well as wield her influence in the calculus of European power politics. Both the English and the Dutch could see that this dynastic change affected their own abilities to use maritime trade routes, to obtain markets, and to maintain their own political independence and economic growth.

William III was deeply involved in diplomacy to establish a stable and lasting peace on a basis acceptable to the Maritime Powers. In at least one instance during this period, the English navy played a key role in maintaining the peace of Europe.

In 1700 a crisis in northern Europe threatened to disturb the balance that William was trying to establish in regard to the Spanish succession. The Great Northern War was just breaking out, and William III feared that this war among the northern powers could interfere with the settlement of the Spanish succession issue. To resolve the matter, William dispatched a joint Anglo-Dutch squadron under Sir George Rooke to the Sound in the summer of 1700 where, in concert with the Swedish navy, they bombarded Copenhagen and facilitated the Swedish army's invasion of Denmark. By doing this, the allies successfully forced the Danes to withdraw and confined the conflict to northern Europe.

The death of King Carlos in Madrid a few months later meant that the entire Spanish empire passed to Philippe of Anjou, the younger son of the heir to the French throne. Within two years, international tensions increased following French military movements into Italy and the Spanish Netherlands. Shortly thereafter, in September 1701, the Dutch, the English, and the Austrians agreed to a new Grand Alliance. Following William's own accidental death in March 1702, the government of his successor, Queen Anne, continued his policies with the allies, jointly declaring a new war against France in May 1702.

THE WAR OF THE SPANISH SUCCESSION

In their conduct of this war England's (or, after the 1707 Act of Union, Britain's) leaders employed many of the same strategic ideas that they had tried to initiate in the Nine Years War, using the lessons they had learned there. At the outset they had clear offensive and defensive objectives. On the one hand, England sought to achieve her own safety, to prevent foreign interference in the Revolution settlement, and to protect and to expand her trade abroad. In order to do this, she sought a balance of power among the European states that would allow her economic growth at the expense of France and Spain. The basic strategic idea behind England's conduct of the war was to engage France's superior strength on as many fronts as possible, forcing France to divide and thus to weaken her forces. Since no nation could do this alone, the Alliance was an essential feature of English strategy, allowing her to attack from the Mediterranean, Germany, and the Low Countries, as well as in Spain. At sea, England sought to use the Alliance not only to support the military operations in these widespread theatres, making amphibious landings and coastal attacks, but also to isolate the French economy through privateering and naval attacks on French trade and to cut off the flow of silver and gold from Spanish America—something Englishmen of the day perceived as vital to the French and Spanish war effort. At the same time, the navy served to protect allied trade and colonies abroad. Ideally, England wanted to have the main naval squadrons in the North Sea, the western Channel approaches, and in the Mediterranean, while other vessels served as convoys in dangerous waters.

The Campaigns of 1702–1704 The Grand Alliance lacked the strategic position to carry out England's basic offensive conception of the alliance strategy. She fought the campaigns of 1702, 1703, and 1704 to achieve the required position, while at the same time carrying out what other tasks she could.

In terms of naval operations, the thought of cutting off the enemy's seaborne source of silver and gold was something that English councils had considered

Sir Cloudesley Shovell (1650–1707), by Michael Dahl (1650–1743). Joining the navy as a rating, Shovell was noted for his seamanship and daring, earning him an officer's commission. He was knighted and promoted to rear-admiral after Bantry Bay in 1690. After fighting at Barfleur, Gibraltar, and Malaga, he became Commander-in-Chief of the fleet in 1705. Returning from the Mediterranean in 1707, he and his flagship, HMS *Association*, were lost on the rocks of the Scilly Isles.

from the moment war was declared. A squadron under Sir Cloudesley Shovell was poised in the Soundings, the western approaches to the Channel, waiting to receive the latest intelligence and to intercept the richly laden Spanish galleons before they could reach a safe port. In the event, however, the intelligence came too late, although an Anglo-Dutch squadron under Sir George Rooke was diverted to attack the galleons at Vigo harbour in north-western Spain in October 1702. Though the attack was hailed as a great Anglo-Dutch naval victory at the time, modern historical research has shown that it made little impact on the French war effort.

Equally important in strategic terms, the English sought to secure the practical means to conduct naval operations in the Mediterranean. Facing an even greater problem than the Maritime Powers had faced in 1695, they lacked not only a port to support large warships within the Mediterranean, but also one near the entrance to the Straits of Gibraltar from which operations could be carried out. In the Nine Years War, Spain had been an ally and Cadiz had been the staging-point for Mediterranean operations. At the outset of the war in 1702, the closest friendly port to the Mediterranean was Portsmouth. Thus one of the first major actions of the war was Admiral Sir George Rooke's attempt to land forces and to seize Cadiz in September 1702.

This attempt failed, however, and the next closest harbour to the Straits was Lisbon, 350 miles further away. The diplomats set to work at once, and by May 1703 had arranged Portugal's accession to the Grand Alliance. Lisbon provided the safety of a harbour as well as a secure base for fleet supplies and repairs that made it possible for the fleet to operate in the western Mediterranean. In the summer of 1703, an Anglo-Dutch squadron under Sir Cloudesley Shovell entered the Mediterranean and, although the 1703 campaign at sea was not a major one, it bore fruit. Within a few months, the duke of Savoy broke with Louis XIV and, joining the Grand Alliance, completed the strategic ring around France.

In 1704 the Maritime Powers planned a major naval campaign in the Mediterranean to secure their position, parallelling the allied army's campaign against French-supported Bavaria. Before moving to attack the main French Mediterranean base at Toulon, Admiral Sir George Rooke planned, once again, to take Cadiz, but, finding it too heavily defended, he seized Gibraltar instead. In response to these operations, the French mobilized both their Atlantic fleet at Brest and the Mediterranean fleet at Toulon. This force met Rooke's Anglo-Dutch fleet in mid-August off Malaga, east of Gibraltar. While both sides claimed victory in a naval battle that occurred at almost exactly the same time as the battle of Blenheim, there had been no clear-cut, tactical decision at sea. Nevertheless, France refused ever again to oppose the allies with a battlefleet at sea during the remaining seven years of the war. The Maritime Powers had clearly opened the Mediterranean for their own use while holding Gibraltar for future use.

The Campaigns of 1705–1710 The middle period of the war marked a significantly different approach to the use of the navy. While the first three years had been marked by the attempt to establish the Alliance's strategic position and had ended by securing dominance at sea, in the remaining, major part of the war, the allies exercised their control of the sea. It was a period in which there were no great sea battles, but the navy was fully occupied in the protection of trade, in maintaining the sea lines of communication from England to the war theatres in the Low Countries, on the Iberian peninsula, and in supporting military operations from the sea. Among these operations in the Mediterranean, the navy assisted in an unsuccessful amphibious assault on Toulon in 1707, and following this established a more secure base from which the allies could conduct military operations in the western Mediterranean. In this the most important operation was the capture of Port Mahon on the island of Minorca in 1708.

In the colonies, Rear-Admiral Charles Wager attacked the Spanish plate fleet in the West Indies in 1708 while in both the North Sea and the Mediterranean the navy attempted to enforce a blockade of grain, as means of forcing the French to the peace table in a period of famine. At the same time as they attacked French trade, the navy needed to protect that of the allies. In a political controversy which surrounded the Admiralty in 1706–7, an Act of Parliament required even more convoy protection as the French continued to promote their privateering attacks on English trade. At the same time, the navy was held on the alert to forestall any attempt to invade Britain, as the Jacobites had attempted to do, with French naval support, in their March 1708 landing in Scotland. In this phase of the war there were repeated calls to launch expeditions against the French and Spanish in the colonies. Most of the plans, however, came to little, as the main priorities remained in the European theatre.

The Final Campaigns, 1710–1712 The terms of the Portuguese treaties were high for the allies, but essential to allied Mediterranean naval operations in the war. Nevertheless, it proved impossible to succeed in Spain, as the treaties required. Following a change of ministry in London, the government continued after 1710 to follow the basic strategy that had been laid out years before, while making clear its desire to end the war. One change, however, was a shift of emphasis in Britain's maritime position. As the Dutch Republic increasingly felt the financial strain of the war, it began to restrain its joint operations with the British. Strain became apparent in the discussions over naval priorities, as the Dutch stressed their priority of protecting trade rather than undertaking other types of actions. The inability of the Dutch to meet all the demands made upon them resulted in the increasing use of privateers when naval forces were under pressure. By this time, the navy had been able to clear the enemy from the sea, capturing 2,200 enemy ships.

Consonant with Britain's initial objectives, the government sought to lead the peace, as William had led the entry into the war. The new ministry continued the old ministry's strategy, but at the same time it looked to securing Britain's commercial and political position after the war. Along with a new-found ambition to retain Gibraltar and Minorca, there was a plan to attack Canada in 1711, removing the French threat to the other colonies on the American mainland. Despite the failure of this expedition, owing largely to poor navigation in the St Lawrence River, the expedition had played a role in the preliminary peace negotiations with France.

THE PEACE OF UTRECHT   With the peace negotiations in 1712–13, Britain had risen to the position of a great power. The treaty itself was an epoch-making agreement that laid the basis of an ambitious plan for general European peace. Through it Britain became a guarantor of that plan, marking the completion of the revolution in foreign policy and English attitudes toward Europe and the world that had begun in 1688, and completing the turn towards involvement and collective co-operation initiated by William III. The agreement was a complex one, but it clearly recognized the maritime character of Britain's role. The annexation of St Kitts in the Caribbean with Minorca and Gibraltar in the Mediterranean, as well as Britain's first legal right to trade with the Spanish Indies through the *Asiento*, recognized her position as a commercial and oceanic nation as well as a European power. These were key factors which allowed Britain to continue to improve her naval position and to develop her trade in America and in the Mediterranean.

The long period of general European stability that followed, between 1714 and 1739, saw the Royal Navy active in a number of ways. First, ships helped to implement the agreements made at Utrecht, removing Austrian troops from Spain, installing the duke of Savoy's forces in Sicily. In other areas, British naval forces were involved in maintaining the general settlement of European affairs.

Through the Triple Alliance of 1717, Britain and France became partners in collective European security over the next 20 years. The most dramatic naval incident involved Spain's attempt to break the Utrecht agreement and seize territory she had lost in Italy. Through the agreement of the European powers, a squadron under Sir George Byng defeated this attempt in a peacetime attack on the Spanish fleet off Cape Passaro in 1718.

Following this, Admiral Sir John Norris took a squadron to the Baltic in 1719, 1720, and 1721 using the navy as an instrument in diplomacy and diplomatic balance in the settlement at the final stages of the Great Northern War involving Sweden and Russia. Sir Charles Wager took another squadron to the Baltic in 1726. In that same year, Vice-Admiral Francis Hosier sailed for the West Indies. There, through a peacetime blockade of the Spanish galleons at Puerto Bello, he was able to prevent Spain from obtaining the means to provide a subsidy to her ally, Austria, and thereby prevented the two countries from undertaking military actions that would upset the balance of power in Europe.

THE WAR OF 1739–1748   The war that broke out in 1739 marked the beginning of a new cause for warfare. Unlike the conflicts in the previous half-century, this war grew directly out of colonial trade, which in turn led to conflict in Europe.

The War of Jenkins's Ear   The root cause of the war was the rising friction between the Spanish and English over trade rights in Central America. Irritated over illicit

British trade and smuggling connected with the *Asiento*, the Spanish authorities tried to discourage it through harsh measures. The attempt to repress these British intrusions was taken up by privately owned coastguard vessels, whose crews caused heavy British losses and occasional atrocities. One of these allegedly involved Spaniards slicing off the ear of one Captain Robert Jenkins. When this was blown out of proportion in public discussion, parliament declared war. The ministry, which had seen every hope of negotiating a settlement of the mutual grievances with Spain, sincerely regretted parliament's action. This phase of the war concentrated on the now traditional objective of stopping the Spanish galleons in their passage from America to Spain and interfering with Spanish trade. To do this, the navy made attacks on Cartagena, Puerto Bello, and Santiago in the Caribbean, and blockaded Cadiz in Spain. At the same time, the Royal Navy took defensive measures to protect British trade and overseas possessions. In particular, Britain took precautions in the Channel and at such vulnerable outposts as Minorca and Gibraltar against possible French involvement in the war, invoking her stated opposition to British conquests in the West Indies.

Admiral Edward Vernon (1684–1757), by Charles Philips (1708–47). Nicknamed 'Old Grog' for his grogram coat, Vernon ordered the rum ration diluted with water in 1740, thereby giving it a name. A captain at age 21 and a rear-admiral at 24, he captured Porto Bello in the West Indies in 1739, but his attempt on Cartagena failed. During the '45, his fleet dispositions helped prevent a French invasion.

The War of the Austrian Succession, Phase I In October 1740 Emperor Charles VI of Austria died and, in accordance with the Pragmatic Sanction, his daughter, Maria Theresa, succeeded him. The Elector of Bavaria attempted to seize Austrian lands in Hungary, and Spain supported his claims, making her own plans to seize Austrian lands in Italy. In December, the new king of Prussia, Frederick II, invaded Silesia for his own purposes. Reacting to these incursions, Maria Theresa called upon England and France, among the guarantors of the Pragmatic Sanction, to defend her.

In this situation, Britain was best able to support Austria with a strong squadron in the Mediterranean. Using Minorca as a base, the British fleet interrupted coastal supply lines and also blockaded Toulon, where a Franco-Spanish fleet capable of opposing its operations lay at anchor.

Pursuing earlier ideas, the ministry sent a major expedition to the West Indies for a renewed assault on Cartagena. Under the command of Admiral Edward Vernon and Major-General Thomas Wentworth, who showed poor leadership, the expedition was plagued by delays in outfitting as well as numerous technical problems in conducting amphibious operations in the tropics. More than 10,000 men died, mainly through disease.

The War of the Austrian Succession, Phase II The war entered a new phase in October 1743, when France entered the 'Second Family Compact' with Spain and decided to intervene actively in the war. In February 1744, even before a formal declaration of war, France ordered the fleet at Toulon to sea and engaged the British fleet under the command of Admiral Thomas Mathews and Vice-Admiral Richard Lestock. The British gaining some success, the Franco-Spanish fleet retired, but Mathews was reluctant to pursue them as they attempted to draw him away from his duties in protecting the Italian coast. A major controversy arose over British conduct in this indecisive battle, along with a public dispute between Mathews and Lestock. This led to a series of charges and counter-charges, involving both admirals and their subordinate captains, followed by courts-martial and an inquiry by the House of Commons. Stripping away the acrimony, one can conclude that Mathews had insufficiently prepared for the battle, that he did not fight hard enough, and that he failed to take every step that he could have taken to destroy an enemy that he had been in contact with for several days. In the end, Mathews was cashiered and Lestock justified, but Lestock's conduct in blindly adhering to the letter of his instructions was no better. The failure was one of

Admiral Lord George Anson (1697–1762), by Sir Joshua Reynolds (1723–92). Promoted to rear-admiral for capturing a richly laden Spanish galleon during his circumnavigation of the globe 1740–4, Anson defeated the French off Cape Finisterre in May 1747. Well connected politically, Anson became the key figure in naval administration from 1748 to 1762.

command based on a tactical system dependent on instructions rather than individual initiative.

While it was still necessary to maintain a strong fleet in the Mediterranean, Britain faced demands in a variety of other areas. In the Channel, France began to prepare an invasion force for the Young Pretender. Although a February gale dispersed this, the king's announcement of the attempt in parliament brought an immediate declaration of war against France. The threat was renewed in the spring of 1745, with the French army's victory at Fontenoy and the subsequent capture of Brussels. French command of the Low Countries renewed the traditional threat from any hostile power controlling the Scheldt, and it soon emerged with the rising of 'The '45'. Squadrons in the North Sea and the Channel were on guard to prevent further French reinforcements arriving in Scotland and to protect shipping around Britain, while France returned once again to the strategy of *guerre de course* after the battle of Toulon.

With the need for a strong defensive strategy in European waters, the Royal Navy had few resources to send to distant stations. In India, Britain was at a disadvantage while the French maintained a permanent squadron at Mauritius. At the same time, the Royal Navy had a similar advantage over the French in North America, through the availability of their West Indies-based squadron. With these facts in mind in the spring of 1745, there was a new departure in naval strategy involving the North American colonies. Admiral Sir Peter Warren, then on the Leeward Islands station, took command of naval forces north of Carolina, including the colonial privateers who wished to serve under him and the forces on the Newfoundland station. Joining a colonial expedition mounted on the initiative of Massachusetts in June 1745, Warren succeeded in taking the harbour and fort at Louisbourg, from which the French had controlled the mouth of the St Lawrence, supplied their Newfoundland fishing fleets, and threatened both New England and Nova Scotia.

In the following years of the war, while Britain maintained superiority in the Channel, France backed away from further invasion threats based at Dunkirk. At the same time, she was unable to support the Jacobites. Unable to maintain a concentration in the Mediterranean, Franco-Spanish naval forces became less of a threat. As a result, Britain could move some of her fleet to the Bay of Biscay, while at the same time bringing more ships into service.

By 1747 the Royal Navy maintained a strong squadron to the west of the Channel approaches. In May, a squadron under the command of Vice-Admiral George Anson intercepted a French convoy off Cape Finisterre which was taking troops and supplies to India and Canada. In October, another squadron under the command of Rear-Admiral Edward Hawke destroyed a large French convoy bound for the West Indies in almost the same location. Bereft of protection, French maritime trade slowed, tightening the financial strain on France. Within a year, France agreed to the peace at Aix-la-Chapelle in 1748.

By the end of the war, the Royal Navy had performed some very distinct functions, complementing allied military, diplomatic, and financial strategies. It had helped protect allied territories from sea attack, moved and supported military operations by sea, interdicted France's communications with her colonies, and helped to undermine French finances.

Unlike the Peace of Utrecht, the Treaty of Aix-la-Chapelle provided no long-term basis for the structure of international relations. Nevertheless, the dispute between Spain and Britain, over which the war had begun, was resolved and an equilibrium was reached in Italy. As she had in 1712, France again agreed to dismantle her base and fortifications at Dunkirk, and withdrew her support for the Pretender, marking the full recognition of the Hanoverian succession and ending more than half a century of conflict involving active French support for the Jacobite cause. For France and Britain, the main unresolved issues for the future turned on imperial and overseas trade ambitions. In this respect there was a stale-mate which foretold a future war.

In 1749, mindful of the importance of British North America, the government established a major naval base at Halifax, the new capital of Nova Scotia, and settled 2,600 British settlers along with additional German and Swiss Protestants in the colony. No similar move was made in India, and the rival East India Compa-nies remained hostile. Moreover, the task of maintaining British naval superiority over the combined naval strength of France and Spain now seemed particularly burdensome. Her interests stretched from one side of the globe to the other, Britain depended upon her capacity to move troops and to maintain trade and communications by sea, while preventing interference from enemies.

In fact, the practice of maintaining British trade while denying the benefits of trade to opponents had become a central object of British war policy. It repre-sented not only a naval challenge, but raised a practical legal problem of neutral rights. This was a particular issue in this period when several wars occurred simul-taneously; a belligerent party in one war could well be neutral in another. In the years between 1702 and the 1750s, belligerent powers sometimes found themselves in the unusual situation of having to consider their neutral rights at the same time as they planned attacks on enemy trade. In the years leading up to the Seven Years War, this problem gave rise to a legal principle that came to be called 'The Rule of the War of 1756'. Illustrating the uncertain nature of maritime law, it did not in fact begin as a rule of law, but as a British government policy to relieve some earlier restrictions on captures. This rule stated that neutrals lost their immunity to capture when they undertook with an enemy trade that was not normally permit-ted in peacetime.

Prelude to War In the spring of 1754 the slow westward growth of the English colonies in North America came into direct conflict with the interior regions controlled by the French. In a pitched battle, the French defeated and repelled a small militia force from Virginia under Major George Washington that had attempted to uphold British claims in the Ohio valley. Reacting to this in January 1755, ministers in London ordered General Braddock with British regular troops to Halifax, while Admiral the Hon. Edward Boscawen took a squadron to protect the colonies by sea. The French responded in kind, sending a force to Quebec. The two rival fleets did not meet, although Captain Richard Howe, with several other British ships, took two French warships. News of this capture caused a great public outcry in France, fanning war hysteria. At the same time, on the other side of the Channel, French naval preparations had created the fear of an invasion. Despite large naval mobilizations and rising tensions, no war was declared.

By March 1756 repeated reports had reached London that the French planned to seize the British-held island of Minorca. It was a position that was not highly valued in London, and the ministry moved slowly to respond to these reports, more concerned over the possibility of an invasion. Finally, they ordered Vice-Admiral John Byng to prepare a small squadron for the Mediterranean, still doubting that Minorca was under a real threat. Arriving in the Mediterranean, Byng found that the French had already besieged the British garrison at Fort St Philip on Minorca and were apparently supported by a superior fleet under the marquis de la Galissonière. Since the island was already under attack in Byng's view, an attempt to relieve it would necessitate leaving the garrison at Gibraltar open to French attack. Although Byng engaged the French indecisively, he declined to relieve Minorca and it fell to the French in June. He was court-martialled and later executed on the quarterdeck of his own flagship; as Voltaire satirically noted in *Candide*, the English 'find it pays to shoot an admiral from time to time to encourage the others'. In fact, the Admiralty and the ministry itself was at fault for failing to provide a fleet of sufficient strength.

War Declared The loss of Minorca was only part of a series of set-backs that created a political crisis, forcing a struggle for political power and a series of changes in the ministry. Following the conclusion of a defensive alliance between France and Austria, completing the diplomatic revolution in Europe, Britain declared war on France on 18 May 1757. The new ministers had no great plan to alter the conduct of the war and began with a defensive strategy. In America, Britain sought only to recover British rights as defined at Utrecht possessing Nova Scotia, a frontier along the St Lawrence River, and control of the Great Lakes, following the French capture of Fort Oswego on Lake Ontario. In India, the ministry planned to assist the East India Company's efforts under Clive to ward off French incursions. At sea the Admiralty planned to defend against invasion and to protect trade and overseas colonies. To support the American colonies an expedition under Admiral Francis Holburne sailed to recover Louisbourg, but called off the attempt in the face of a superior French fleet.

On the Continent, the ministry's only object was to defend the electorate of Hanover from France, yet this posed the most urgent problem. The ministers in London took up a suggestion from King Frederick II of Prussia to make a diversionary attack on the coast of France, which would supposedly force the French to recall their troops from Germany. Casting about for the most important and vulnerable target, they settled on the French dockyard at Rochefort. Arriving on

Admiral Edward Boscawen (1711–61), by Sir Joshua Reynolds (1723–92). Known as 'Old Dreadnought' for the ship he had commanded 1742–4, Boscawen was promoted to rear-admiral following Anson's action off Cape Finisterre, when he was wounded. During the Seven Years War he commanded the successful expedition against Cape Breton, but his greatest achievement was defeating the French Toulon fleet at Lagos Bay in 1759.

the scene, the joint amphibious expedition commanders, General Sir John Mordaunt and Admiral Sir Edward Hawke, found that they had made plans on inaccurate information and could not carry out an assault; they withdrew in failure after 36 hours.

This was not the end of the set-backs, as General Montcalm led French forces in Canada in capturing Fort William Henry and threatened to invade New York's Hudson Valley and to split the British colonies in two. In India, the French threatened Madras from their base at Pondichéry and remained firmly in position in the Carnatic. Only the small squadron under Admiral Charles Watson saw success as it gave some support to the East India Company's efforts at sea, while Clive won victory ashore at Plassey and recovered Calcutta.

A New Beginning At the end of 1757 the ministry began to look for a fresh start after its series of set-backs. The major lesson it had learned was that it should not have staked everything on a single seaborne attack. In developing new ideas William Pitt, as a secretary of state working with his colleagues, had considerable influence in choosing new commanders and making arrangements for the following year. Orders had already been sent in the autumn to station eight ships at Halifax over the winter and to establish a naval base on Cornwallis Island in Halifax Bay. At the end of the year, plans were made for a series of three assaults in North America: (1) An expedition of provincials and regular troops under General James Abercromby up the Hudson Valley and across Lake George, Lake Ticonderoga, and Crown Point against Montreal or Quebec; (2) an assault on Louisbourg under General Jeffrey Amherst supported by the fleet under Admiral Edward Boscawen; and (3) a land attack on Fort Duquesne in western Pennsylvania under Brigadier John Forbes.

In European waters, the ministry continued to resist repeated requests from Prussia to maintain a squadron in the Baltic, needing ships in the Mediterranean, America, the Channel, the western squadron in the Bay of Biscay and in blockading Brest. In discussions with the Prussians, they agreed to resume European coastal operations, sending Commodore Charles Holmes to patrol the River Ems and assist in taking Emden, thereby covering the Prussian flank.

At sea, Admiral Henry Osborne had chased part of the Toulon fleet into Cartagena on the southern coast of Spain, then defeated the squadron sent to relieve it. This was followed in April 1758 by Hawke's attack on Basque Roads, destroying more ships loading to reinforce Canada. Together these actions temporarily stopped the French from diverting the British plans to attack Louisbourg. While these events were taking place, Pitt laid plans for a series of coastal operations, including another amphibious landing in France. This time the ministry agreed on Saint-Malo, where the duke of Marlborough's forces did considerable damage ashore from ships commanded by Captain Richard Howe, before moving on to capture Cherbourg in August.

The diversionary intentions of the British coastal raids on France failed to have any effect on French military deployments in Germany. In their view the decisive military theatre was on the Rhine, and victory over Hanover and Prussia would allow them to concentrate later on the maritime war.

Taking the next step in creating a more offensive maritime war, Pitt took up suggestions for assaults on the French colonies at Gorée in West Africa and the rich West Indian island of Martinique. The more important of these was the expedition to the West Indies, which eventually altered its objective and succeeded in capturing Guadeloupe in May 1759.

From the winter of 1757 British leaders felt that they faced a serious threat of invasion from France, while their forces at sea and ashore were in a delicate strategic balance. The various expeditions had not yet borne fruit, and there were those who doubted whether the navy could, simultaneously, prevent invasion. The first hopeful sign came when Boscawen, after detecting the French Toulon fleet under the marquis de la Clue trying to escape from the Mediterranean and get past Boscawen's fleet at Gibraltar, brought it to battle in early August, off Lagos on the coast of Portugal. Boscawen took three French ships and destroyed two others, forcing the remainder of the fleet into the Tagus at Lisbon, where the British blockaded them. Yet, despite this set-back, the French continued to prepare troopships at Bordeaux and Nantes for an invasion attempt.

In North America, British military operations against Niagara, Ticonderoga, and Crown Point in the summer of 1759 were successful. Consolidating British control on the Great Lakes, these actions allowed British forces to command the principal lines of communication into Canada from the south and, provided they could take control of Lake Champlain, would establish a clear land route to advance on the St Lawrence River to join up with Wolfe's seaborne forces. In a risky operation, Wolfe had not waited for Amherst to control Lake Champlain and move up from the south. At the end of the season, Wolfe and Vice-Admiral Sir Charles Saunders gambled on their own. Sailing up the difficult waters of the St Lawrence River, facing the constant threat of fireboats and shore artillery, Saunders landed Wolfe's army just below Quebec, where Wolfe could base himself and find a weak point in the heavily defended city. Eventually, Wolfe assaulted the Heights of Abraham overlooking Quebec, supported by naval gun-fire from improvised gunboats that could negotiate the shoals and ledges of the river. Wolfe died heroically as his forces, supported by Saunders and the fleet, overwhelmed the French, forcing them out of the city.

At the end of the campaigning season in America, forces returned to Europe. Admiral Bompard returned from the West Indies to Brest, providing much-needed reinforcements for France. The success of operations in North America could still be offset by the French forces remaining in Canada and by the French counter-strategy of invading Britain. To deal with that particular threat, Hawke's western squadron had cruised for an extraordinarily long time off Brest in 1759, preventing the French from leaving port. However, taking advantage of a momentary absence of the British blockading squadron off Brest, the French court decided to strike England. They ordered the marquis de Conflans to sail from Brest, gather the transports from Nantes and Bordeaux at Morbihan, and sail for

Scotland along the western coast of Ireland. Although Conflans had a two-day head start, a squadron under Sir Edward Hawke caught up with the French on 20 November in Quiberon Bay, where they had taken refuge. Fighting in bad weather and in confined waters studded with rocks and shoals, the action became a mêlée. With only 'the space of a short winter's day', Hawke had been unable to do much more than destroy a few ships, letting the others escape. His strategic success was far greater than his tactical achievement suggested. Much of the French fleet took refuge far up the bay in the River Vilaine for the remainder of the war. In French politics, Hawke's victory gave a political edge to the party that favoured land war over maritime war. For Britain, this change removed the threat of invasion. For the remainder of the war, France made no further major attempts with her navy, leaving Britain to use the seas as she wished.

French fireships attacking the English fleet off Quebec, 28 June 1759, by Dominic Serres the Elder (1722–93). Following the initial landing on the Isle of Orleans, just downstream from Quebec, the French launched seven fireships, attempting to destroy the British warships at anchor. As the furthest upstream, HMS *Centurion* sighted them first, cut her cable and alerted the others, giving time for them to launch boats, grapple the fireships and tow them clear. Subsequently, naval forces under Admiral Saunders and army units under General Wolfe collaborated successfully in the combined operation which resulted in the capture of Quebec.

The Final Phase Following the victories of 1759 Britain needed to maintain her position until a satisfactory peace could be negotiated. There was little immediate prospect of peace and Britain was faced with using her navy in an entirely different manner than she had up to this point in the war. The British still needed to watch the French fleet bases at Rochefort, Toulon, and Brest, guarding against the possibility of a new attack and preventing them from joining with their Spanish allies and renewing a threat of invasion. She needed to protect her trade while continuing to attack French maritime trade, support troops in Canada, fend off Spanish claims to the Newfoundland fisheries, and support Gibraltar. In delicate circumstances and not wanting to offend Spain and bring her into the war, the Admiralty was directed to respect Spanish territorial waters and avoid direct confrontation with Spain.

The death of the king in 1760 and the accession of George III brought some unexpected changes, one of which was the new king's interest in amphibious expeditions. A series of plans was laid for assaults on Belle Isle, Martinique, and Mauritius. Commodore Augustus Keppel and Major-General Studholme Hodgson took Belle Isle on 8 June, just as peace discussions were getting under way in Paris. This added to French apprehensions that Britain might inflict a damaging peace and in August 1760 France signed the 'Second Family Compact' with Spain.

The nature of the strategic situation changed once again as the ministry in London haggled over its response, breaking off peace negotiations with France. Unable to get support for a pre-emptive strike against Spain, Pitt resigned in 1761. Nevertheless, Spain declared war in December, and Britain reciprocated in January 1762. At this point, Admiral Saunders was in the Mediterranean with a fleet of 30 ships-of-the-line based at Gibraltar to prevent the Toulon and Brest fleets from joining in another invasion attempt. To ensure that the French did not try to repeat their 1744 attempt in the Channel, Hawke readied the western squadron to fall back from the Bay of Biscay to Plymouth. Admiral George Rodney and General Robert Monkton took Martinique in February 1762, as Pitt had planned. Pursuing their success, they continued and occupied the nearby islands of St Lucia, St Vincent, and Grenada. Following these events, Admiral Sir Edward Pocock and General the Earl of Albemarle took Havana on the Spanish island of Cuba in August, and the forces under Rear-Admiral Sir Samuel Cornish in the East Indies took Manila in October.

Admiral Lord Edward Hawke (1705–81), by Francis Cotes (1726–70). As a captain Hawke had distinguished himself in the battle of Toulon, and as a rear-admiral he had defeated a French squadron in the second battle of Finisterre in October 1747. In 1757, blame for the failure of his unsuccessful joint expedition to Rochefort fell on his army colleague. Two years later Hawke chased a French squadron in a full gale and defeated it among the rocks and shoals of Quiberon Bay.

As these events were taking place, France had opened secret peace talks and was ready to negotiate the final articles of peace in September 1762, signing the final treaty at Paris on 10 February 1763. Both politicians of the day and historians have continued to debate the strengths and weaknesses of the treaty's terms. Britain had clearly won the war and defeated French imperial aims in North America and India. At the beginning of the century she had stressed the importance of trade; by mid-century she had acquired a position of dominion, with the right to expand her colonies as far west as the Mississippi, and retaining Canada, Grenada, St Vincent, Dominica, and Tobago. This was done, however, at the expense of

THE PEACE OF PARIS

The Bombardment of Morro Castle, 1 July 1762, by Richard Paton (1717–91). Three weeks after the siege of Havana began, HMS *Cambridge* (80), *Dragon* (74), and *Marlborough* (70) bombarded the castle at the harbour entrance. After more than six hours the ships were so badly damaged, with 42 killed and 140 wounded, that they withdrew, leaving further bombardment to shore batteries. The castle eventually fell on 30 July.

returning Gorée, St Lucia, and Martinique, along with limited fishing rights on the Newfoundland banks, to France. At the same time, Belle Isle was exchanged for Minorca and Britain acquired Florida in exchange for Havana and Manila. With the signing of the Peace of Paris, Britain rose to new heights of global power and prestige, but in the years that followed she sank to a new low.

For in 1763 Britain found herself isolated. The old pattern of alliances, around which William III and Queen Anne had structured England's position on the Continent, was gone; so was Prussia's friendship, a casualty of Britain's desire for limiting obligations on the Continent. In the years between 1763 and 1778, Britain found herself challenged at a variety of points. For the navy, the most dramatic incident was the crisis which erupted in 1770 over a Spanish attempt to claim the Falkland Islands. Britain armed her navy and made preparations for war, but no fleet put to sea. As the years passed, it became increasingly clear that the Bourbon powers of France and Spain intended to use any weakness in the British empire as a wedge to regain their lost imperial positions. Unexpectedly, the weakness that hurt Britain most was the open revolt of the American colonies.

The first naval operations in America began in 1775 and were devoted directly to the problem of quelling colonial revolt. In the beginning it seemed a minor affair. Captain Charles Douglas's arrival with a squadron of ships in the St Lawrence on 12 April 1776 was enough, in itself, to dissuade a small group of Americans from continuing their blockade of Quebec, and by the end of the year had completely eliminated the tiny American flotilla on Lake Champlain.

THE WAR FOR AMERICA

In most respects the naval forces sent out from Britain to deal with the rebellion were too small to provide a crushing blow and were restricted in their use of force. Several squadrons sailed with troops to quell the rebellion; among them Vice-Admiral Lord Howe sailed for New York, which he took in the operations during the summer and autumn. At the same time, Commodore Sir Peter Parker made for Charleston, South Carolina, but diverted to New York after being repulsed at Charleston. By late 1777, British forces had succeeded in establishing themselves at Philadelphia, New York, and Newport, Rhode Island, but in larger terms they had neither inflicted a significant defeat on the small American army nor cut off its vital supply of arms and ammunition by denying the use of the sea to American and neutral vessels carrying war supplies. The surrender of the British army at Saratoga in October 1777 was the signal, drawing France into the war and transforming it from a local rebellion to a broader geo-strategic problem for Britain.

The War against France Immediately following the signing of the Franco-American treaty in March 1778, a French fleet under count d'Estaing sailed from Toulon to the American coast, where he was directed to engage the fleet under Lord Howe. With the French arrival in American waters, the naval war changed from an attempt to contain a revolt and support British troops in North America to a war of global dimensions in which Britain had to contend with major battle-fleet operations. It was a change from a focus on amphibious warfare, emphasizing the support of military operations along the coast with frigates and smaller vessels, to one requiring ships of the line.

The first fleet action took place in late July 1778 off Ushant, when Admiral Keppel encountered d'Orvilliers's Brest squadron on a training cruise. After an inconclusive engagement in which both sides were heavily damaged, the French outmanoeuvred the British and slipped back into the safety of Brest. For his part, Keppel was court-martialled and charged simultaneously with the charges Mathews had faced in 1744 and those that Byng had faced in 1756: engaging the enemy before forming a battle line and failing to do his utmost against the enemy. In this instance, however, the court martial exonerated Keppel, emphasizing an admiral's responsibility to choose his own tactics and not blindly follow the *Fighting Instructions*.

Not long afterwards, the opposing fleets met for the first time in North America. In mid-August, the French fleet under d'Estaing attempted to make a landing on Rhode Island, where the British army occupied Newport. D'Estaing put to sea to meet Lord Richard Howe's squadron; the course of the engagement was determined not by tactics, but by an approaching hurricane that scattered both fleets.

The war quickly spread to the West Indies; in September 1778 it took on the character of imperial rivalry when a French force from Martinique captured the British island of Dominica. A series of engagements with British fleets followed, with Vice-Admiral Samuel Barrington taking St Lucia in December, and the full fleet under Vice-Admiral John Byron engaging d'Estaing's French fleet in a disastrous action off the island of Grenada on 6 July 1779. Despite Byron's tactical defeat, the French victory brought them no results.

In June 1779, when Spain entered the war, Britain faced a heavy threat in the West Indies, Gibraltar, and Minorca, where Spain wanted to regain her old losses. Moreover, a combined Franco-Spanish force could overwhelm the Royal Navy in the home waters of the Channel, putting Britain at risk.

Troops quickly gathered at Le Havre in preparation for such an invasion, and in 1779 a Franco-Spanish fleet of 66 ships cruised unopposed in the Channel. In the West Indies, the French added to their conquests, taking Grenada and St Vincent. After relieving Gibraltar in January 1780, Admiral Sir George Brydges Rodney fought two battles with the French under de Guichen in the West Indies. The first took place on 17 April, and the second on 15 May 1780. Together, the battles helped to prevent further French incursions in the West Indies, but made no decisive action.

Reinforcing the French naval presence, Admiral de Ternay took advantage of the British evacuation of Newport, arriving in July with Rochambeau's troops. Rodney might have sailed northwards to New York and joined with Arbuthnot and others to dislodge the French from Narragansett Bay, but they failed to agree on a plan. Returning to the West Indies, Rodney found that a hurricane had destroyed two ships-of-the-line, several smaller vessels, and fleet supplies at Barbados, St Lucia, and Jamaica. Trying to use the disaster to good effect, Rodney attempted to reconquer St Vincent, but on landing there he discovered that the reports of hurricane damage to the French fortifications were inaccurate and he was forced to withdraw.

At the end of 1780 the European diplomatic scene was further complicated by Russia, Sweden, and Denmark creating the League of Armed Neutrality against Britain. In order to pre-empt the Dutch from joining it, Britain declared war on the Republic on 20 December 1780, and immediately sent orders to take Dutch possessions in the West Indies. On receipt of the orders, Rodney at once seized the Dutch islands of St Eustatius, St Maarten, and Saba on 3 February 1781. In the Channel, the fleet under Sir Peter Parker engaged the Dutch under Johan Arnold Zoutman off the Dogger Bank in August 1781.

The French had sent Count de Grasse from Brest with reinforcements for the West Indies, where they arrived safely at the end of March, after Rodney and

Hood failed to intercept and stop them. Shortly afterwards, the French naval forces in the West Indies were able to join Destouches' squadron from Newport off Chesapeake Bay, where Arbuthnot engaged them in March. During the following months, Lord Cornwallis's army had advanced from South Carolina as far as Virginia, where it had made camp near Yorktown, awaiting reinforcements by sea or co-ordinating operations from the northern colonies.

In September Graves engaged the French fleet under de Grasse off the Chesapeake Capes, where they arrived with troops to support military operations under Generals Washington and Rochambeau against Cornwallis. Graves's failure to defeat the French fleet and dislodge them from the Chesapeake to allow the British to bring reinforcements from New York was, in Washington's words, 'the pivot upon which everything turned', forcing Cornwallis to surrender his army.

Admiral Lord George Brydges Rodney (1719–92), First Baron Rodney, by Sir Joshua Reynolds. In the last great portrait of his working life, painted in 1787, Reynolds depicted Rodney worn by age. Rodney had joined the navy in 1732 and become a post-captain ten years later. Distinguishing himself as a flag officer in the Seven Years War, his greatest victory was over De Grasse at the battle of the Saintes in April 1782.

While the battle of Yorktown in September 1781 virtually decided the issue on the North American continent, the war was not over. The allied French and Spanish forces began to extend their conquests in the West Indies, taking St Kitts and Nevis in January, despite the presence of a squadron under Rear-Admiral Sir Samuel Hood. Arriving on the station shortly thereafter, Rodney took command of the fleet and prepared to stop the French plan to take Jamaica. When the French began to move out of Martinique for this purpose, Rodney intercepted them in

Major naval operations, 1688–1815.

Major battle or siege

1 North America
War of the Spanish Succession:
Annapolis Royal (1710)
Quebec (1711)
Seven Years War:
Louisbourg (1758)
Quebec (1759)
War of American Independence:
Chesapeake Capes (1781)
War of 1812 (1812–14)

2 West Indies
All wars of the century:
Cartagena (1708)
Landings,
Porto Bello (1739)
Cartagena (1741)
Havana (1762)
Martinique (1780)
Saintes (1782)
Landings, numerous
French possessions (1808–10)

3 South Atlantic
Porto Praja (1781)
Cape of Good Hope (1806)
River Plate (1806)

4 Convoy Landfall Area
Smyrna Convoy (1693)
Vigo (1702)
Finisterre I–II (1747)
Glorious 1st of June (1794)

5 Baltic
Expeditions (1700), (1719–21)
Napoleonic wars:
Copenhagen (1801), (1807)
Expeditions (1807–13)

6 Home Waters and Channel
Nine Years War:
Torbay Landing (1688)
Beachy Head (1690)
Barfleur/La Hougue (1692)
All wars:
Counter-invasion, control
and trade protection
Napoleonic wars:
Camperdown (1797)
Walcheren (1809)

7 Brest/Ushant and the Western Squadron
All wars:
Blockade
Blockade of Brest and
protection of trade
Seven Years War:
Quiberon Bay (1759)
War of American Independence:
Ushant (1778)

April between Dominica and Guadeloupe, near the small island group called the Saintes, where Rodney defeated de Grasse and forced him to abandon the planned attack on Jamaica.

In the Mediterranean, a Franco-Spanish force seized Minorca in February 1782, but the final Spanish attack on Gibraltar was repulsed.

In the East Indies, Rear-Admiral Sir Edward Hughes's small squadron had been unopposed in the years between 1779 and 1781. With the Dutch entry into the war, Commodore George Johnstone attempted to capture the Dutch colony at the Cape of Good Hope, but this expedition was prevented by the French under Suffren in the action at Porto Prayo in April 1781. After securing the Cape,

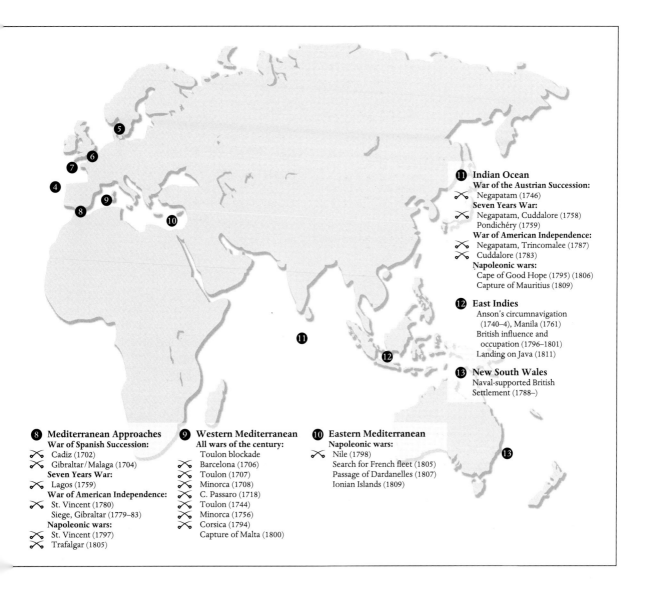

11 Indian Ocean
War of the Austrian Succession:
⚔ Negapatam (1746)
Seven Years War:
⚔ Negapatam, Cuddalore (1758)
  Pondichéry (1759)
War of American Independence:
⚔ Negapatam, Trincomalee (1787)
⚔ Cuddalore (1783)
Napoleonic wars:
  Cape of Good Hope (1795) (1806)
  Capture of Mauritius (1809)

12 East Indies
  Anson's circumnavigation
    (1740–4), Manila (1761)
  British influence and
    occupation (1796–1801)
  Landing on Java (1811)

13 New South Wales
  Naval-supported British
  Settlement (1788–)

8 Mediterranean Approaches
War of Spanish Succession:
⚔ Cadiz (1702)
⚔ Gibraltar/Malaga (1704)
Seven Years War:
⚔ Lagos (1759)
War of American Independence:
⚔ St. Vincent (1780)
  Siege, Gibraltar (1779–83)
Napoleonic wars:
⚔ St. Vincent (1797)
⚔ Trafalgar (1805)

9 Western Mediterranean
All wars of the century:
  Toulon blockade
⚔ Barcelona (1706)
⚔ Toulon (1707)
⚔ Minorca (1708)
⚔ C. Passaro (1718)
⚔ Toulon (1744)
⚔ Minorca (1756)
⚔ Corsica (1794)
  Capture of Malta (1800)

10 Eastern Mediterranean
Napoleonic wars:
⚔ Nile (1798)
  Search for French fleet (1805)
  Passage of Dardanelles (1807)
  Ionian Islands (1809)

Suffren moved toward India, where he engaged Hughes in a series of four actions, defeating him at Negapatam in July 1782. A fifth, off Cuddalore in January 1783, marked the last naval action of the war, five months after the preliminaries of the peace were signed. Both tactically and strategically, these actions were French victories in which the smaller French forces maintained themselves on the French coast of India, took Trincomalee, reinforced Hyder Ali, and retained Cuddalore.

The Peace of 1783 Despite the political difficulties that parliamentary ratification of the peace involved, from a strategic standpoint the Peace of Versailles was not the disaster that many expected. In most theatres, French moves had been stalemated

107

and there were no grounds for overturning the basic principles of the 1763 peace. France made some gains when Britain ceded Tobago, restored Sénégal, obtained some minor areas in India, and enlarged French fishing rights on the Newfoundland banks, but otherwise the two countries exchanged captured territories. France's refusal to allow partitioning of the Dutch empire enabled them to resist most British demands to extend control in India, except for giving Britain the small Indian enclave of Negapatam and the right to navigate (but not trade) in the Dutch East Indies. While Britain lost the thirteen American colonies, she was assured of Canada. After the hard-fought defence of Gibraltar, the British public refused to allow it to be given up and Britain gave up Minorca and Florida instead. Significantly, the treaty was a substantive agreement among the western European states of Spain, Britain, and France. For the first time, the eastern and central European states played no determining role in a major settlement, something that confirmed the increasing division between the interests of eastern and western European powers.

By 1788 Britain had clearly emerged from the isolation of the previous twenty years, joining in a series of diplomatic alliances in the period up to 1791. Taken in the context of Britain's role in the Triple Alliance, her preparations for a naval war in 1790 with Spain in the Nootka Sound crisis, which involved mercantile rights on the Pacific coast of North America, demonstrated Britain's diplomatic and naval recovery. Admiral Lord Howe put to sea in August and September 1790, cruising European waters to back up British diplomatic demands on Spain with a fleet of 29 ships-of-the-line and 17 smaller vessels. In the following year, during the Ochakov crisis with Russia, the Admiralty commissioned a fleet of 36 ships-of-the-line and 10 other vessels under Vice-Admiral Lord Hood, to serve a similar diplomatic purpose as Russia's victories in the Black Sea threatened the security of Turkey and the sea route to India. However, parliament's refusal to support this move, during a period of general peace, ended its employment, and resulted in further reductions in naval spending, capability, and readiness.

THE WAR OF THE
FRENCH REVOLUTION

As revolution swept across France in 1789, the British government explicitly refrained from any involvement in the internal affairs of France. London abandoned its aloof detachment only after Louis XVI's execution in 1793. The French declaration of war on Britain and the Dutch Republic on 1 February 1793 confirmed a course which had already been clear after a similar declaration against Austria and Prussia nearly a year earlier.

The War of the First Coalition, 1793–1798 Joining as partner in the First Coalition with Austria, Prussia, Holland, and Spain, Britain chose a diffuse strategy, aimed at achieving several different objectives at once.

Following precedents from the days of William III and Anne, the Admiralty sent orders to stop all French merchant ships wherever they were, while at the same time stopping neutrals carrying grain to France at the time of a great famine. It was this order that sent Howe looking for Villaret-Joyeuse convoying

American grain from Chesapeake Bay. Engaging him on the 'Glorious First of June' 1794, Howe defeated the French warships, although the grain reached its destination safely.

In the West Indies, Vice-Admiral Sir John Jervis, with Lieutenant-General Sir Charles Grey, set about capturing the exposed French colonies. They moved rapidly, and Martinique fell in February 1794, followed by Guadeloupe, which the French quickly retook. In 1796, Rear-Admiral Sir Hugh Christian continued British success and captured St Lucia, St Vincent, and Grenada. In February 1797 Rear-Admiral Henry Harvey and Lieutenant-General Ralph Abercromby succeeded in taking Trinidad from the Spanish.

One of the first events of the war came almost by chance, when Hood's squadron in the Mediterranean took advantage of local discontent with the revolutionary regime in Toulon and seized the city and base. From a broader strategic perspective it was not a position which British forces could maintain, and they soon withdrew and took a more careful, defensive stance in those waters, moving back as far as Lisbon and the refuge of the Tagus to find a safe base of operations. As before, operations in the western Mediterranean required a closer base to be effective. The revolt in Corsica, led by Paoli, offered the possibility that that island could serve as an effective substitute for Minorca, a base from which British naval forces could try to blockade Toulon. Hood's amphibious assault on Corsica in 1794 was designed to achieve that end. Although British forces succeeded for a time, the French prevailed and retook Corsica.

The Royal Navy retired from the Mediterranean in 1796 to await another opportunity. As had happened a century before, Britain's withdrawal from the Mediterranean was interpreted by some allies as a failure to support the Alliance. Using this as an excuse, Austria quickly withdrew from the coalition war in 1797, joining Holland, Prussia, and the Dutch Republic, who had made peace in 1795. From 1797, Britain remained the only opponent.

British prospects at this point in the war were not good, but the Royal Navy continued to use the Tagus as an advanced base. It was there that Sir John Jervis learned that the Spanish fleet was sailing northwards to join the French at Brest for a possible invasion attempt. Sailing to prevent that union, Jervis encountered the Spanish fleet under Don José de Cordova off Cape St Vincent on the Portuguese coast, defeating it on 14 February 1797 and forcing it into Cadiz. Coming at a low point in British fortunes, news of the battle raised morale at home, despite the difficult situation, which was further complicated for the navy when mutiny broke out in the Home Fleet in spring 1797, leaving momentary doubts as to whether or not British seamen would continue to fight if they had to face the Dutch or French fleets.

After the French conquest of the Dutch Republic in 1795, Britain had begun to seize Dutch shipping in an effort to prevent France from utilizing Dutch resources. This policy began to bear fruit as Commodore Peter Rainier took Amboyna and the Banda Islands in the East Indies in 1796 and Vice-Admiral Sir George Elphinstone seized the Dutch colony at the Cape of Good Hope in

Rear-Admiral Sir John Jervis (1735–1823), First Earl St Vincent, by Sir William Beechey (1753–1839). Entering the navy in 1749, Jervis first distinguished himself in the Quebec expedition in 1759. Later, he fought off Brest in 1778 and captured the 74-gun *Pégase* in 1782. For his victory off Cape St Vincent in 1797, he was created earl St Vincent. After commanding the Channel Fleet, he served as First Lord of the Admiralty 1801–4, making numerous reforms.

Lieutenant's Log of HMS *Culloden* for the Battle of St Vincent, 14 February 1797. Captain Thomas Troubridge, commanding the 74-gun ship *Culloden*, led the British battle line as it tacked in succession to pursue the Spanish fleet under Cordova. Following Jervis's signal to engage, *Culloden* improvised a line of battle with the other lead ships, supporting Commodore Nelson's initiative in the *Captain*.

September 1795. In October 1797 a Dutch squadron under Vice-Admiral J. W. Winter left the Texel and sought an engagement with Admiral Adam Duncan, who defeated them off Camperdown on the Dutch North Sea coast.

By 1798 Austria seemed ready to rejoin the war against France and pressed Britain to return to the Mediterranean. In the intervening years France had consolidated her position in that region, and by that phase of the war Britain was faced with a wide range of demands for her naval resources in other theatres.

Britain therefore took a great strategic risk when the Admiralty ordered Lord St Vincent to detach a squadron under Rear-Admiral Horatio Nelson to return to the Mediterranean in 1798. This concern for what the French were doing at Toulon deprived the Home Fleet of its strategic reserve, leaving nothing to meet the French fleet if it should break out of the blockade at Brest. It left nothing to deal with a Spanish attack on the West Indies, or a Dutch attempt to dominate the

The Inshore Blockading Squadron at Cadiz, July 1797, by Thomas Buttersworth (1768–1842). Illustrating the general difficulties and dangers of maintaining a close, inshore blockade with large ships, Buttersworth shows Rear-Admiral Sir Horatio Nelson's squadron blockading the Spanish fleet in Cadiz, just after the Battle of Cape St Vincent.

North Sea. After long and anguished debate, the Cabinet in London took this strategic risk and sent the fleet into the Mediterranean in order to protect Naples and to influence Austria into joining a new coalition against France. It was a time, as the historian Piers Mackesy has described it, of 'great conceptions and disappointed hopes'. The idea of sending the fleet into the Mediterranean could easily have led to failure, but with great good fortune and after a long, frustrating, and misdirected search, Nelson found and defeated the French fleet at the Battle of the Nile in Aboukir Bay on 1 August 1798.

The War of the Second Coalition The Battle of the Nile was not enough, in itself, to mobilize a European coalition, but it played a role. After the British succeeded in taking Minorca in November 1798, they once again had the base they needed to conduct operations in the western Mediterranean and control a revived threat from Toulon. For a moment, when the French fleet under Bruix based at Brest slipped past Admiral Bridport's blockade and entered the Mediterranean in May 1799, it seemed as though the British advantages would be quickly nullified, but

The Battle of the Nile, 1 August 1798, by Nicholas Pococke (1740–1821). Anchored in Aboukir Bay, Admiral Brueys's French squadron seemed secure from seaward attack. Nevertheless, Nelson approached at dusk, as this painting shows, and fought the battle by night, with some vessels rounding the French line on the west, with others from the east, bringing overwhelming gunfire to bear in a tactical concentration. Only two of the 16 French ships-of-the-line, with two of the four frigates, escaped to Malta afterwards.

the French soon left, taking the Spanish fleet with them. Throughout 1799 allied British, Austrian, and Russian diplomats worked to forge a policy for an offensive war strategy against France.

The new strategy called for the encirclement of France, with concentric attacks from the Channel, the Alps, the north, and from the Mediterranean. However, the plan for Lord Keith's fleet to lead a sea attack from the Mediterranean was soon diverted amidst a debate within the ministry in 1800 following Bonaparte's June victory at Marengo. The issues debated in London reflected a shift in Britain's strategic stance from all-out support for offensive, coalition warfare to a more guarded, careful position which would allow her to preserve her options should the coalition fail. With this new stance, military operations in Holland and amphibious attacks on Brittany became more important, while the Mediterranean became an area for British defensive action. In this context the capture of Malta became an important strategic goal; like that of Minorca, it played an important role in consolidating Britain's defensive position in the western

Mediterranean. In March 1801 Lord Keith landed Abercromby's expeditionary force in Egypt, contesting the French presence in Egypt without trying to conquer the whole area. By seizing Alexandria, Britain hoped to be in a stronger bargaining position to achieve her goals in the Levant in a future peace negotiation.

Nothing, however, could diminish France's domination of the Continent. Russia abandoned the coalition and formed a League of Armed Neutrality with the Baltic powers. The Tsar's assassination in March 1801 and Nelson's carefully planned attack on Copenhagen in April destroyed the League, but Bonaparte's continued success forced both Austria and Naples to agree to peace.

The Peace of Amiens Following Pitt's fall from power, Addington's new ministry negotiated the British preliminary agreement on 1 October 1801. When the final

Admiral Adam Duncan (1731–1804), First Viscount Duncan, by Sir William Beechey (1753–1839). In defeating the Dutch off Camperdown, 12 October 1797, Duncan approached in two squadrons: half the British broke through the Dutch line, raking them as they passed, before engaging the Dutch from leeward. Refusing to take the Dutch admiral's sword in surrender, Duncan told De Winter, 'I would much rather take a brave man's hand than his sword'.

peace was signed at Amiens in March 1802, Spain regained Minorca, the Knights
of St John recovered Malta under Russian supervision, while Bonaparte evacuated
Naples, the Papal States, and Egypt. France retained most of her continental
conquests and gave up none of her overseas gains, while Britain retained only
Trinidad and Ceylon, yielding the Cape of Good Hope, Egypt, Malta, Tobago,
Martinique, Demerara, Berbice, and Curaçao.

For Britain, the Peace of Amiens had been a peace of necessity rather than
advantage. In the uneasy period that followed, it soon became clear that
Napoleon intended to dominate the Mediterranean, exclude Britain from the
Levant trade, capture Russian trade, and threaten the British position in India.
Britain had returned Minorca to Spain. Napoleon now proceeded to annex
Leghorn and Elba, appearing only to be waiting for Britain to surrender Malta, as
required by the peace treaty, before proceeding further.

The War against Napoleon Malta, in its role as the key position to command the
Mediterranean, was, as Talleyrand said, *'la vraie pomme de discorde'*. The ministry's
decision in 1803 not to evacuate that island in accordance with the treaty of
Amiens became the immediate cause for renewing the war. In reality, it was more
than just a renewal of the previous decade's fighting. Britain had fought the war
between 1793 and 1802 to contain the French Revolution. From 1803 to 1815, she
fought to defeat Napoleon's attempt to unite continental Europe under his own
control, building up French maritime strength in the process. As Napoleon moved
into Germany, Britain watched carefully. She returned a strong squadron under
Nelson to the Mediterranean to watch Toulon, another expedition sailed under
Sir Samuel Hood for the West Indies, where it took St Lucia, and a squadron under
Commodore John Loring co-operated with black troops on Santo Domingo.

During 1804 the French began to make preparations for an invasion of Britain
which continued into the following year. Simultaneously, Spain joined France in
the war, allowing France the possibility of achieving a naval superiority in the
Channel as well as helping her recover her position in the Mediterranean. As
Britain worked to form a Third Coalition, the French naval forces at Toulon grew.
Nelson watched, not knowing where they would strike: Sardinia, Sicily, Naples,
Egypt, or elsewhere. For this reason, he followed the French when they left
Toulon, sailing as far as the West Indies and back, to try to engage the fleet in
battle and remove it as a threat. While the squadrons commanded by Sir Robert
Calder in the Bay of Biscay, by Cornwallis in the Channel, and by Nelson pursu-
ing the French Mediterranean fleet across the Atlantic jointly prevented the vari-
ous Franco-Spanish naval forces from combining in a superior force to support the
invasion, Napoleon cancelled his plans and moved his army against Austria,
defeating her at Ulm before she could become an effective member of the Third
Coalition. Thwarted in its original purpose, the Franco-Spanish fleet at Cadiz
sailed east in order to enter the Mediterranean and, in co-operation with the
French armies in Italy, help recover Sicily and re-establish the French position.
Shortly after the Franco-Spanish fleet set sail for this purpose, Nelson intercepted

Rear-Admiral Sir Horatio Nelson (1758–1805), by Lemuel Francis Abbott (1760–1803). This heroic portrait captures the legendary Nelson as Captain A. T. Mahan would later describe him: 'The Embodiment of the Sea Power of Great Britain.' Based on a 1797 study from life, the artist supplied this updated version to Nelson's first biographer, John M'Arthur, in 1800, to show recent decorations. Among them are the artist's fanciful representation of the diamond 'Chelengk', that Nelson reportedly was wearing in his hat at the time, a gift the sultan of Turkey had taken from one of his own turbans.

A small man, Nelson's physical appearance played a role in creating the legend surrounding him, having lost the sight in his right eye during an attack on Calvi on Corsica in 1794 and his right arm while attempting to capture a treasure ship at Santa Cruz de Tenerife in 1797. His death at the battle of Trafalgar in 1805, the moment of his greatest victory, gave poignancy to his reputation. In addition to his bravery, Nelson's remarkable ability as a leader, marked by the then unusual method of sharing his innovative tactical thoughts with his trusting subordinate 'band of brothers', marked him as the greatest fighting officer in the history of the Royal Navy.

it off Cape Trafalgar, before it could enter the Mediterranean, and defeated it on 21 October 1805. By that victory, Britain prevented Napoleon from controlling the Mediterranean, defending her own presence in that sea. Nevertheless, Napoleon continued his military success, decisively defeating the Austrians at Austerlitz in December, forcing her out of the war and thus ending the Third Coalition.

The Naval War after Trafalgar From 1804 to 1806 Pitt's war strategy had been focused on the coalition, but 'The Ministry of all the Talents' which followed in 1806–7 took a different approach, seeking independent action and attempting to use overseas and distant actions as a means of influencing the war.

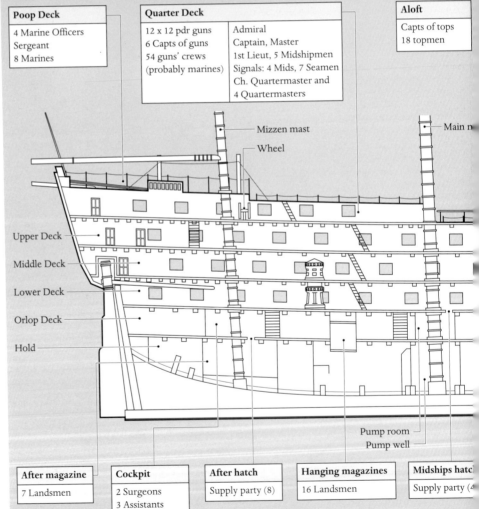

Poop Deck

4 Marine Officers
Sergeant
8 Marines

Quarter Deck

12 x 12 pdr guns	Admiral
6 Capts of guns	Captain, Master
54 guns' crews	1st Lieut, 5 Midshipmen
(probably marines)	Signals: 4 Mids, 7 Seamen
	Ch. Quartermaster and
	4 Quartermasters

Aloft

Capts of tops
18 topmen

Mizzen mast

Wheel

Main m

Upper Deck

Middle Deck

Lower Deck

Orlop Deck

Hold

Pump room
Pump well

Note: 1. The chain of ammunition supply is not fully established but probably each charge passed through several pairs of hands before reaching the gun.

2. Guns' crews were only sufficient to man one broadside. If fire was shifted from one broadside to the other, guns on the disengaged side were left loaded where possible.

3. Men in the guns' crews were pre-detailed by name in the quarter bill for emergency tasks as follows: 'F', fire-fighting; 'P', pumps; 'S', seamanship tasks; 'B', boarders. Generally there were 1 'F'; 1 'P'; 2 'S'; 1 or 2 'B' per gun.

After magazine

7 Landsmen

Cockpit

2 Surgeons
3 Assistants

After hatch

Supply party (8)

Hanging magazines

16 Landsmen

Midships hatch

Supply party (4

Source information supplied by Peter Goodwin, I. Eng. Curator, HMS *Victory*

HMS VICTORY (1805)

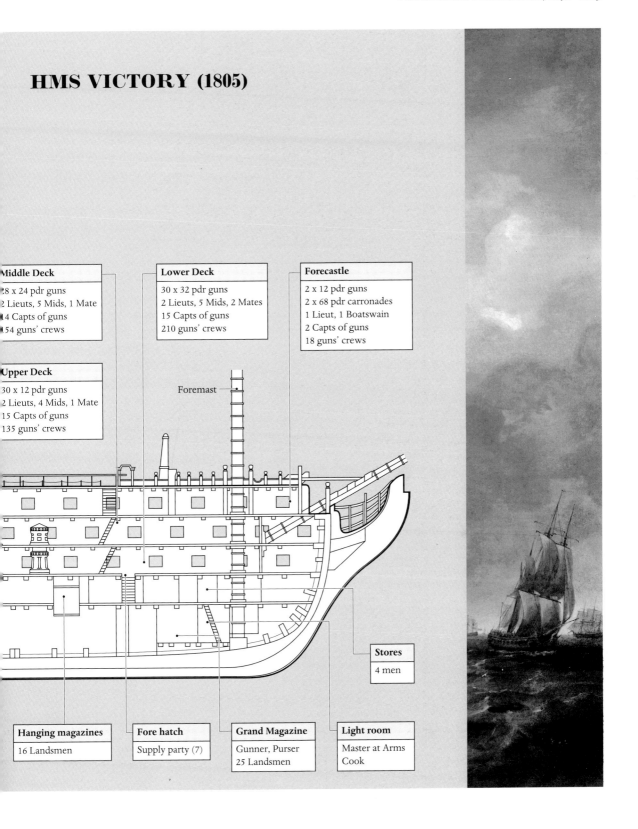

Middle Deck

28 x 24 pdr guns
2 Lieuts, 5 Mids, 1 Mate
14 Capts of guns
154 guns' crews

Upper Deck

30 x 12 pdr guns
2 Lieuts, 4 Mids, 1 Mate
15 Capts of guns
135 guns' crews

Lower Deck

30 x 32 pdr guns
2 Lieuts, 5 Mids, 2 Mates
15 Capts of guns
210 guns' crews

Forecastle

2 x 12 pdr guns
2 x 68 pdr carronades
1 Lieut, 1 Boatswain
2 Capts of guns
18 guns' crews

Foremast

Stores

4 men

Hanging magazines

16 Landsmen

Fore hatch

Supply party (7)

Grand Magazine

Gunner, Purser
25 Landsmen

Light room

Master at Arms
Cook

In the wake of Trafalgar, the weakest point of British Mediterranean strategy was Sicily. Britain could neither afford to give it up voluntarily nor could she defend it adequately. It remained a shackle on the Royal Navy, as it lay under constant threat from southern Italy and from Toulon. The blockade of Toulon remained a key undertaking, for by keeping the French squadron there Britain limited what France could do in other areas in the Mediterranean. To prevent France from dominating Turkey, Vice-Admiral Sir John Duckworth backed diplomatic negotiations in the Dardanelles and attempted to obtain the surrender of the Turkish fleet in 1807; Captain John Spanger conquered the Ionian Islands in 1809.

In 1807 a new ministry returned to the earlier focus on European affairs and began to work towards a new coalition. In this context, the Baltic became an area of concern. Fearing that Napoleon might take control of Denmark and the Baltic approaches, Admiral James Gambier attacked Copenhagen in September 1807, seizing the Danish naval ships and supplies. Following this, Lord Saumarez sailed in early 1808 for the Baltic to support Sweden against Napoleon. The Franco-Russian Treaty of Tilsit changed the strategic situation for Britain, and shortly afterwards she began to focus her attention on southern Europe, beginning with intervention in Portugal as France attempted to occupy that country. Concerned about Napoleon seizing the Portuguese naval assets, Rear-Admiral Sir William Sidney Smith's squadron intervened, escorting the Portuguese royal family and its navy to Brazil. By August 1808, the navy had landed troops in Portugal under Sir Arthur Wellesley. As these operations were just beginning, the main thrust of British strategy in 1809 focused on a large, amphibious assault on Walcheren Island in the Scheldt estuary, with the aim of removing the threat from Antwerp, reducing the resources for French maritime power, and paralleling Austria's military campaign against France in the Danube valley. Involving some 44,000 men and 235 armed vessels, the expedition proved a disaster, with bad weather, widespread illness, poor planning, and ineffective leadership.

Another new ministry came to power in autumn 1809, and strategic changes followed. In 1810 Napoleon made his last serious attempt to shift the balance of power in the Mediterranean when he attempted to seize Sicily, but for the Royal Navy the task of defending the position was further complicated by the need to support the new focus of the war in the peninsula. The revolt in Spain offered the opportunity to expand operations outwards from Portugal, and with persistence this theatre became the scene of the major British effort against Napoleon.

In other areas, the Royal Navy provided differing contributions to the war effort. Small warships and privateers were the key elements in the war involving attack on and protection of trade, as Napoleon's continental blockade threatened to damage the British economy and the basis for her financial support. British expeditions captured French Guiana, Martinique, Guadeloupe, and Santo Domingo between 1808 and 1810, while other expeditions sailed from India to take Mauritius and Réunion, removing the French privateers from those islands. This was followed in 1811 by the East India Company's seizure of Java. By this point, all of the bases which Napoleon might use outside Europe had been seized, and Britain was consequently able to reduce her forces in overseas areas.

However, Britain's emphasis on stringently controlling trade and maintaining her rights as a belligerent in the war had its effect on other nations. Between 1812 and 1815, it led Britain, while still fighting Napoleon, into a war with the United States over neutral trading rights and the impressment of seamen from American vessels. This conflict was maintained at a low level while active fighting against Napoleon continued, but, despite this, the Royal Navy maintained a blockade of the eastern coast of the United States and had to increase convoy protection. As the war in the West Indies against France ended, British troops were transferred to fight in North America. In addition to convoying these troops, the navy supported British military operations along the US–Canadian border, in the Chesapeake Bay area at Baltimore and Washington, and in Louisiana. The war at sea was marked by American successes in single-ship frigate actions and in small vessels quickly built to sail the Great Lakes. In the peace that was agreed at Ghent in 1814, neither side won the objectives for which it had gone to war.

As Napoleon's authority began to crumble following his disastrous campaign in Russia, the ministry began to develop the opportunity for forming a new coalition in 1813–14. Naval activity increased in northern European waters to support military activity there, but the emphasis on defence in the Mediterranean and support for the offensive in the peninsula remained until Napoleon's abdication in the spring of 1814.

The peace lasted for nearly a year, but Napoleon broke it with his escape from Elba and The Hundred Days. Just before Napoleon's defeat at Waterloo and his final abdication, the British squadron in the Mediterranean laid plans to support a rebellion in Provence. At the same time, the ministry chose to concentrate its forces in the Low Countries, the traditional area of key strategic concern.

Looking back across this 126-year span of warfare at sea between France and Britain, one can see repeated strategic patterns emerging as Britain became involved in alliances and used her navy to support them, while at the same time protecting herself and her trade and extending her colonies and her markets. Battles between fleets were the most dramatic events in a pattern that increasingly emphasized the importance of general naval supremacy over France and bringing pressure to bear on her adherents. It was supremacy in naval strength that allowed Britain to challenge France effectively. Throughout the period the French navy was, with relatively few exceptions, reluctant to challenge British squadrons to battle. Battles figure prominently in the illustrations of these pages only because their drama lends itself to artistic representation, not because they were the sole essence of naval power. The British navy's capacity to win large battles was ultimately essential to the continuance of naval operations, mainly employing smaller warships, which sought to deny use of the sea to opponents and to allow the ships of Britain and her allies to undertake the myriad and often unpicturesque workaday tasks of marine transportation and communication.

5

THE EIGHTEENTH CENTURY NAVY AS A NATIONAL INSTITUTION 1690–1815

DANIEL A. BAUGH

*S*amuel Pepys took a dim view of the notion that the English people had always looked to the sea and he fondly collected evidence to make his case. He explained the recent rise of the Royal Navy in terms of dynastic good fortune: 'If it had not pleased God to give us a King and Duke that understood the sea, this nation had 'ere this been quite beaten out of it.' Pepys, in retirement, witnessed the vast amounts of money that William and Mary's parliaments allocated to the navy and could not have been blind to William's strategic uses of the navy, but he clung to his prejudices: the new king's genius seemed 'bent to land-action only'. As to parliament, Pepys's view reflected his own bitter experiences: the House of Commons was more interested in censuring administrators than providing a fleet.

By still supposing in the 1690s that the fate of the Royal Navy lay in a monarch's palm and that parliament, being a body of landed men, would fail to understand its needs, Pepys showed how thoroughly his mind was rooted in the past. Public and parliamentary support for a strong navy came forward with a surge in 1690. The débâcle off Beachy Head in late June had exposed the kingdom to invasion. When parliament assembled that autumn the temper of the House of Commons was such as Bishop Burnet claimed he had never seen in his whole life. All the members he had spoken with said that 'they dare not go back into their country if they do not give money liberally . . . we seem now not to be the same people that we were a year ago'. In the year following Beachy Head, a special fund was voted for building 27 men of war, a new dockyard at Plymouth was authorized, and sweeping investigations of naval administrative problems were begun. From this time forward the constancy of public and parliamentary concern for the navy was a prominent fact of English political life. It did not matter whether the nation's monarchs were naval enthusiasts, military enthusiasts, or neither.

One result was massive growth. The number of seamen and marines borne on the books of the active fleet reached a wartime peak of 48,000 in 1713; 85,000 in 1762; and 142,000 in 1813. The list of lieutenants in 1715 showed about 360 names; there were 640 in 1748; 880 in 1763; 1,350 in 1783; and 3,270 in 1813. The number of captains did not grow so greatly, from about 260 in 1715 to 800 in 1813, but there were 600 commanders in 1813, a rank which did not exist a century earlier. The surges of growth came, of course, in time of war, and during each major war the average yearly cost of the navy reached a new plateau, as the figures in Table 5.1 show.

Table 5.1 Average annual spending on the navy in wartime, 1689–1815

Date	£ millions
1689–97	1.8
1702–13	2.4
1739–48	2.9
1756–63	4.6
1776–83	7.8
1793–1815	15.2

Source: B. R. Mitchell, *Abstract of British Historical Statistics* (Cambridge, 1962).

What is especially interesting, however, is the rise in peacetime naval expenditures. Table 5.2 shows average annual spending between major wars and its relationship expressed as a percentage to that during the preceding war. Some of this spending reflected occasional spurts of mobilization. Especially during the 'peacetime' intervals of the period from 1715 to 1775, battle squadrons were frequently readied for purposes of deterrence. Yet most of it represented the growing cost of stationing more cruisers, keeping up a larger fleet in reserve, administering an elaborated permanent service, and improving the efficiency and global reach of the shore establishment.

Table 5.2 Average annual spending on the navy in peacetime, 1715–92

Date	£ millions	%
1715–38	1.0	43
1749–55	1.3	45
1764–75	2.0	42
1784–92	2.1	34

Source: B. R. Mitchell, *Abstract of British Historical Statistics*.

As of old, the Commissioners of the Navy (the Navy Board) continued to bear the main burden of naval administration. They still supervised shipbuilding and repair, the purchase or hiring of vessels, the purchase of naval stores, and all aspects of dockyard administration. They controlled every aspect of naval expenditure, auditing the accounts of officers serving at sea as well as departments ashore. They had financial control of both the Victualling and the Sick and Wounded Commissioners.

The victualling service began a long course of improvement in 1700. Traditional naval histories have focused on the occasional failures, but in fact the Victualling Office's record in supplying warships on ever more distant stations was the envy of other navies. The care of sick seamen sent ashore, however, saw little progress before the 1740s; in that decade the Sick and Wounded Board (also charged with the care of prisoners of war) became permanent and work was begun on permanent naval hospitals.

But the key administrative development of the period was the rise of the Admiralty. During the 20 years after 1689, the office of Lord High Admiral became a permanent bureaucracy under commissioners who comprised the Board of Admiralty. Its location became fixed and its records ceased to be regarded as the personal belongings of its chief or his secretary. The secretaryship became a stable position, especially under Josiah Burchett, who held it for 44 years. In 1699 the first Admiralty building was erected in Whitehall. Although this structure's foundations were unsound and it was torn down, the current Admiralty building, built in 1725, stands on the same site.

All other branches of naval administration reported to the Board of Admiralty and came under its direction. When important issues of policy or sudden emergencies arose, the board in whose province the business lay—Navy, Victualling, or Sick and Wounded—was asked to come to the Admiralty Office for a joint-meeting. Otherwise, everything was done by letters and messengers. But the speed of communication with the outports was revolutionized in 1796 by the construction of hilltop telegraphs running from the roof of the Admiralty building to Sheerness, Chatham, Portsmouth, and, in 1806, Plymouth.

The most substantial business of the Admiralty concerned the fleet and its personnel. The board not only wrote orders but dealt with the variety of problems disclosed in a vast flow of letters from all ranks, especially from flag officers and captains. Extra clerks were hired in wartime, as they were in all the branches of naval administration, but by modern measures the Admiralty staff was small—two dozen established clerks and perhaps four dozen temporary hires even at the height of the Napoleonic wars—and the latter were dismissed when peace came. Considering the mass of surviving papers, and the volume of hand-copied orders and letters, one can only be amazed that it was all done on so modest an establishment.

The First Lord of the Admiralty was a member of the cabinet. His position was inevitably political as well as administrative, whether it was held by a sea officer or a civilian. Admirals such as Sir Charles Wager and Lord Anson understood this completely. All First Lords met with the narrow circle of ministers who conducted strategy whenever the use of the navy might be involved; if the First Lord happened to be a civilian, an admiral was brought into the deliberations.

The real importance of the Admiralty, however, lay in the sum of the parts. Its capacity to establish the means of naval effectiveness and to co-ordinate all aspects of naval preparedness was of immense administrative value. Yet even more important was its position as the permanent, authoritative centre of accumulated naval expertise, in regard to both naval needs and naval strategy. The eighteenth-century French navy had no equivalent and felt the want of it repeatedly and severely.

Ships were the sole instruments of naval power in the age of sail; their numbers are the starting-point of any assessment of naval strength. The figures in Table 5.3 refer to warships on official navy lists, whether laid up in reserve or commissioned for active service. What is represented, therefore, is the mobilizable force, not the force fitted out and manned. In examining Table 5.3 it will be noticed that the number of British ships of the line of battle (ships of the top four rates, carrying from 100 to 60 guns) did not increase very much during the period. Although, in

SHIPS, DOCKYARDS, AND BASES

Table 5.3 Ships and tonnages, 1680–1815

Year	Numbers		Displacement tons (000s)		
	Line of battle	Cruisers[a]	British	French	Dutch
1680	95	20	129	132	64
1690	83	26	124	141	68
1695	112	46	172	208	106
1700	127	49	196	195	113
1710	123	57	201	171	119
1715	119	63	201	108	98
					Spanish
1720	102	53	174	48	22
1730	105	49	189	73	73
1740	101	58	195	91	91
1745	104	98	235	98	55
1750	115	108	276	115	41
1755	117	108	277	162	113
1760	135	172	375	156	137
1765	139	136	377	175	124
1770	126	112	350	219	165
1775	117	115	337	199	198
1780	117	187	372	271	196
1785	137	181	447	268	211
1790	145	180	459	314	242
1800	127	264[b]	546	204	227
1805	135		569	182	139
1810	152	390[b]	725[b]	194	100
1815	126		616	228	60

[a] Fifth- and sixth-rate frigates, and sloops of war.
[b] Rough estimates which include cruisers displacing less than 500 tons, ships which are omitted in the other post-1790 figures.

Source: Compiled from figures in Jan Glete, *Navies and Nations: Warships, Navies and State Building in Europe and America 1500–1860*, 2 vols. (Stockholm, 1993).

A sixth-rate frigate ready for launching at a merchant builder's yard, 1758. Cruisers were essential to the policy of comprehensive sea control toward which Britain moved decisively from the 1690s onwards. Nearly all cruisers were built in private yards to meet surging wartime needs. Note (foreground) that this yard, on the Thames, had a dry-dock capable of receiving a small frigate.

fact, the total tonnage of this category grew substantially, the main reason was that between 1710 and 1810 the average size of a ship of the line doubled.

The growth of Britain's cruiser force must therefore command particular attention. Cruisers (frigates of the fifth and sixth rates, sloops of war, and armed brigs) accounted for 15 per cent of British warship tonnage in 1710, 30 per cent in 1760, and 43 per cent in 1810. Table 5.3 shows that their numbers compared to 1710 tripled by 1760 and rose almost seven times by 1810. Although frigates and sloops of war of each class were built larger as the century progressed, the British navy's insatiable demands required all sizes. In the decade after 1801, for instance, armed brigs of about 350 tons (equal to the displacement of a 20-gun frigate a century earlier) became the most numerous class of ships in the fleet.

The table shows how important the allied Dutch navy was in helping to offset French naval strength in the 1690s. Older histories held that the Royal Navy was in decline during the period 1700–40, but the error of this idea emerges instantly when British tonnage is compared with that of the Bourbon powers, Britain's usual opponents. It was the French navy that fell into decline, and the recovery of Bourbon naval strength after 1720 depended considerably on Spain's naval awakening. The really remarkable fact concerning this period is Britain's evident determination *not* to reduce her fleet regardless of the obvious decay of the French navy.

From about 1730 onwards one may reasonably discuss the British naval situation in terms of a modern concept, a 'two-power standard'. Having come close to parity of tonnage in 1740, the Bourbon powers fell behind during the wars of the 1740s. During the peace of the early 1750s—notwithstanding the presence of Lord Anson at the Admiralty—Britain's preponderance over the two rivals came to an end: by 1755 the House of Bourbon reached parity. In the Seven Years War British preponderance was impressively re-established, but the combined French and

Spanish navies again achieved parity of fighting units during the subsequent peace. (Their combined tonnage was actually 18 per cent greater than Britain's in 1775, but, rate for rate, Spanish and French ships of the line were about 20 per cent larger.) The American Revolutionary War was the only eighteenth-century war in which the British navy did not wind up with greater tonnage. Finally—a little known fact—during the peace after 1783 the Bourbon navies, notwithstanding their governments' financial crises, substantially augmented their fleets, and thus, though the British navy was well cared for in the 1780s, it did not enjoy a clear preponderance in 1790. To sum up, during the first half of the century the mobilizable force of the British fleet remained stronger than that of its combined rivals in peacetime, but during the second half it became relatively weaker.

May it be concluded that Britain prevailed by outbuilding her opponents in wartime (except in the case of the American Revolutionary War)? This seems plausible. In time of war Great Britain enjoyed far superior financial resources and better access to essential materials. Yet, as the figures in Table 5.4 reveal, the war of the Spanish Succession and the war of 1739–48 were the only ones during which Britain built significantly more tonnage than her opponents. In the Seven Years War the exertions of the two sides proved about equal. Actually, Britain's most intensive shipbuilding effort of the whole century occurred between 1778 and 1783, but it did not manage to produce parity. This helps to explain why British wartime governments were so intent on using sea control to deny passage of timber, masts, and vital stores to enemy dockyards.

The other point revealed by Table 5.4 is the importance of captures. Although some of this captured tonnage was taken from a third party, one may best appreciate the weight of this factor by adding *and* subtracting the last column as if the

Table 5.4 Tonnage of fighting ships built and added by captures, 1691–1815

| | Displacement tons (000s) | | | | | | |
| | Built by | | | | Captured and added by | | |
	Britain	France	Spain	House of Bourbon	Britain	France & Spain	Britain net
1691–1700	115	132			9	13	−4
1701–15	156	62			28	18	+10
1716–40	193	82	149	231		peacetime	
1741–50	186	111	40	151	39	7	+32
1751–5	17	71	77	148		peacetime	
1756–65	201	142	55	197	69	6	+63
1766–75	103	80	96	176		peacetime	
1776–85	267	223	77	300	96	32	+64
1786–90	69	95	60	155		peacetime	
1791–1815	600			739[a]	451	84	+367

[a] Includes the French Empire with its satellites and allies.
Source: Compiled from figures in Jan Glete, *Navies and Nations*.

tonnage were simply transferable from one side to the other. Admittedly, this would produce rogue statistics, but it exposes a fundamental truth, and in the wars of 1756–63 and 1793–1815 the consequences were as sensational as they were real. This is the main reason, in fact, for the huge preponderance of the British fleet by the end of each of those wars. Clearly, the Bourbon naval efforts were overcome only by a combination of wartime shipbuilding, sea control, and effectiveness in combat.

Table 5.4 also suggests that the British navy's superiority in combat increased as the century wore on. This was broadly the case, but one must remember that the figures only count ships which were actually taken into the captor's navy; they do not encompass those rejected and, more important, those destroyed. For instance, only one ship of the line was captured at La Hougue in 1692, although a huge quantity of French fleet tonnage was destroyed. Whatever doubts must be allowed, there can be none concerning the Napoleonic era. During the long struggle of 1793–1815, the British captured or destroyed 139 enemy ships of the line and 238 cruisers (of 28 guns or more), of which 83 of the line and 162 cruisers were added to the Royal Navy, whereas the total number of British warships destroyed or taken as a result of enemy action was twenty-four.

Nevertheless, Britain had to build an immense number of warships. Throughout the period, 'Great Ships' of 90 and 100 guns were built in royal dockyards and nearly all frigates and sloops were built by contract in private yards. The navy would have preferred to build all its warships in royal dockyards, but although from about 1700 to 1755 ships of the line were generally government-built (or rebuilt), at other times ships of less than 90 guns were commonly built in private yards. The massive building programme of 74-gun ships during the American Revolutionary War was almost entirely carried out in private yards. This practice continued in wartime down to 1815.

Shipbuilding was not the primary function of an English dockyard. In time of peace, caring for the ships in reserve ('Ordinary') by ventilation, regular inspection, and timely repair held priority. Granted, the Ordinary was not looked after as it should have been. The warrant officers—boatswains, gunners, carpenters, pursers—who were supposed to reside on board and look to their duties were mostly long-service men in the condition of pensioners. The salvation of these ships was really the 'triennial trimming', which placed them in dry-dock and got them properly inspected and repaired. Well-constructed ships, well cared for and given timely repair, lasted for decades, but in most cases hurried construction, peacetime neglect, and wartime wear and tear took their toll; toward the end of the century longevity was unlikely to exceed ten years without a costly major repair.

In wartime the focus of dockyard attention was on the active fleet: cleaning, refitting, and performing small repairs. The figures in Table 5.3 measure potential sea power. Actual sea power depended of course on the numbers in active service. For instance, 95 of the listed 123 ships of the line in 1710 were in commission; 108 of 152 were in active service in 1810; and practically all British cruisers fit for service

were in commission on those dates. The royal yards of France and Spain rarely shouldered anything like the burden of these numbers.

In 1690 all the royal dockyards were in the Thames basin and orientated towards war with the Dutch, save Portsmouth. Shallow water and distance upriver rendered Deptford and Woolwich yards unsuitable for quick refits, and during the eighteenth century their work was focused on shipbuilding and preparing specialized equipment. Sheerness, near the Nore, became the cleaning and refitting station for east-coast cruisers. Chatham, with no fewer than five dry-docks, was by far the largest yard. It came to be used chiefly for seasonal (winter) overhauls and major repairs. As already noted, a new dockyard at Plymouth was established in the 1690s; its growth was phenomenal, accounting for over 80 per cent of the navy's added shipwrights between 1711 and 1772. The facilities at Portsmouth as well as at Plymouth were greatly improved and enlarged during the century. The crucial role of these south-coast dockyards in our period is exhibited by sample statistics for 1774–83 when Portsmouth and Plymouth accomplished the docking and refitting of 87 per cent of the active 74-gun ships (161 dockings).

The launching of the *Cambridge* at Deptford dockyard, 1755. She was the last of the 80-gun three-deckers, a class not well suited to all-weather operations. But the *Cambridge* remained on the fleet list for almost 40 years, testimony to the quality of unhurried shipbuilding in royal yards.

Lower masts could be replaced by means of a sheer hulk. The men at the hulk's capstan are providing the lifting power. Warships no longer fit for service were converted to hulks at overseas bases as well as at home.

England's dry-docks gave her a considerable naval advantage. The growing French arsenal at Brest did not have dry-docks until the 1740s, and Toulon did not have a dry-dock until the 1770s (partly because of the narrow range of tide in the Mediterranean). Docking put less strain on a hull than careening, and except for the two largest rates it was possible on a spring tide to bring a ship of the line into dock without clearing her lower hold and ballast. Even so, refitting a 74-gun ship took an average of 14 weeks, five in preparing for the dock, three in dry-dock, and six in re-rigging and stowing for sea.

The use of private contractors for shipbuilding (and sometimes repairs to frigates and sloops) in the second half of the century was partly dictated by a skilled-labour problem. After 1714, shipwrights were no longer impressed for dockyard service yet there was a tremendous wartime demand for them. Aware of the high wages paid by merchant builders, the navy's shipwrights insisted on overtime wages when wars broke out and went on strike to secure these plus other special allowances, all of which had the effect of raising their incomes significantly. The authorities talked of discipline but yielded to market realities and the shipwrights' organized pressure. The result of working 'Two Tides Extra' (five hours) was a nominally 14-hour day, very useful in emergencies such as an invasion threat, but conducive to desultory work habits at other times. Reforms were attempted, and in some degree task work was actually introduced, but there was always the difficulty that the hardest-working shipwrights were ready to go to the merchant yards. Expanding the number of apprentices in the dockyards was an obvious remedy; an apprentice shipwright was often a valuable worker after just three years. The failure to add apprentices between 1775 and 1778, and the deep cuts in dockyard personnel made by order of the earl of St Vincent in 1802, were probably the two greatest administrative follies of the whole period: both preceded serious invasion threats. The most astonishing statistical fact is that, notwithstanding the increase in ships and tonnage, the number of journeymen shipwrights employed in the royal dockyards hardly changed: 3,100 in 1743; 3,200 in 1772; 3,300 in 1781; 3,100 in 1792; and 3,300 in 1804. (In 1804 there were 5,100 shipwrights in the British Isles.)

Britain, nevertheless, enjoyed some formidable advantages with regard to the upkeep of the fleet. One, as already noted, was access to proficient contract builders. A second was the ease with which most non-shipwright workers could be added (for instance, in the six main yards employment rose from 9,000 in 1792 to 15,000 in 1813).

A third was superior access to vital naval materials. It used to be thought that there was a serious timber problem. Very large timbers and certain oddly shaped pieces for knees and standards were, indeed, always scarce. When old ships were broken up, salvageable pieces of this sort were always saved for re-use. Also, iron was sometimes substituted. When large lower masts became scarce after 1775, 'made masts' became more common. All navies experienced these problems, but with respect to lesser timbers and masts it all came down to money and access.

Port Royal, at the entrance to Kingston Bay, had suffered so much from an earthquake and three hurricanes by the late 1720s that the Admiralty chose to build its main naval installation on Jamaica elsewhere, at Port Antonio. That place proved too rainy and unhealthy, and Port Royal became the focus of development after 1735. The 1782 plan emphasizes the base's military defences. The careening wharves were located on the north-east side.

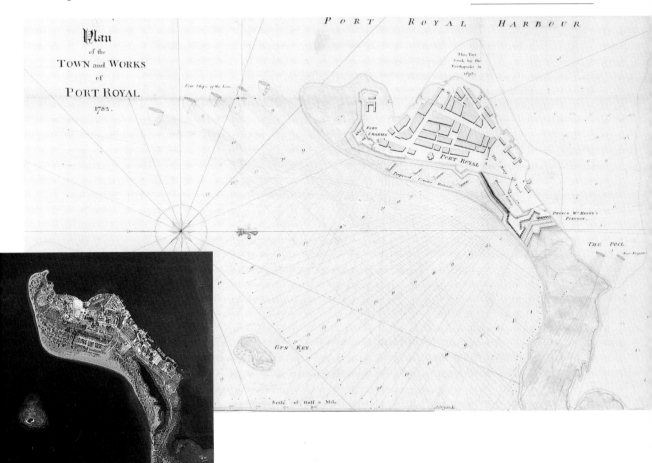

English Harbour, Antigua. Easily defended and the best hurricane refuge in the West Indies, this small harbour was developed from 1729 onwards as a refitting base. Eventually, stone careening wharves and storehouses were created on filled land in the centre of the inner harbour. Before copper, wood sheathing was used as a barrier to the shipworm, *teredo navalis*; it had to be replaced periodically, especially in tropical waters, lest the hull become fatally infested.

Here the British navy enjoyed great advantages. It could pay the market price, get its bills readily accepted, rely on British Baltic merchants who were connected to the best sources of supply, and interrupt supplies to enemy yards. These factors applied to naval stores in general; most hemp and tar came from the Baltic countries, where supplies were greatest and the quality highest.

Overseas bases conferred a fourth great advantage. Naturally their locations followed strategic needs. Gibraltar, captured in 1704, helped to transform the advantage France gained by Toulon's distance from England into something of a liability. Minorca was acquired, in 1708, because of its superb harbour, Port Mahon. The island base was twice lost to French assaults, in 1756 and 1782. At the end of the century, the British took Malta, and from 1800 onwards its excellent, fortifiable harbour became almost as important to the navy as Gibraltar. Hostile relations with Spain in the later 1720s prompted plans for a permanent, well-equipped base at Jamaica. Initially, the strategically superior location of Port Antonio, on the north-east coast, was chosen. It was only after extensive works were completed there that it was found that the constancy of rain seriously hampered work and also caused the place to be extremely unhealthy for ships' companies. As a result, the wharves and storehouses were built at Port Royal, despite its recent history of natural disasters. English Harbour, Antigua was established because the Leeward Islanders wanted better trade protection; it was found that the security of the harbour rendered it a useful place to develop as a base. The stationed ships on the North American coast could rely on the main colonial ports for support. Halifax was first established as a base in 1749 in response to the heightened Anglo-French contention in the region. Its subsequent development was a response to the alienation and loss of the Thirteen Colonies.

On the other side of the world, in the Indian Ocean, a significant presence of

Royal Navy ships was not seen until the later 1740s. Not one of Britain's overseas bases had a dry-dock, but at Bombay the East India Company built one in 1750 (and another in 1806 that could dock a 74-gun ship). As the British admiral who had been on that station in the early 1780s reported, 'at no other port or place in our possession could the ships of the Squadron be even properly refitted much less repaired'. The well-stocked East India dockyard at Bombay, with its own native shipwrights, was not strategically in the best location, but it gave the British navy an immense advantage in maintaining its squadrons in the Indian Ocean.

Any navy would use a port facility available to it, as had the British navy when it based a fleet on Cadiz in 1694–5. British warships were refitted from time to time at Lisbon throughout the period. What is remarkable about British naval bases is the manner in which they were quickly looked upon as permanent and were progressively developed. The only exception to the pattern was Gibraltar where investment was held up by the possibility that British diplomats might give it back to Spain. A base was always a refreshment point for the crews, with a victualling agent and storehouse present or contractor nearby, and perhaps there was a hospital too. With regard to ships the most common tasks were to clean the bottom and replace damaged masts. Often it was also necessary to replace worm-eaten sheathing and make minor repairs. A hulk could be positioned for re-masting, and a ship could be cleaned and re-sheathed by makeshift methods of careening. But a careening wharf, with nearby storehouses and purpose-built capstans for easier heaving down, promoted safety and efficiency. It was the key investment. Careening was essential to restore speed, but in tropical waters it was also necessary for combating the shipworm which attacked hull timbers.

Did the Royal Navy suffer from any serious disadvantages with respect to ships? It is commonly believed that British warships suffered from inferior, unscientific design and that this had serious operational consequences. During the first half of the century this was probably the case; French ships on the whole were superior. To the British navy the seventeenth century bequeathed tough, compact, heavily armed, and seaworthy ships which were well suited to a navy that kept its ships at sea, but they were not as fast as the ships of the reborn French navy and did not carry their guns so well, especially the 80-gun three-deckers. It appears that British ships should have grown larger sooner, but a combination of factors—the need to economize, the sanctity of the 'Navy List', the quest for uniformity within rates by establishments of dimensions, the practice (and mentality) of rebuilding—all conspired to prevent bold increases in ship size. The human embodiment of this conservatism was the opinionated, arrogant, and seemingly immortal Surveyor of the Navy, Sir Jacob Ackworth, who held the job for 34 years (1715–49).

After his retirement, however, British ship design caught up. In 1755, under Anson's influence, the Deptford Master Shipwright, Thomas Slade, was made Surveyor, and he fully deserved his subsequent knighthood. Within 20 years new designs permeated the fleet; the two-decked 74 in particular was an extremely efficient ship, combining strength, firepower, swiftness, and good sailing qualities. British ships were further improved by copper sheathing, which was aggressively installed throughout the fleet between 1779 and 1781. It inhibited hull fouling and thus enabled ships to sail near their best operational speed for a year or more without cleaning. During the 1780s the carronade was introduced. Stubby but effective at short range, the larger version could fire the same ball as a 32-pounder even though it weighed about the same as a 6-pounder long gun; it was ideal for mounting on an upper deck. With respect to fighting ships, these almost simultaneous technological innovations were the most notable of the century, and Sir Charles Middleton (later Lord Barham) is to be credited with pushing hard for both of them.

Sir Charles Middleton, Controller of the Navy, 1778–90. A vigorous, reforming head of the Navy Board, he played key roles in expediting the coppering of the fleet in 1779–81, overcoming objections to carronades, and promoting the readiness of ships in Ordinary. As Lord Barham, he was First Lord of the Admiralty in 1805–6.

In fact, hull contours and 'scientific' naval architecture were not nearly as important as is commonly supposed. If British ships, unit for unit, were smaller and tougher, and therefore less expensive to deploy, that made sense in terms of Britain's vast naval needs. As for speed, very slight alterations in ballasting and rigging—to say nothing of variations in wind and sea conditions—could nullify the advantage of an inherently swifter hull. As for general capabilities, British ships of the 1740s (their moment of greatest inferiority in terms of design) managed to take or destroy 20 French ships of the line and 15 Spanish, losing one. A retrospective study of the 1793–1815 period has recently analysed 58 chases in which a ship or squadron was clearly faster than an opponent. The key comparisons involved 40 British-built ships (frigates and larger) manned by British crews, eight captured

French-built ships manned by British crews, and six French-built ships manned by French crews. The findings: French-built ships manned by British crews were only slightly faster than British-built ships manned by British crews; French-built ships manned by French crews ran a very poor third. Each of these chases resulted in combat, in which the advantage of being manned by a British crew was equally evident. In some measure these outcomes may be explained by the higher quality of British ordnance, but it was well known that British crews, especially in this period, were capable of a much higher rate of fire than their opponents. As far as speed was concerned, seamanship clearly mattered more than hull design. British superiority at sea during the eighteenth century rested on men more than ships.

The Royal Navy had access to the largest reservoir of seamen of any nation. SEAMEN AND MANNING Skilled seamen working as a team were essential to combat effectiveness. Aloft they enhanced speed and manœuvrability; at the guns they enhanced rate of fire. They were also essential to a ship's survival. Anticipation and quickness of response not only prevented catastrophe but minimized damage to masts and rigging; moreover, a good crew could repair their own ship (spare materials were carried aboard) to an extent which the modern age can scarcely comprehend. Without a body of proficient petty officers and seamen an eighteenth-century warship was not worth much.

Yet our modern age has formed a general impression that the sailors who manned eighteenth-century British men of war were all virtually kidnapped, nameless to their commissioned officers, callously given bad food, often whipped, and ever of a mind to mutiny. This impression has proved indelible, regardless of the obvious record of success in combat and sea-keeping. As a result, the superior performance of British seamen in the line of duty is either left unexplained or implicitly, sometimes explicitly, attributed to the officers—as if all those boyish lieutenants and 26-year-old commanders could have instilled energy, skill, and teamwork into a personally unknown, miserable, sullen crew by means of the lash.

Plainly there is a puzzle here. In trying to solve it one must begin with correct facts as to how most naval seamen were recruited. What steps did the Admiralty take to improve the means of recruitment? Were most men, in fact, 'kidnapped', that is to say impressed? These questions will be addressed in this section. In the section to follow, the Admiralty's attention to matters of health and provisioning will be examined. After that, it will be time to reopen the puzzle involving discipline, morale, and shipboard performance.

For all the failings of its mode of recruitment, the British navy solved its problem of wartime shortage better than its opponents solved theirs. At the height of a war it could claim a preponderance of skilled manpower even when it did not, on some occasions, possess a preponderance of ships. It is equally the case, however, that manning the fleet was its most intractable problem—one that generated other administrative problems in turn and, above all, seriously affected naval readiness in the early months of every war.

Robert Williams, boatswain's mate of the *Venerable*, a drawing by Philippe Jacques de Loutherbourg. Petty officers were keys to a ship's effectiveness. Their wages were higher and their prize-money shares larger than an able seaman's, but the navy had no method for giving them an established claim to their rating; the matter was entirely in the hands of the captain.

The Admiralty recognized its severity. During the first half of our period it tried to create a registry system along the lines of the French *inscription maritime*. A scheme was adopted by the English government in 1696: it was voluntary, entailing a payment of £2 per year to those who chose to register, and it failed. In 1720 a bill was introduced into parliament by the government for a compulsory register and was speedily rejected. In 1740 the administration of Sir Robert Walpole supported a bill urged by the Admiralty for a mandatory register of all seafaring men, but again parliament turned it down. Ideological concerns about Crown power and Englishmen's liberties were voiced, but as will be seen in a moment there were other concerns.

A registry might have helped speed mobilization, but it could not have done much more than that. France's system worked fairly well so long as her naval mobilizations were seasonal or modest in scope, but certainly by the 1740s her system of 'classes', which allowed for a rotation of service among the maritime population, was overwhelmed by the scale of France's naval needs. In the case of Britain the maritime population had grown rapidly in the seventeenth century and continued to grow in the eighteenth, but in the 1690s, when seasonal manning ceased to be commonplace and naval needs swelled (by early 1692 four times as many men had been mobilized as in the autumn of 1688), the problem took on a new character. The most important fact, however, was that the navy's wartime requirements for seamen fell on top of existing demands which did not slacken, because Britain—this was less true of France—had to try to maintain the nation's merchant shipping as if war made no difference. Moreover, privateering started up and there was always a huge government demand for transport vessels to carry troops and supplies. In 1690, even before the navy's demands escalated, a pamphleteer observed: 'That there are not seamen sufficient in England and the rest of their Majesties dominions to man the royal fleet and to drive the trade thereof, experience tells us.'

Recent research reveals some precise dimensions of the problem. Table 5.5 provides a comparison of average employment levels of British seafaring men, naval and civil (the latter including seaborne, coastal, and inland navigation) in three wartime and pre-war periods. It shows the wartime imbalance of supply and demand growing worse as the century progressed. Because each figure represents an average for a span of time the table does not reveal the full extent of wartime expansion; peak years show much higher figures for both naval and total seafaring employment.

There was never a moment during the whole century and a quarter when a merely administrative remedy could have solved the manning problem. Although proponents of a registry often envisioned a general solution, the real benefits would have been confined to the initial months of mobilization. All the same, why did the shipping interest so pertinaciously oppose a registry? Because—rhetoric about Englishmen's liberties aside—if the government had commanded a systematic power to transfer large numbers of seamen from merchant ships to the navy, the main burden of the shortage would have fallen on the merchants. An admin-

Table 5.5 Average annual seafaring employment, 1736–83

Years		Naval	Privateering	Civil	Total
Peacetime	Wartime				
1736–8		14,845		35,239	50,084
	1739–48	43,303	2,602	30,392	75,997
1753–5		17,369		40,862	58,231
	1756–63	74,771	3,286	37,584	115,641
1773–5		18,540		50,903	69,443
	1775–83	67,747	3,749	44,947	116,443

Source: David J. Starkey, 'War and the Market for Seafarers in Britain, 1736–1792' in *Shipping and Trade 1750–1950*, ed. L. R. Fisher and H. W. Nordvik (Pontefract, 1990).

istered system, though it would have benefited the navy and the seamen, was thus unlikely to benefit the merchants and they had sufficient influence in the House of Commons to prevent this.

The modes of recruiting seamen therefore remained unchanged and amounted to a contest in which the merchants relied chiefly on paying higher wages while the navy relied ultimately on physical compulsion, which was exercised through the inequitable and erratic practice of impressment. This applied only to 'seamen, seafaring men and persons whose occupations or callings are to work in vessels and boats upon rivers', but included anyone who had ever been so employed.

Although the navy had no legal right to impress landsmen (invariably called 'landmen' in the eighteenth century), quite a few men who had never been to sea were caught up in the navy's clutches. Magistrates sent men and did not much care whether they had ever been to sea or had the slightest potential; most were rejected when they reached the ports. Criminals were also rejected, except for smugglers (good seamen usually) and debtors. A good deal of evidence suggests that most landsmen were released by the naval authorities as soon as their status was discovered.

Still, pressing ashore invited violence. Gangs and their lieutenants often got into trouble with local authorities, and one readily understands why Regulating Captains were instituted, first in London and later in provincial ports. Captains on half-pay were thus called upon, for an additional £5 a week, to regulate the activities of press gangs. In 1793 the Impress Service ashore was formalized as a separate recruiting arm and headed by an admiral; by 1797 it employed 80 lieutenants, and 47 captains and commanders.

Whatever the mode of administration, impressment ashore was not a pleasant assignment for an officer. Mayors of the ports could be expected to flaunt their hostility to it: lord mayors of London often did, mayors of Bristol almost without exception. In 1756 the mayor of Liverpool gave out that any officer who pressed a man in his city would be thrown in prison, and in an ensuing riot the regulating captain was nearly killed. The attitudes struck by civic authority served not only to invite prosecutions but mob actions, and regulating captains and lieutenants had to make quick decisions in dangerous situations, as if in combat with an enemy. Assigned to the Impress Service at Hull in 1803, Lieutenant William Dillon was courteously enough received by the city's élite, but not by some others. He later remembered, 'a volley of either musket or pistol balls was fired into my room one evening as I was reading at my table.' He luckily escaped injury.

Most men—especially the best seamen—were obtained afloat, and in this sphere serious eruptions of violence were less frequent and relatively invisible. Here one can detect a pattern of conventions. The 'rules' of this game were shaped so that each side—the merchant interest and the navy—would get a share of the available trained seamen. The key 'rule' was that the navy must not take men from outward bound vessels. It was too easy to do, thus ending the game before it started, for no seaman who hoped to avoid naval service would risk a voyage under such a threat; he would abandon the sea for some other occupation or perhaps find foreign employment. The navy's focus was therefore on incoming vessels. Upon mobilization for war, cruisers were ordered to blanket the coast. Tenders were hired, and newly commissioned ships were ordered to send them out as fast as they could be manned. One difficulty was that pressing at sea necessitated extra men who could be substituted for those removed to see the vessel safely to her destination. Because such men had to be reasonably competent and also trusted to come back, usually overland, they had to be volunteers. Fortunately for the navy, quite a few seamen, perhaps long out of work, were willing to enter the service voluntarily at the beginning of a mobilization, and though these were seldom the most skilled of their profession, they were indispensable.

George Morland's 'Jack in the Bilboes' (1790) portrays a London waterman being pressed. The power of impressment, which resisted all legal challenges during the century, extended to anyone who worked on the water.

As a supplement to these efforts, cruisers bringing in convoys were routinely ordered to press seamen out of the ships under their care. When convoys got close to home, some ships, their masters urged on by the crew, broke away and fled. When incoming ships reached the Downs, Folkestone cutters received high fees for rushing seamen ashore. Homecoming East India captains sometimes put into Irish ports, ostensibly for other reasons but probably to enable their men to flee ashore. In fact, as a category, East India crews were likely to offer the most violent resistance—understandable when one remembers how long they had been away. For instance, they fired upon approaching warships' boats in the Downs in September 1740, and in 1743 there was a bloody fray at the entrance to the Channel in which six men were killed and many wounded.

In every considerable mobilization there was a moment when the navy chose to invoke the extreme measure of an embargo. A manning embargo forced shipowners to disgorge a portion of their seamen in return for protections given to the rest and permission to sail, an arrangement known as compounding. Everyone understood that this could not be used long or often—otherwise British maritime trade would be ruined by its own navy. And if trade were ruined Great Britain could not financially sustain a war nor could merchant ships, transports, colliers, and fishing vessels train up more seamen. Another extreme measure was a general press, or 'hot press'. This was a comprehensive surprise sweep of the ports. Secretly arranged, and scheduled to occur after dark, it gathered in every seaman, afloat or ashore, regardless of whether he carried a protection; the selection of those to be released was worked out later.

Protections were formal documents. They were issued by officials in all branches of the navy and many other government offices. An account drawn up for the House of Commons showed that 50,000 protections were circulated during the year 1757. This does not mean that there were 50,000 protected seafaring men; many protections were written for a limited time-period and covered men who were not really seamen. By an act passed in 1740 no landsman could be pressed during his first two years at sea. Shipowners were thus encouraged to train up landsmen, but this was open to substantial abuse; the navy sometimes countered with another extreme measure, namely a temporary order to press one out of every five or six protected men.

Britain's effective seafaring population was increased by the service of foreigners; in wartime eighteenth-century governments suspended the clause in the navigation laws which required merchant crews to be composed of British nationals. (In the Napoleonic era the Royal Navy enlisted many foreigners as seamen—up to 15 per cent of ships' complements toward the end.) The pool was also enlarged by youngsters who were trained up in the merchant service in wartime, but the main contributor was probably the navy itself.

The navy trained up many more landsmen than has commonly been recognized. Admirals and captains repeatedly complained of their want of true seamen, and perhaps later generations have been led by such evidence to assume that landsmen were unwelcome in the eighteenth-century navy. This was not the case.

A protection certificate. Even dockyard labourers were vulnerable to impressment because they often worked afloat. All protections included a personal description to inhibit fraudulent use. This one, with its narrow time-limit, was issued to ensure safe travel from Deptford to Chatham.

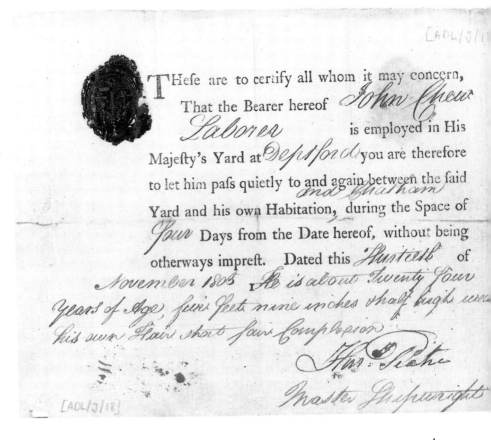

Nine ships stationed to receive men at the Nore and in the Downs reported on their recruiting during five weeks in July–August 1739; they counted 2,452 (59%) volunteer seamen, 837 (20%) pressed men, and 877 (21%) landsmen. Obviously, even at a time such as this—the very outset of mobilization, when it was possible to recruit volunteer seamen—able-bodied landsmen were not refused. In 1755, Admiral Edward Boscawen was happy enough to bring 55 Cornishmen—his own 'country'—aboard his newly commissioned ship: 'stout fellows but all landmen'. One captain commented in 1760 that, rather than keep a dissatisfied seaman who was the best in the world, he would prefer 'a willing and contented landman, who with a little time, and his own endeavours, I could make a seaman of'. This would have been widely endorsed by his colleagues.

How many men volunteered? At the beginning of a war, quite a large number. The above sample for summer 1739, indicating that pressed men accounted for only 20 per cent of the recruitment, was typical only of an early phase of mobilization. In answer to a parliamentary request, the Navy Board came up with figures spanning the first three years of mobilization for the Seven Years War, 1755–7. Of a total of 70,566 recruits, 20,370 (29%) were entered as volunteers—how many of these were landsmen was not reported—and 16,953 (24%) as pressed. Nothing was specified about the remaining 47 per cent. We are not in a position .

to assume that they were all pressed; nor can we assume that the 29 per cent labelled volunteers were truly willing volunteers. The problem arises because most volunteers were administratively distinguished by their having been paid the bounty, which was established as a cash inducement. (At this particular time its amount for an able seaman—three years or more of experience at sea being the rough-and-ready qualification—was £3 and for an ordinary seaman £1 10s.) Once men saw they were trapped by a press gang, especially at sea, they commonly volunteered for the bounty. The port admiral at Portsmouth in 1741 put it thus to the Admiralty: 'I should be glad to know if the Impress'd men out of Merchant Ships, upon their declaring themselves Voluntiers, are entitled to the King's Bounty, for they are all Voluntiers as soon as they find they can't get away.' Generally speaking, the Admiralty was willing to allow it because it eased the problem of transferring the men without their escaping (the bounty was not paid until after the second muster aboard ship). As a consequence, however, historians will probably never know how many men on board the king's ships were true volunteers. Still, the figures above suggest that the numbers were quite considerable around mid-century, especially during the earlier portion of a war, and random samples of recruiting returns during the great contest of 1793–1815 indicate that the ratio of volunteers to men recruited under compulsion in the earlier portion of that struggle was about 50:50 and later on about 25:75.

Marines were a special category of volunteer landsmen. The marines before 1740 were really designed to become sailors. After 1740, they were given a more permanent form of military organization, but their officers were forbidden to discourage them from going aloft and becoming seamen if that was their desire. After 1755, the marines came firmly under Admiralty control. Formed into independent companies, they were sent aboard ships in proportion to complement; even small sloops of war had a few marines on board. Although their supposed speciality was small arms, they usually came aboard as raw recruits untrained in musketry. Most sea fights saw them working alongside seamen at the great guns, but in close engagements they commonly lined up on the quarterdeck to pour musket fire on to the decks and into the tops of the enemy. It was not until the 1750s that their specialized role in beach assaults began to be studied and developed. In respect to filling out ships' complements, their importance grew enormously in the Napoleonic wars; by 1810 the Royal Marines accounted for 30,000 of the 145,000 men borne.

In every war the navy alone at its peak strength had more men aboard than the *total* number of pre-war British seafaring men. Although there was a serious problem of getting the fleet manned when war broke out—French naval strategists knew this and it often influenced their plans—the numbers gradually increased. Whatever source they came from, British ships' companies attained through experience at sea a standard of proficiency which other navies could not match. This growth in numbers and skills was managed in spite of the huge disparity of naval and merchant wages in wartime, the inherently demoralizing effects of impressment, and serious losses to disease.

In the Seven Years War about 1,500 men of the Royal Navy were killed in combat; at least ten times as many died from disease. During 20 years of war between 1793 and 1813, when the causes of naval fatalities were recorded more carefully, approximately 100,000 naval personnel died of all causes: 6.3 per cent by enemy action, 12.2 per cent by shipwreck or other disasters, and 81.5 per cent by disease or accident. It is likely that more than three-quarters of the last group died of disease.

The navy felt the ravages of disease most severely at the beginning of a war, and the reason was typhus, commonly known as 'ship fever' or 'gaol fever'. The worst naval typhus epidemic of the century occurred between August 1739 and October 1740. After 15,000 men were quickly recruited by November 1739, only 1,000 more could be mustered six months later. No fewer than 25,000 men fell ill and were sent to hospital ships, sick quarters, and hospitals; of these, 2,570 died and 1,965 deserted. In late 1755 and early 1756, the various squadrons, one of which was designated for the defence of Minorca, sent thousands ashore to hospitals and sick quarters, and over 2,000 men died. Once again, at a crucial time, mobilization was seriously retarded.

These figures do not tell the whole story because more men died after the ships put to sea. Sailing under such circumstances may sound cruel but it had its advantages: new sources of infection were stifled, bedding could be better aired, and decks washed with vinegar. In the 1750s, ventilating apparatus to force fresh air below decks began to be installed.

In the wake of the terrible experience of 1740 the Admiralty formally asked the king to provide hospitals at the leading naval ports. Nothing happened, but in 1744, the Admiralty asked again; this time funds were authorized and foundations for the royal hospital at Haslar (Portsmouth) were laid in 1746; the first patients were admitted in 1754. Plymouth's hospital was started later and began to receive patients in 1761. At the equivalent cost of building two or three ships of the line, these hospitals were worth every penny to the navy—not only because they provided better care and prevented drunken relapses, but also because the men could not easily run away upon convalescence, both of which objects the Admiralty had in mind when it asked for them.

The best way to defeat typhus, however, was to prevent its spread. It was noticed that men caught the fever while huddled together on tenders. The disease was commonly conveyed by lice, and it is not irrelevant that the winter of 1740 was one of England's coldest on record. In 1757 Dr James Lind pointed out the disease's infectious nature and the chief source: 'The fatal mischief lurked in their tainted apparel.' He also pointed out the remedy: special receiving vessels where the men would be scrubbed, their clothes washed, and their health evaluated before they were allowed aboard a man of war. In 1780, at last, a vessel of this description was established at Portsmouth. Bad as this delay was, it nevertheless appears that matters of hygiene were far better ordered in the British navy than in the French, especially aboard ship. Time and again, the ravages of disease prevented both the French and Spanish fleets from remaining at sea long enough to exploit a naval opportunity.

Tropical diseases were a well-known scourge, feared as much by officers as by the seamen. During the half-century after 1690, British warships sent to the West Indies suffered dreadful losses. Vice-Admiral Neville's small squadron, sent out in 1697, lost half of its men and all the captains; Neville himself died. Admiral Francis Hosier's force, in 1726, was larger, and the mortality in that squadron was forever made famous by the ballad, 'Admiral Hosier's Ghost'. The cause in both cases was the virulent form of malaria known as 'yellow fever'. Those whom tropical fevers and dysentery did not kill were reduced to helpless weakness. In early 1742, only 3,000 of 6,600 on board Sir Chaloner Ogle's large West India squadron were fit for duty.

Both Neville's and Hosier's squadrons arrived in the West Indies in May, the first month of the rainy season when these fatal diseases, especially mosquito-borne yellow fever, were most to be feared. The best remedy was to stay out of port during the dangerous summer months. It helped that the British bases at Port Royal and English Harbour were on dry south coasts; as noted earlier, Port Antonio's location made it a death trap. For whatever reasons, shipboard losses to tropical disease in the West Indies seem to have diminished substantially as the century progressed, though the death toll remained high for soldiers, marines, and sailors who went ashore to fight on tropical soil.

When Commodore Anson led a small squadron against the Spanish in the Pacific Ocean and ended up voyaging round the world (1740–4), he lost 1,050 of the 1,955 men who started out with him to the most famous sea disease—scurvy. He noticed that the surviving sick men recovered miraculously when given oranges on the island of Tinian. East India Company ships took on board citrus fruits where they could, and so did many Royal Navy ships when cruising in the Mediterranean. Lind conducted an experiment in 1747 which plainly revealed the efficacy of lemon juice. Yet scurvy continued to give serious trouble to the navy almost until the end of the century.

Merchant crews coming in from a long voyage had a special dread of the press because it might kill them. They knew they had to get ashore soon—they were not sure why. One of Lind's recommendations in 1757 was that all pressed men 'just arrived from a long and sickly voyage' should be 'allowed fresh provisions, and especially a quantity of greens . . . for at least three weeks'. It thus appears that Lind knew exactly what to do and that the navy's failure to adopt the necessary measures bespeaks neglect and complacency. Nothing could be further from the truth. Even Lind himself did not regard his now-famous experiment as decisive, and, contrary to popular belief, the views of Captain Cook did not help. Expert advice from distinguished London physicians was worse than useless. In short, reputable sources provided numerous plausible theories, and the authorities were lulled into thinking that the vexations of trying to supply fresh vegetables, fruits, and juices could be avoided.

At length, the navy focused on the right solution. In the mid-1790s, scurvy was tamed by bottled lemon juice, issued to ships' surgeons according to Admiralty orders (under urging from the Commissioners of Sick and Wounded) as a curative

James Lind, MD, naval physician and first head of Haslar Hospital. Though famous for his experiment with lemon juice, reported in *A Treatise on Scurvy* (1753), he unfortunately did not single-mindedly push that remedy. His discerning *Essay on the Most Effectual Means of Preserving the Health of Seamen in the Royal Navy* (1757) laid out effective methods for curbing typhus.

DANIEL A. BAUGH

The ghosts of Admiral Hosier and his men are portrayed as haunting Admiral Vernon's squadron in 1740 (when the ballad, 'Admiral Hosier's Ghost', was written). Hosier's blockade of Porto Bello in 1726 succeeded in its purpose but at the cost of 4,000 lives:

Sent in this foul clime to languish
Think what thousands fell in vain,
Wasted with disease and anguish,
Not in glorious battle slain.

for those who fell ill. Subsequently, lemons were added to the regular shipboard diet. Without this breakthrough the blockading squadrons of the Napoleonic era could hardly have kept their long vigils off the ports of Europe, as they did in some cases for two years.

The delays in adopting proper remedies for typhus and scurvy came not from complacency but from confusion. To the sea officers and everyone at the Admiralty the health of seamen was a serious matter in the eighteenth century—not only because of humanitarian concerns, but also because of the manning problem, and, sometimes quite directly, in respect of the exercise of sea power. There is statistical evidence that deaths from disease were cut at least by half between 1780 and 1813, and it is quite possible that similar progress was made between 1740 and 1780.

Obviously scurvy persisted because the standard shipboard diet provided by the Victualling Office was mainly designed so that the food would keep—salt beef and pork, Suffolk cheese which might be as hard as beeswax, a large daily ration of unleavened bread, a gallon of beer a day per man, plus other less consequential items. Not only was this diet entirely deficient in vitamin C, it was also, as Lind remarked, 'extremely gross, viscid and hard of digestion'. Sick men were taken off it. Certainly the allowance was more copious than the competition's. Early in the century it was observed by an expert, 'Our Seamen will do more bodily labour on their Ships, than the common Seamen either of Holland or France, as being better fed, and really stronger.' Throughout the century the attitude of the Victualling Commissioners towards the sea diet was excessively conservative. The question of its healthiness, they claimed with perfect administrative rectitude, was not their business. Their business was to try to provide whatever the Admiralty stipulated. Yet they habitually offered administrative objections to proposed innovations and on that account their performance cannot command admiration.

In their main sphere of responsibility, however, their achievement was remarkable. Victualling lapses had been frequent in the war of 1689–97. Immediately afterwards the department was put through an administrative revolution. In the seventeenth century the main victualling concerns had been wise purchasing, sound accountability, and avoidance of profiteering; after 1700, quality and distribution held priority. The eighteenth-century popular press continued to cherish lurid stories about spoiled food as if nothing had changed since the old days, but spoilage rates in fact became very low—less than 1 per cent for all items except stockfish and perhaps beer. Yet the most impressive achievements were in the realm of distribution.

The commissioners were assiduous in anticipating requirements. Even in the 1702–13 war they had agents and contractors in place in the Mediterranean and the West Indies. In the latter region they enjoyed a key advantage: regular access to supplies from North America. French *munitionnaires* did not begin to match the British level of service in the West Indies, and French warships, unless lucky enough to capture a vessel carrying a cargo of food, were sometimes forced to buy up available local supplies, thereby throwing a small island like Martinique into famine. Otherwise they had to go home. One measure of the department's achievement was the drop in Short Allowance payments (cash compensation to seamen when short supplies forced a cut in rations). The amounts paid during the war of 1739–48 were one-tenth of those paid in the 1702–13 war. This radical reduction occurred in spite of the fact that the Victualling Commissioners' repeated requests during the peace after 1715 for better facilities for storing, baking, and brewing at Portsmouth and Plymouth were turned down. In this aspect of their business the Victualling Commissioners were far from displaying complacency and conservatism.

From about 1740 the department was increasingly required by the Admiralty to meet a new challenge, the supplying of fresh meat to Channel and Western Squadrons. In 1747, fresh greens were specified along with the fresh beef for the

Livestock at sea, a commonplace on long voyages. This view of the quaterdeck of the *Deal Castle* from aft, painted in 1775 by Thomas Hearn, shows a goat and coops for keeping edible fowl in the foreground. In the background noon sun sights are being taken.

ships that took turns coming in to Plymouth. A decade later, with Anson as First Lord, a mighty effort was made to supply the Western Squadron *at sea* with these items, an effort which led to Sir Edward Hawke's triumph at Quiberon Bay in 1759. The bullocks had to be carried out live on the hoof and hoisted aboard in slings from vessels bumping alongside in open ocean. Hawke welcomed the vegetables which thus came out to his blockading squadron, but his attention seems to have been chiefly fixed on the fresh beef.

His attention was also fixed on beer. Although the official sea allowance was a gallon of strong beer per day per man, the men actually had unlimited recourse to it. Resupply of beer was always urgently demanded by admirals and captains at sea, and the Victualling Office (reluctantly) provided it even though, if it had to be brewed in summer, it tended to go bad. In parts of the world where beer was out of the question, regulations called for wine, and in the West Indies when the wine was exhausted, rum was issued. Rum being too powerful taken neat, Admiral Vernon, in 1740, ordered that it be mixed with water, and thus began the issuance of 'grog', which by the end of the century had become an optional naval issue even in the British Isles.

Any attempt to supply fresh vegetables and fruit in those days was bound to be difficult and wasteful. But the same may be said of supplying live animals and summer-brewed beer, and although fresh beef was not useless against scurvy—it can, in fact, deliver small amounts of vitamin C—it is easy to see in retrospect that a different emphasis would have been greatly beneficial to seamen's health. It seems reasonable to conclude that the extraordinary efforts to supply fresh meat and beer, though consciously made in the interests of health, were also responses to the known desires of the men of the lower deck—in other words, for the sake of morale.

Any attempt to enquire into shipboard discipline and morale in the eighteenth-century navy confronts baffling contradictions. At one extreme there are reports of captains who ordered their topmen flogged simply because another ship in the squadron was observed getting her topgallant yards up first. There were captains like Hugh Pigot whose cruelty turned the *Hermione* into a perfect hell. In this case the men, in 1797, rose up; they murdered him and all the officers, and carried the ship to a foreign port in the West Indies. At the other extreme, a whole year sometimes passed on board the flagships of Nelson or Collingwood without a single sailor being subjected to the lash.

Then there is the question of how complaints were dealt with. The Admiralty kept on the Hon. William Hervey as captain of the *Superb* in October 1740, even though an episode prompted the port admiral to recommend that he should be removed from command; seventeen months later this insanely brutal captain was cashiered by a court martial (convened by an admiral in the West Indies) for ill-treatment of his officers and barbarity to his crew. Yet it appears that this instance of Admiralty leniency toward a cruel captain was an aberration: complaints of ill-treatment were normally taken very seriously. If a seaman or group of them thought that the normal course of complaining to the captain was out of the question, they might go directly to the port admiral, or make a written complaint to some other flag officer or to the Admiralty. The Admiralty did not reject a letter because it failed to come 'through channels'. Letters signed in the form of a 'round robin'—in a circle so that no one could be deemed a ringleader—were particularly attended to. Anonymous letters were not ignored. Investigations were ordered, and in general there was no disposition to favour the captain or his officers. In 1809, Commander William Dillon was incensed when he suddenly learned that a Court of Inquiry (which in this case consisted of three senior captains) was coming aboard his ship in consequence of an anonymous letter written by one of his crew—incensed because he had been given neither notice nor opportunity to respond before the three captains appeared on his quarterdeck, not because someone higher up had decided to pay attention to an anonymous letter; he expected that.

During the century it appears that shipboard discipline grew tighter. In the 1750s a few ships began dividing the crew into divisions, each under cognizance and care of a lieutenant; before the century ended the divisional system was adopted generally, and would long endure. This introduced better order, but whether it increased the commissioned officers' authority much is doubtful. It is certain, however, that the use of the marine detachment to support officers' authority over the lower deck had such an effect. This became a common practice toward the end of the century. One possible indicator of a more formalized shipboard discipline is the fact that toward the end of our period quite a few captains in home waters brought their wives aboard for extended periods of time.

Whatever the style of discipline, administrators had every reason to think that lots of seamen would try to run away. In terms of economic incentives alone, the navy deserved a high rate of desertion in wartime. The naval wage for an able seaman was set at 24s. (£1 20) per lunar month in 1686 and remained unchanged until 1797. Yet at the height of the Seven Years War, for instance, wages in merchant ships reached £3 or more. The average pay in the port of London was £1 10s. per month on the eve of war in 1793; it leapt immediately to £3 15s. upon the outbreak of hostilities and stood at over £5 in 1804.

Although ministers and parliaments instituted measures (in 1728 and 1758) which required the navy to pay wages more regularly, thus preventing the Navy Board from indulging too freely its habit of placing the seamen last in the queue of creditors, nothing was done to raise wages. For all the glowing rhetoric about naval seamen being the true saviours of the nation, this policy was abandoned only under pressure. Early in 1797 the seamen at Spithead clandestinely concerted a petition for higher pay and put it in the post. It mentioned the rapid advance of prices and the fact that other branches of the armed forces had recently been granted higher wages. When their petition was rejected the men mutinied. All officers were forced ashore, where they stayed from mid-April to mid-May. It was really a strike and the government wisely yielded and brought the mutiny to an end by granting a 23 per cent increase. (A further increase in 1806 put wages 40 per cent above the old level.) Not a single man in the Spithead mutiny was punished.

(A month later mutiny spread through the ships at the Nore. It differed from the Spithead mutiny in respect to its leadership and range of demands. The Nore mutiny was put down by force and the supposed ringleaders, some of whom were landsmen, were hanged.) In this manner, the wages of naval seamen were finally raised.

There is no question that seamen were motivated by the prospect of prize money. For crews of large ships, however, the chances of a profitable voyage were slim. Therefore prize money was especially useful as a counter to the lure of privateering and a means of making service in frigates and sloops—an arduous, crowded, dangerous, hard life—more appealing. By established protocol, non-rated seamen and marines divided one-quarter of the condemned value (the captain alone received one-quarter). In a 32-gun frigate close to the size of the *Pallas* that meant dividing it among about 150 men. Sailors in privateers, however, rarely enjoyed larger shares, because the owners generally took half.

In view of the possibility of suffering under harsh discipline in the navy and the fact of higher pay in the merchant service, it seems incredible that any seaman would volunteer for naval service in time of war, and, indeed, not many able seamen did volunteer. But some did, for there is another dimension to naval recruiting and retention.

Whereas an officer may be said to have joined the *navy*, as a profession, a volunteer seaman joined a *ship*, as a community, engaging himself to serve under a particular captain or some other officer of the ship—for whom he felt respect, with whom he could claim some connection, and from whom he might expect preferment in rating. One sees the importance of the captain in a request made by Admiral John Balchen, in March 1729, for clarification of policy. He asked whether he could allow men to be discharged from one ship to another to follow a captain who had been transferred:

The captains tell me that some of them are people that have been with them in other ships and are uneasy they can't go with them again . . . I am likewise informed that several men that are entered in town [London] decline coming down to the ship they entered for upon hearing the captain was removed from that ship and believing they should not have the liberty to go with the captains they entered for.

Balchen suggested that the Admiralty should make its guarantee of this liberty clear by publication in a newspaper. To all of this the Admiralty agreed.

Did captains know their men by name? After extended cruises they probably did. A long voyage to the Cape of Good Hope enabled Captain Thomas Pasley to know all his men well. The divisional system helped to bridge the gap between quarterdeck and lower deck, but was no substitute for confidence in the commanding officer. Shortly after the mutiny at the Nore, Collingwood was offered the command of a larger ship, which meant a higher salary. He declined the offer, 'for I know and am known here, which, in these ticklish times, I hold to be of much consequence'.

Under good officers life with one's mates of the lower deck developed into a

GOD save the KING.

Doublons.

SPANISH
Dollar Bag
Consigned to Boney.

My LADS,
    The rest of the **GALLEONS** with the TREASURE from
LA PLATA, are waiting half loaded at CARTAGENA, for the arrival
of those from PERU at PANAMA, as soon as that takes place, they are
to sail for PORTOVELO, to take in the rest of their Cargo, with Provisions and Water for the Voyage to EUROPE. They stay at PORTO-
VELO a few days only. Such a Chance perhaps will never occur
again,

THE FLYING
PALLAS,
Of 36 GUNS,
At PLYMOUTH,

is a new and uncommonly fine Frigate. Built on purpose. And ready
for an EXPEDITION, as soon as some more good Hands are on board;

Captain Lord Cochrane,

(who was not drowned in the ARAB as reported)
Commands her. The sooner you are on board the better.

    None need apply, but SEAMEN, or Stout Hands, able to rouse
about the Field Pieces, and carry an hundred weight of PEWTER,
without stopping, at least three Miles.

To British Seamen.   **COCHRANE.**

BONEY's CORONATION
Is postponed for want of COBBS.

J. BARFIELD, Printer, Wardour-Street.

Rendezvous, at the White Flag,

Recruiting poster of early 1805.
This was not false advertising:
Cochrane's initial orders allowed him to cruise for profit,
and near the Azores he took
four richly laden vessels from
the Spanish West Indies. Everyone on the 'Golden Pallas', as
she came to be known, got rich
from the prize money.

tolerable existence, and perhaps something better than that. There are figures for
the mid-century period indicating that desertions were most common from ships'
companies newly formed. Another indication of the importance of personal relations is found in a dozen or so seamen's memoirs for the Napoleonic period—
scarcely any genuine ones have come down to us from the preceding
decades—which testify to the ill-treatment the men suffered at the outset, before
settling into a ship's company.

All this made 'turn-overs' devastating to morale. Where the whole crew were turned over with the captain as he moved to a new ship, it might be considered a 'good' form of turn-over. But turning-over parcels of men in order to get certain ships manned in a hurry was as administratively deplorable as it was operationally tempting. Early in the century the Navy Board set forth a catalogue of what grievous results had arisen from an unwise use of turn-overs during the 1690s. Instead of the men being paid off,

when the Ships come in to be laid up or refitted, [they were] kept in floating Prisons, and turn'd over, and lent from Ship to Ship for several Years together, by which means they were detained from their Families, as well as from their Election of serving under such Officers as they had Expectation from, and had no sooner got Acquaintance onboard one Ship, but were removed to another, many times absent when the Ships they served in came to be paid, and consequently left to the Recalls, often rated one way in one Ship, and another in another, and qualified as [petty] officers in one, and only as Able or Ordinary in another.

Thus, 'the commanders lost their Interest with the Seamen, and the Seamen with them'. Involuntary turn-overs of this injurious sort continued throughout the century, though most Boards of Admiralty tried to restrict their use to true operational emergencies.

Even a 'good' form of turn-over kept men from going home to their families. The extreme duration of wartime service was indeed a serious drawback. After particularly long service abroad, many captains beseeched the Admiralty to allow them to offer their men leave of absence, trusting to their loyalty to watch newspaper announcements and return when summoned to serve under them again, often in a new ship. It should not be imagined, however, that ships in harbour were socially isolated; women—as has been said, some of them wives—were permitted on board. Theatricals were sometimes produced by crew members, and, in harbour, guests as well as the ship's company formed the audience. Bacchanalian shipboard revels were to be expected at times. At sea, when work and weather did not exhaust all energies, music and dancing were commonplace.

A ship's efficiency and fighting capacity depended upon camaraderie and trust as well as discipline, for the skills that were most crucial to success—alert sailing and rapid reloading—were team skills. Two days after leaving Spithead with a newly formed crew in March 1781, Captain Pasley noted: 'We made but a poor figure in Manœuvering today and setting our Canvas; by and by things may go more to my satisfaction—we are yet strangers to each other.' In his journal he did not record that there were any special training measures to be undertaken, nor does he mention gunnery drills, though there must have been some. In any case, after the first engagement with an enemy, a month after leaving Spithead, he could jubilantly take notice of the 'intrepid and Cool Gallantry of my Ship's Company; the fire they kept up was past all conception—not one accident.' Nothing like this could have been achieved without men already familiar with their business.

Women were regularly allowed aboard during recreation times in port, as this scene on the forecastle of the *Prince George* (90) in 1779 shows. At sea, especially during long blockading stints, entertainments were essential to morale. 'We have lately been making musical instruments,' wrote Collingwood from off Cadiz in 1798, 'and have now a very good band. Every moonlight night the sailors dance; and there seems as much mirth and festivity as if we were in Wapping itself.'

Good performance of young men is obtainable under harsh discipline, especially where the tasks evoke a group pride, but there is a question of whether first-class seamanship could have been elicited in this manner. The Royal Navy enjoyed huge advantages in the alertness and practised teamwork of its able seamen. While detained as a prisoner in France, Commander Dillon interviewed a captain who had fought in the French line at Trafalgar. Dillon asked him one last question: 'What was the act on the part of the British Fleet that made the greatest impression on your mind during that battle?' The French captain's reply:

The act that astonished me the most was when the action was over. It came on to blow a gale of wind, and the English immediately set to work to shorten sail and reef the topsails,

149

with as much regularity and order as if their ships had not been fighting a dreadful battle. We were all amazement, wondering what the English seamen could be made of. All *our* seamen were either drunk or disabled, and we, the officers, could not get any work out of them. We never witnessed any such clever manœuvres before, and I shall never forget them.

An eighteenth-century man of war carried what would strike a modern observer as an unusual number of boys. On a 74-gun ship the captain alone was permitted to carry 20. Most of these boys, their parents hoped, would become sea officers. On one occasion the highest-ranking parent in the kingdom, George III, asked various sea officers about the proper starting-age for a youngster who intended a naval career and was informed that 'fourteen is as late as so hardy a profession can be embraced with the smallest chance of success'; the future William IV went off to sea shortly before his fourteenth birthday. George Anson went to sea at 14, Edward Hawke and Samuel Hood at 15, but most began before that: Samuel Barrington at 11, Peter Warren, Hugh Palliser, and Horatio Nelson at 12. The largest single category of entrants consisted of sons and nephews of captains; the rest were youngsters recommended by friends, relatives, and persons of political or social consequence. It is clear that entry to the naval profession was basically in the hands of individual captains and admirals.

Every boy was expected to become a sailor. Prince William, like the others, was made to learn the ropes, climb the rigging, and take the physical risks. No captain would wish to have aboard a lieutenant who lacked this sort of first-hand experience in seamanship; it was the most essential element of officer training. Admiral Collingwood's comment on one poorly performing youth who had been recommended to his care went straight to the point: 'As to his being an officer, it seems entirely out of the question. That would be sporting with men's lives indeed.'

There was, however, a serious drawback. Boys who went to sea at so young an age might well learn seamanship and how to adjust to sea life, but were deprived of that gentlemanly course of education which the navy had come to value and also of the mathematics needed for conquering celestial navigation. Shipboard schoolmasters, established in 1702, proved to be a feeble remedy. The Admiralty demonstrated its concern by establishing the Royal Naval Academy at Portsmouth in the 1730s; cadets were to spend three years there (two of which counted as sea service) before going aboard ships. Yet in the early decades of its foundation the academy lacked applicants and barely survived. Parents preferred to send a boy immediately to a known captain who could be trusted to look after him. Besides, everyone realized that a youngster needed to be brought under the eye of a sea officer who could attest to his performance at sea and try to assist his professional advancement.

Although the Admiralty was not in a position to select the entrants, it did uphold the minimum requirements for a lieutenant's commission that were first established in the 1670s and 1680s. As for the examination in seamanship and navigation, conducted by three captains under the directions of the Navy Board, system and continuity seem to have been lacking; stories from across the centuries suggest that its chief benefit lay in reminding candidates of what they were expected to know. The minimum age for lieutenants remained 20, but this was not strictly enforced. The minimum sea-service requirement was lengthened in 1728 to six years, two of which had to be served in the rating of midshipman or mate. Muster and pay books made this requirement almost impossible to evade, and it was the best guarantor of minimum competence.

Satisfying the requirements, however, did not automatically yield a lieutenant's commission, for that could only come through appointment to a lieutenant's berth in a commissioned ship. In peacetime such positions were scarce; one needed connections and strong recommendations. In wartime, on the other hand, lieutenants were in demand—a shortage actually developed in the mid-1740s—and this was the easiest rank to attain.

During much of the century the next rank above lieutenant was captain, or 'post captain'. Taking post, achieved by commanding a ship of 20 guns or more, was of crucial significance because after about 1700 the captain's date of rank, rather than the size of ship he had commanded, came to confer both precedence in command and an inviolable position on the ladder for promotion to rear-admiral. An officer who failed to take post early in life appeared to have little chance of flying a flag before physical decline set in; one sees here a motive for sending boys to sea so young. Seniority did not guarantee promotion to every captain who lived long enough, yet for decades the Admiralty had no regular method of dealing with those deemed undeserving or unsuitable. The problem was solved in 1747 when such captains were given the title and half-pay of rear-admiral on the understanding that they could never expect active employment at sea. The Admiralty at that moment was desperately in need of some younger admirals.

To become first lieutenant of a ship, especially a flagship, was to be singled out for early promotion to captain, but early in the century a trend toward recognizing service in the capacity of 'master and commander' (of a ship of fewer than 20 guns) as a preliminary step toward captain began to emerge. In due course 'commander' became a rank. Jane Austen and her family were ecstatic when they learned in 1799 that her brother, after six years' active service as a lieutenant, was to be appointed commander of the *Scorpion* sloop. They knew what it meant, and in fact Francis Austen was elevated to captain two years later.

Time served at sea as a lieutenant or commander was commendable, but promotions went to officers with good connections regardless. A progression from first appointment as lieutenant to post captain in four to seven years was typical for men favoured both by luck and connections; Anson, Warren, John Byng, Hawke, and John Jervis are examples. Some lieutenants were never promoted and grew old in rank, especially in times of peace. The system, which remained decentralized and personal rather than centralized and bureaucratic, obviously gave rise to enormous disparities. It might have become centralized if the power to promote had rested entirely in the hands of the Admiralty.

Promotion to the ranks of lieutenant, commander, and captain could, however, come from commanders-in-chief on distant stations who had the power to fill

James Ramsay, able seaman, drawn by Loutherbourg. A sailing warship's fighting effectiveness and safety depended upon having enough brave, well-trained seamen. Ramsay was part of Nelson's assault party when he boarded two Spanish first-rates during the engagement off Cape St Vincent in 1797.

vacancies. To be sure, the lion's share of vacancies belonged to the Lords of the Admiralty; they controlled appointments to ships preparing for sea and in the home fleet. On distant stations, however, direct control was impractical, since deaths, dismissals by court martial, and resignations on grounds of ill-health were to be expected. At times vacancies opened up more frequently in the West Indies than in home waters.

The Royal Naval Academy at Portsmouth was built in the 1730s. It was designed to accommodate 40 boys (over 13) of good family for two to three years of instruction before they went to sea as midshipmen, but its enrolments remained conspicuously below quota until the 1770s. Even then the academy provided less than 3 per cent of the navy's officer corps during the Napoleonic wars.

This explains why the private and unofficial letters that always went back and forth between a First Lord and a Commander-in-Chief abroad were so heavily laden with recommendations regarding young officers. The requests sailed in both directions. Sir Charles Wager was the newly minted First Lord in 1733 when he sent the following to Commodore Sir Chaloner Ogle in the West Indies:

I have writt you severall letters, but I believe they have been generally Recommendations which I could not avoid; my own, was, principally Mr. Knight, who I have had so good a Character of, from everybody he has been with, that I suppose I must have been misinformed, or you must be so from his present Capt. Obrien with whom he went out first Lieut., or else I should have thought you would have taken some notice of him before now.

This is an exemplary fragment. The recommendations which he 'could not avoid' presumably came from politically or socially influential persons and were being dutifully relayed. His personal reminder about Lieutenant Knight was crafted with exquisite professional delicacy. The key point, however, is that the choices lay entirely with the commodore, who had his own followers and career position to consider.

So long as the channels of preferment were dominated by active naval professionals there subsisted a mutually beneficial system of exchange between juniors and seniors, a system whose tendency was to reward professional abilities. It strongly favoured young men who contrived to remain active at sea: they could thus prove their worth, stay close (circumstances permitting) to captains or admirals who were likely to help them, and be at hand in case a vacancy opened. These considerations applied regardless of whether their best connections were aristocratic, parliamentary, or naval.

One readily sees the benefits for juniors, but what benefit did captains and admirals receive from devoting so much attention to the business of observing and recommending younger officers? The answer mainly lies in the nature of naval service: the career was a minefield, perilous both physically and politically. Senior officers therefore needed 'followers'—lieutenants and captains who would exert themselves in combat or other moments of danger with competence and unflinching devotion. Their loyalty would also incline them to give favourable testimony to boards of inquiry and courts martial; in the aftermath of ship disasters as well as strategic and tactical misfortunes such investigations were to be expected. A flag officer might even win the loyalty of officers hitherto unknown to him who happened to fall under his command if they could hope for eventual preferment. Flag officers who were not given authority to promote always protested to the Admiralty that their power of command was thereby weakened: their followers would grow restless, and other officers assigned to their command would see little advantage in taking risks to obtain their good opinion. The power of command and patronage were seen as inseparable.

Historians of the eighteenth century have tended not to notice that this patronage system was operated almost wholly by the navy's professionals. Peers and politicians recurrently applied to the Admiralty for favours, but if they had some connection with an admiral they were better advised to go through him. Even so, professional considerations were never absent, and a senior officer who sought to gratify aristocrats and politicians too readily—to the point of ignoring the competence of the young man in question—could be suspected by his colleagues of looking to a future elsewhere than in the sea service. Certainly his fellow professionals would deem him fully deserving of whatever misfortunes at sea came his way.

One cannot ignore the inherent unfairness of a system that elevated some young men to the crucial rank of captain in less than seven years and consigned others who might be equally deserving to professional oblivion. It should not be imagined, however, that political and aristocratic influences dominated the selections. For one thing, the network of naval patronage was extremely broad. It involved the entire naval organization—civil and warrant as well as sea officers. A master shipwright's son might be looked after by the local dockyard commissioner, a warrant officer's son by a well-remembered captain, a purser's by an Admiralty Secretary who knew his father. Such occurrences were not rare and the British navy avoided, mainly for this reason, the dreadful rivalry of sword and pen which so injured the French navy. As is well known, the system was capable of identifying youngsters of great merit, such as Horatio Nelson and Cuthbert Collingwood, who started out with nothing more than the fact that a near relative commanded a ship in commission. This huge domain of naval patronage accounted for the social comprehensiveness of the officer corps—the feature that the snobbish Sir Walter Elliott, in Jane Austen's *Persuasion*, considered to be the great defect of the navy as an institution, 'as being the means of bringing persons of obscure birth into undue distinction'.

Was this naval network any less powerful than the network of court and government? Extreme cases of political assistance—clearly evident in the naval career of the Hon. George Clinton, later governor of New York—may be matched by the tolerant attitude of captains and admirals toward each others' sons, the bad effects of which were seen in the cases of Richard Norris and John Byng. In fact, purely naval connections could advance a young man as fast as any other kind. Practically no one matched the velocity and youthfulness of Augustus Keppel's promotions: he moved from acting lieutenant to captain in three years, taking post two months before he was 20. Yet Horatio Nelson was under 21 when he took post, and this occurred only two-and-a-quarter years after he became a lieutenant. Keppel possessed the best court connections imaginable. Nelson was the son of a penurious Norfolk clergyman. Both gained their promotions in hazardous circumstances—the South Seas and the West Indies—and both were admirable sea officers.

For all its imperfections, the system of naval patronage had the capacity to harmonize individual officers' interests in a manner that tended to uphold professional standards. For this to prevail, however, two conditions had to be met. First, there had to be ample opportunity for service at sea. The frequency and length of wars, and the character of British naval strategy (which entailed sustained cruising), provided this; even in peacetime there were enough ships in active service to keep the system from collapsing. Second, naval patronage had to be shielded from excessive political pressures. It helped considerably if the First Lord was a sea officer himself. It helped enormously if the First Lord, whatever his background, was known to favour professional competence in any quarter where it might be found, since the system as described above was bound to generate rival coteries.

At its best the system was capable of encouraging professional standards in all ranks, but unpleasant realities abounded. The Admiralty issued a steady stream of admonitions—hurrying commanding officers aboard ships fitting out, urging them to dispatch press gangs, to put to sea, to cruise rather than linger in port, and so forth. Captains complained of bad lieutenants. Lieutenants and warrant officers complained of inhuman treatment by their captains. Too often, commanding officers neglected shipboard administrative duties and manipulated muster books and purchasing accounts. (Captains' salaries for the smaller ships scarcely covered active-duty expenses; they had to pay for their own spyglasses, charts, navigational instruments and books, and uniforms.) The varieties of neglect and fraud led the Admiralty to supplement the navy's *Articles of War* by publishing *Regulations and Instructions Relating to His Majesty's Service at Sea*, a book first issued in 1731 and revised and enlarged frequently thereafter. The Admiralty did not lack means of disciplining officers who failed to comply. It could call for a court martial or board of inquiry, stop pay, terminate half-pay, and—perhaps most effective—threaten to show disfavour in regard to future appointments and assignments.

Low salaries invited fraud, but the government rejected Admiralty petitions of 1730 and 1754 for higher pay for captains. Low salaries also amplified a huge defect in the system itself, namely the too-distant prospect of promotion to rear-admiral. Officers who failed to make post in their twenties could scarcely hope for a flag, and the more fortunate could do nothing to hasten the day of promotion. Only long wars and deaths of seniors could do that. Consequently, upon making post an officer's ambitions abruptly shifted: locales and types of assignment, rather than promotion, now dominated his thoughts.

There were very few shore-duty billets, and if a captain accepted one of them—the Controllership at the Navy Board or a dockyard commissionership, for instance—while still desiring to remain in good standing for promotion to a flag, he was well advised to settle that point with the Admiralty beforehand. For almost everyone, therefore, active service meant going to sea, but once an officer made post the need for employment became less urgent. A period ashore on half-pay was, for many, a welcome respite—a time to marry and start a family.

In peacetime there were more captains seeking employment than could be accommodated. The Admiralty's problems arose in wartime. Captains often tried to refuse appointments not to their liking and wait for something better. The Mediterranean was relatively pleasant; the West Indies were hazardous to health and life but possibly lucrative. Captains mobilized what influence they had chiefly for the purpose of avoiding assignments where service was likely to be more hazardous than profitable, and profit came much in view. The main source was prize money. One-quarter of the value of a rich cargo (three-eighths if not under command of a flag officer) could yield an instant fortune; smaller takings could make a man financially comfortable. Another source of income was 'freight money'. This was a payment made by merchants (often foreign merchants) for a sort of armoured-vehicle service—coin, bullion, plate, or jewels forming the only category of 'freight' which regulations permitted a navy ship to carry. At 1 or 2 per cent of the value it could come to a handsome amount (and was not shared with the rest of the ship's company). As is well known, when ships were deployed the lure of prize money could interfere with proper execution of orders, and some captains were inclined to linger at a port like Lisbon hoping for a chance to earn 'freight money'.

Portrait of Lord Collingwood. Admiral Lord Collingwood possessed every quality one could hope to find in a sea officer. He also embodied the sustaining role of sea power. After Lord Nelson's death, Collingwood presided over the containing of the French Empire on its Mediterranean flank, a role involving astute intelligence evaluation and adroit diplomacy as well as years of patient cruising.

Refusals of assignments and deviations from orders for the sake of profit could damage one's reputation with the Admiralty. In wartime a captain who refused an assignment injured his 'pretensions to a flag'. Yet he could always hope for a shift in national politics which would sweep a more favouring set of men on to the Board of Admiralty. The Admiralty sometimes responded to reluctant captains with a carrot: more attractive duty next time round was promised, but there was of course no guarantee. To the entrenched hold-outs their Lordships applied the stick: the captain was dropped from the half-pay list. This order could be, and often was, later reversed. In 1749, after experiencing numerous frustrations in dealing with its captains (and admirals too) during the war just ended, the Admiralty, under the urging of Lord Anson, undertook to revise the Articles of War. Among the proposals was one stipulating that an officer who refused an assignment in wartime should risk not only loss of half-pay but be court-martialled. When this measure came to a vote in the House of Commons it was seen as giving a government department too great a power and was rejected. Officers who felt themselves to be out of favour were in the forefront of the opposition's pamphleteering against it. 'Do officers then know', asked Captain Augustus Hervey, 'that there are now two sort of Services widely differing, the one Honourable and Advantageous, the other Distressful and Severe, and do they know that if they are without Interest or Friends, they cannot hope the former, and therefore, if they are employ'd it can only be to execute the latter?'

His opposition to the measure was justified, but no one should suppose that a simple division existed between captains with and without 'Interest or Friends'. Nor should one forget the unpredictability of a ship's assignments in wartime; after the amazingly lucrative first cruise of the *Pallas* Lord Cochrane took his turn at convoy duty, and her next assignment, on blockade, might have been equally tedious to a less adventurous captain. In the final analysis, all creditable service entailed some risk and privation, and those officers who were prepared to endure sea service in all forms generally fared well in the profession.

THE NAVY AND THE NATION

By mid-century the Royal Navy as an institution had acquired almost all of the attributes that would carry it through the Napoleonic era. It had become a 'standing navy', a fulfilment of the scheme of permanent service which was laid down in the Restoration period. After 1689, the most important developments were the rise of the Board of Admiralty and its busy secretariat, the establishment of a truly efficient victualling department, the development of overseas bases, and the elaboration of a system of seniority, half-pay, and superannuation for sea officers. For the navy, the 1749 revision of the Articles of War may be seen as marking the end of an era of administrative reform. During this era the Admiralty even tried to institute a system for coping with the problem of manning, but its key proposals were rejected by parliament.

A registry of seamen would have initiated a permanent connection between a body of seamen and the navy, as half-pay did for the officers, and thus have made possible such things as pay differentials for length of service and official recogni-

tion of a petty officer's rating. None of this occurred, the result being a huge gap in the institutional framework. A standardized uniform for officers was established in 1748, but nothing was ever done to render the 'distinctive dress' of an eighteenth-century seaman distinctively naval. The seaman's contribution for Greenwich Hospital also manifested the situation. Although its benefits were reserved for men who had served in the navy, the sixpence per month for its support was deducted from the wages of all seamen of the British Empire, naval or otherwise.

Throughout the naval organization wage rates remained unaltered, and certainly from the later 1760s onwards, when price inflation commenced (it surged in the mid-1790s), this created problems in almost every branch of the service—problems which far outweighed the money conserved. The inadequacy of seamen's wages represented the greatest sphere of exploitation because, except for highly unpredictable prize money, naval seamen did not enjoy the opportunities available to other naval employees for bringing their wartime incomes closer to market levels.

These were blemishes upon an otherwise impressive record of parliamentary support. Roughly 60 per cent of Great Britain's budget for armed forces during the century went to the navy. The cost was met mainly by a heavy incidence of taxes on consumption items, the yields of such taxes growing with the growth of trade. The levies on malt, tobacco, and candles bore heavily on the poor, but rates were actually higher on items favoured by the affluent. At all social levels the people of Britain paid a higher percentage of their wealth in taxes than their opposites in

The Royal Hospital at Greenwich, a home for invalid naval seamen, was established in 1694 on land donated by Queen Mary. A significant portion of its ongoing support came from unclaimed prize money and the sixpences deducted from seamen's wages. The fourth wing of Christopher Wren's magnificent design was not finished until after 1763, but the first pensioners arrived in 1705. By 1738 there were 1,000; by 1782, 2,350.

France did. In time of war expenditures soared and were chiefly met by government borrowing. Britain's advanced financial system, which emerged in the first two decades of our period, was everywhere recognized as one of its greatest weapons of war. Although in peacetime the sea service was always cut back sharply and Britain failed to match its rivals in new ship construction, the navy benefited immensely from the consistency of the support it received. A considerable cruising force was kept in sea pay and shore facilities were improved. To be sure, the victualling department should have been given the buildings it needed sooner, but the investment in new West Indian naval bases in the 1730s shows that the nation was contemplating a permanence of global reach.

In return for its money the public demanded efficient and effective performance. Although the Augustan age was ready enough to regard supremacy at sea as glorious, this was not a prestige navy; it was a functioning instrument for the purpose of maintaining Great Britain's safety, wealth, and strength.

Far from inhibiting the enthusiasm of the populace, these practical objects were causes of the navy's popularity. It is not accidental that 'Rule Britannia' was first performed, in 1740, at a gathering of gentlemen and politicians from opposition groups—people who were inclined to play to the populace's known enthusiasm for the navy. Great sea victories always set the church bells ringing. Successful admirals were placed in nomination for popular constituencies such as Westminster. Lloyd's, in 1803, set up a Patriotic Fund, both to honour heroic deeds in all ranks and to provide aid for sailors who had been crippled by wounds or slain, leaving widows and orphans.

The British people were captivated not only by heroism and sacrifice but also by prize money. At every social level a huge haul was regarded as big news. Lucky ships' companies ceremonially paraded chests full of treasure in London's streets

The Royal William Yard. This huge victualling facility at Plymouth was laid out in 1822 and completed a decade later. Throughout the eighteenth century, in spite of government procrastination in providing proper facilities, the superiority of the navy's victualling service afforded significant operational advantages.

in the manner of a Roman triumph. Among the élite, landowning families exchanged opinions as to which distressed estate a recently fortunate admiral might buy. Englishmen of the eighteenth century had not yet pretended to that lofty public posture which quarantined commerce, profit, and plunder from imperial and maritime grandeur.

For the admirals and captains, however, the joys of popular acclaim were to be weighed against the intensity of popular criticism. The public seemed to expect not just success in battle but faultless interception of enemy forces, the latter of course being subject to wind, weather, and chancy intelligence. Recalling that in those days the wartime power of censorship was either temporary or nil, one may grasp how immediately an officer's conduct might be censured in the press. Usually the light of publicity fell on the flag officers. Admiral Sir Robert Calder, in 1805, was not alone in resenting newspaper stories which dwelt upon what 'John Bull seems to have wished me to have done', usually written by people who had only the sketchiest knowledge of whether it was at all possible. Any commander-in-chief must suffer such risks, but what was peculiar then was the almost imme-diate circulation of detailed reports and commentaries, and the extension of public scrutiny to individual ship captains, whose names were as available as a sporting line-up is today. Eighteenth-century Britain saw the rise of the newspaper and the demise of censorship, and governments in wartime had not yet developed the means of wartime news control.

To all this must be added criticisms from merchant interests, naturally anxious that commanding officers of ships assigned to protect trade should be fully atten-tive to that duty. Criticisms of particular captains, circulated at Lloyd's, were repre-sented to the Admiralty by delegations from the various trading regions, and readily wound up in the press. The Admiralty, though well aware that it should support its officers unless they were clearly at fault, did not always do so in a full and proper manner; too often there was room for imagining that a particular Board of Admiralty's failure to vindicate an officer stemmed from personal or political bias.

One readily understands why most senior officers thought it necessary, if only for self-protection, to partake of the politics of influential friends and relations ashore. Some allowed themselves to be placed in nomination for 'Admiralty boroughs'. Others spent prize money to gain a seat independently. Admirals who were made peers entered the House of Lords. Twenty of the thirty admirals in 1761 were, or had been, members of parliament. Obviously, naval coteries could become insidiously grouped with political factions ashore.

Disharmony in the corps was greatest when parliamentary divides were most intense, as they were prior to 1714 and during the two decades after 1760. The latter period was characterized by an atmosphere of anxiety and resentment generated by too much ministerial attention to an officer's political affiliation. It culminated in the disastrous Keppel–Palliser court martial of 1778, in the wake of which a number of able admirals refused to serve. During most of the century, however, professional disdain for zealous partisanship prevailed at the Admiralty as well as in the corps. Furthermore, the professional ideal, with its accent on competence in seamanship, bravery, willingness to endure hazards and discomforts, and lengthy service at sea, made the navy an unsuitable choice for aristocratic young men who might be destined for politics, unless the family badly needed money. Three times as many army officers sat in the House of Commons during the eighteenth century as naval officers.

Lord Collingwood's career exemplified the eighteenth-century navy's professional ideal to an extreme degree. After going to sea at 12 he spent 40 of his remaining 49 years in active sea employment. His promotion to captain arose from vacancies created by tropical diseases, to which he was amply exposed himself. He survived three great battles of the Napoleonic wars, receiving a hero's medal for valour in each case. For countless other reasons as well, he was justified in describing his career as 'the precarious and unsteady ladder, by which I have mounted to rank and fortune'. After Trafalgar Collingwood stayed on in command of the Mediterranean fleet, choosing not to resign so long as his services were required. At length illness forced him to quit, but he never made it home. He died in 1810 on board his flagship—in the midst of his men, who revered him, but for whom the choice of going home did not exist.

Facing above: George III presenting a sword to Earl Howe (1726–99) on board HMS *Queen Charlotte*, 26 June 1794, by Henry Perronet Briggs (1792–1844). Recognizing Howe's victory on the Glorious First of June, the king and the royal family visited his flagship, anchored at Spithead, and presented him with a diamond-hilted sword and gold chain. In three days of fighting, Howe had engaged Villaret de Joyeuse's line of battle, with a total of some 7,000 French killed, wounded, or imprisoned and seven vessels taken. The British lost 290 killed and 858 wounded.

Facing below: Manuscript signal book illustrating Lord Howe's code of 1776. Admiral Lord Howe made a series of tactical innovations over a 20-year period between the 1770s and 1790s. A key element in his approach was the use of numbered signal flags, keyed to tactical orders in a signal book. This approach marked the end of the 1673 system and increased the commander's flexibility in tactical control.

Overleaf above: The Bombardment of Algiers, 27 August 1816, by George Chambers, Senior. The centre of the picture is dominated by the explosion of the mole head battery, and the stern galleries of the *Queen Charlotte*. These are framed by the bow of the *Impregnable* and the stern of the *Minden*. The picture also emphasizes the steep rise of the town behind the harbour.

Overleaf below: Chatham dockyard *c*.1800. Its 'monstrously great and extensive' works, as Defoe described them in the 1720s, were a legacy of the era of the Dutch wars. After 1690, the yard grew very little; the emphasis shifted to Portsmouth and Plymouth. The rope yard is at the right rear.

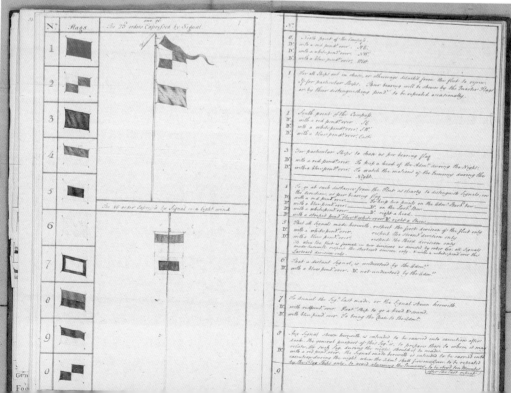

6

THE
SHIELD OF
EMPIRE
1815–1895

ANDREW
LAMBERT

$\mathcal{B}$etween 1815 and 1895 Great Britain was the only world power; her position reflected industrial primacy, economic power, and strategic strength. To maintain this position in a changing world required the constant oversight of clear-sighted statesmen, men who understood the interests and capabilities of Britain, possessed an informed view of the world, and a commitment to sustain national strategy, without bankrupting the Exchequer. As a maritime–economic empire with an insular base Britain alone depended on seapower for security and prosperity; her strategy was dictated by her position. With the technology of the period seapower could only be guaranteed by sea control based on a battlefleet superior to all rivals. This was the foundation of British policy. From it flowed a variety of rewards and options.

In war:

1. The defence of Britain from invasion.
2. The defence of Empire, maritime communications, and the strategic movement of military force in wartime.
3. The defence of oceanic trade.
4. Offensive operations, interdicting trade, amphibious operations, and supporting allies.

In peace:

1. Deterrence, through the possession of overwhelming force.
2. The protection of British commercial interests.
3. The safe use of the seas.

THE STRATEGY OF WORLD EMPIRE

After 1815, diplomacy and war remained vital in Britain's enhanced world position. Despite the illusion of 'Pax Britannica' armed force was in almost constant use against a variety of challenges. Foreign policy attempted to control the international order, but there was an underlying requirement for military force in case diplomacy failed. The instability and hostility of France (there were six changes of regime between 1815 and 1871) sustained defence spending, while Russia and the United States occasionally caused concern. In practice, the major decisions concerned force levels and budgets, where policy-makers had limited freedom of action, rather than wars. Only during the Russian War (1854–6), Britain's only European conflict of the period, was it possible to reconsider the strategic map of Europe.

British policy favoured peace and economic expansion. These were initially secured by supporting the European settlement of 1815, which prevented any one power from establishing a hegemony. In 1817, Castlereagh identified a Franco-Russian combination, 'the only one that can prove really formidable to the liberties of Europe', as the basis for a 'Two-Power' naval standard. In the north the concern was with invasion; in the south Britain upheld the Mediterranean equilibrium, which normally required active support for Turkey. Elsewhere Britain established an extensive system for the protection of trade, including key strategic bases in the Indian Ocean and China Seas.

The dominant issue in British foreign policy remained the stability of Europe and control of Belgium. This strategically sensitive region could not be allowed to fall into the hands of any hegemonic power. French occupation of Belgium in 1793 had been a *casus belli*, while Napoleon's intransigence over Antwerp prevented a negotiated settlement in 1814. The union of Belgium with the Netherlands, as a barrier to France, was settled before the peace in 1814. Palmerston's Treaty of London, 1837, established a Five Power guarantee of Belgian independence and neutrality. That the *casus belli* in 1914 should be the same as that of 1793 reflected the consistency of British policy.

The 1815 settlement reinforced Britain's position in the world with additional imperial bases, including Malta, the Cape of Good Hope, and Mauritius. While the empire was based on the battlefleet there had to be a balance between forces protecting British commercial interests in the wider world, and the battlefleet in home waters and the Mediterranean. Given the central importance of the economy for long-term funding the issue was by no means simple. After 1815 the navy had ever-increasing commitments in defence of British commerce, often at the expense of the battlefleet. The limitations of contemporary ships made sea control a reactive concept: Britain had to meet the strategic programme of her rivals without sacrificing the battlefleet.

Naval decision-making was complicated by economic limitations: the national debt and a House of Commons which would not accept an income tax until 1843 and kept the ministries of the 1870s and 1880s under constant pressure to reduce defence spending. It was no coincidence that Gladstone resigned in 1894 in protest at a Cabinet decision for a large naval programme.

The Role of Force in Diplomacy The role of naval power in diplomacy reflected a perception of wartime potential. A superficial analysis of the short wars of the eighteenth century suggested that Britain required an ally in order to have any influence on the Continent. However, the Russian War demonstrated that naval power was effective against continental states—but only in a long war.

To judge from the money they expended on coast defence between 1815 and 1895 Russia, France, and the United States maintained a healthy respect for British naval power. They had little chance of defeating the Royal Navy at sea, and were conscious of the threat to their interests. In consequence the British used naval forces to signal their intentions. On four occasions British fleets occupied Besika

Bay, outside the Dardanelles, warning Russia of an abiding interest in Turkey. It was only necessary to go to war once. French bluster and threats of invasion in the late 1840s were met by a battlefleet in the Western Approaches and a new strategy. British statesmen made it clear that if diplomacy failed they would commit forces to preserve their strategic interests. The reserve fleet was the key instrument for deterrence diplomacy; mobilizing a large force at short notice gave Britain the power to act. Those who take the 'Pax Britannica' argument to the extreme, suggesting the nineteenth-century navy was merely a collection of gunboats, ignore reality; the Royal Navy could operate as a gunboat force in peacetime, because it could mobilize a superior battlefleet for war. Britain was prepared to fight for her interests. Her response to each new challenge reflected the threat and forces required to ensure a favourable outcome.

Naval Rivals Mahan's argument that battlefleet-based sea control was the key to world power was intended for his fellow countrymen, yet his views have been applied retrospectively to criticize Britain's rivals, without considering how far those powers created their own national strategies. In this period Britain alone depended on sea control for national survival. This forced her to react to alternative strategies adopted by her rivals.

The problem of French naval strategy in the age of sail, choosing between a battlefleet and the *guerre de course*, was partly resolved between 1815 and 1830. The *guerre de course* had been defeated by the strength of British maritime trade and the London insurance market, close co-operation between the Admiralty and merchants, and the convoy system. Convoys could only operate when the British battlefleet had secured a working command of the sea. After 1815 France, unable to challenge British sea control, adopted a fleet-based *guerre de course*, retaining a battlefleet to deal with other powers, to assist frigate squadrons attacking commerce, stage coastal raids, and threaten invasion. Although modified in detail, French strategy followed this pattern until the 1880s.

Some argued that technology changed the balance between France and Britain; that steam might allow an inferior naval force to launch an invasion. This was simply not true; the Royal Navy handled technology with a certain touch, made better use of new equipment than any rival, and relied on the world's largest manufacturing base to outbuild any threat. In addition, steam provided new opportunities for offensive action against French bases. Invasion remained impossible unless France could secure command of the Channel. A superior battlefleet in strategic locations guaranteed British security. The navy could pre-empt any French attempt to use the 'steam bridge' that so alarmed Wellington and Palmerston. Once the fleet was at sea no invasion force could cross the Channel. Steam, often seen as a great leveller, ending Britain's advantage in sailing experience and seamen, was entirely beneficial. For the foundation of seapower lay in political and financial commitment, not in a temporary technological edge. When the steamship became a serious weapon, in the 1840s, Britain rapidly built up a superiority in this class, alongside the battlefleet, to maintain control of the Channel.

Both the Russian and American Civil wars demonstrated that steam favoured the stronger side, making blockade more certain and facilitating coastal operations.

Those who argue that the Royal Navy had been technically or doctrinally weakened by steam ignore the offensive power of the navy, a power that increased after 1815. By 1845 Wellington observed that Napoleonic fortifications were incapable of resisting warships, as Acre demonstrated. After 1840 the navy prepared a 'Cherbourg strategy', to destroy the French base, pre-empting an invasion. The weapons included poison gas, rockets, mortars, and steam gunboats. The development of British strategy in 1853–4, culminating in the raid on Sevastopol, reflected plans for Cherbourg. The fate of Sweaborg in August 1855 demonstrated the offensive power of properly configured naval forces. Enhanced firepower and the tactical mobility of steam made it possible to take the war to the enemy. The 'defensive' measures of the 1840s were dominated by dual-purpose projects; 'Harbours of Refuge' would serve as bases for bombarding flotillas. Similarly, the 'Coast Defence' ironclads of the 1860s and 1870s were designed for offensive operations against enemy bases. As late as 1904, naval arsenals could still be assaulted from the sea.

Invasion was not a real threat; the reservations of the French army, the number of troops required, and the inability of the French navy to contest command of the sea precluded anything more than a raid. Consequently, while fortifying British bases was sound policy, fortifying the country, or building a reserve army, which proved practically and politically impossible, were unnecessary. Although the navy had a sound doctrine to defeat an invasion, exploited the increased offensive power of naval forces, and maintained command of the sea, there was no naval spokesman to set out the case until the 1890s. Invasion was a chimera, employed to talk up army estimates, support flagging political careers, and provide a platform for alarmists of every hue.

During the war of 1812 the United States had been defeated, her trade had been destroyed, and Washington burnt. The small post-war US Navy was only capable of a *guerre de course* strategy, allied to a military threat to Canada, a possibility that troubled British planners. Before 1861 seapower was the counterweight, thereafter the problem was largely ignored. After 1870 the United States posed no naval problems; the Civil War destroyed her merchant shipping industry, while her navy was entirely coastal.

The nineteenth-century Russian navy combined political ambition with administrative incompetence. Tsar Nicholas I (1825–55) admitted that his fleet existed to defend the Russian coast, and overawe Turkey and Sweden; in 1854 his ships were tied up at Kronstadt and Sevastopol. After the war Russia built a coast defence and cruiser navy. Only in the 1880s did she build another battlefleet.

The Defence of Trade The relationship between naval power and trade requires little elucidation. Between 1793 and 1815 new regions were opened for trade, and, more significantly, the number of trading nations increased. In 1815, Lord Melville hoped to get by with *no* battleships outside home waters, but they were soon deployed on the Mediterranean and South American stations. Thereafter demand increased; quite simply the Royal Navy had to support the traders whose taxes maintained the fleet. Stations were reinforced to impress local rulers, or to counter American and French squadrons. The level and disposition of naval forces outside Europe reflected the pressures of the moment, subject to economic constraints. Permanent force levels on stations followed the expansion of the formal empire in the Victorian era. By its flexibility the battlefleet served as both the arbiter of Europe and the shield of British trade. In wartime the battlefleet would blockade hostile warships in European waters. British policy on the protection of seaborne trade was not reviewed until the Carnarvon Commission of 1878.

Steam only gradually affected thinking on trade defence; in 1830 Sir Thomas Hardy favoured a base in the Scilly Isles to protect incoming shipping from French steamships. The Declaration of Paris in 1856 outlawed privateering. This measure had a marked, if usually misunderstood, impact on the defence of maritime trade. It replaced the host of small craft, run for profit, with a handful of warships. Without the number of vessels used in 1803–15 British commerce could not be seriously injured; while if privateers were used, Britain would be free to use the old rules on blockade. Until the 1860s steamships were restricted to short sea routes; auxiliary steamers, such as the American Confederate ship *Alabama*, combined the tactical power of steam with a full rig. Guided by Admiral Milne the Admiralty analysed British commerce, and the key points where it could be defended by cruisers. Only Britain had the bases to operate steamships outside European waters until the end of the century. The Carnarvon Commission, influenced by Milne, adopted Sir John Colomb's vision of a unified imperial structure, based on seapower and linked by defended coaling stations. The navy did not use the report to press for further shipbuilding, for few naval officers shared Milne's interest. The old standard of convoy was seen as 'defensive', an inferior form of war, and was dismissed as irrelevant in the steam age. Offensive measures against enemy ports and patrolling the sea lanes were preferred, although high-value military shipping would be convoyed. During the 1889 exercises raiding cruisers eluded defending ships in the Channel, demonstrating the folly of cruising the sea lanes. One result was a two-power standard, with additional cruisers to defend trade. The following year an Admiralty committee calculated that, allowing for a superior battle-fleet, 106 cruisers were required to patrol the sea lanes, that there were 64 available, and proposed to cover the shortfall of 42 in the remaining programmes of the decade. The new navalist writers reflected these prejudices. Mahan stressed the indecisive quality of the *guerre de course* in the Anglo-French wars of the eighteenth century, and he noted the value of the convoy. With the benefit of hindsight it is clear that convoy was the only answer to the submarine crisis of 1917, but the submarine posed a threat very different from a handful of French and Russian cruisers. The threat to British seaborne trade in the late nineteenth century did not warrant a major switch of resources to convoy defence: it could be handled more economically by blockades and patrols.

After 1815, the Mediterranean fleet remained the most important British force. While the British Isles were secure, the Mediterranean was the most suitable station for the main fleet; favourable weather allowed more cruising, while local politics required frequent intervention. The opening of the Suez Canal in 1869 emphasized that the Mediterranean route to India was the heart of the imperial position.

The first major action of the Royal Navy after 1815 was the bombardment of Algiers, the largest and most arrogant of the pirate cities. Although by 1815 the pirates had long ceased to molest British shipping, they still preyed on weaker nations. The Tsar urged international action, but Britain preferred to act alone. Lord Exmouth was appointed to command; having often visited Algiers he decided that only five battleships could be placed against the batteries, supported by five frigates, seven sloops, four bomb vessels, and rocket boats. The gunners were trained for rapid, aimed fire, while some guns were replaced with heavier weapons. A full-scale rehearsal was conducted at Gibraltar. Exmouth wanted to impress the Algerines with their weakness and destroy their fleet: to do this he had to disable the sea batteries.

The batteries were numerous, but the guns were of many vintages, varied calibre, and inexpertly served. Exmouth arrived off Algiers late on 26 August, his fleet reinforced by a Dutch squadron. He waited until midday on the 27th before coming into range, then, after brief and futile negotiations, the action began at 15.00. Exmouth's flagship, the 100-gun *Queen Charlotte*, moored within 80 yards of the mole-head battery unmolested, an astonishing omission by the defenders. At this range her weight of fire simply overwhelmed the batteries. Once these had been silenced the Algerine ships were burnt. The second flagship, Sir David Milne's *Impregnable*, anchored too far out, and in silencing the Lighthouse battery suffered the heaviest casualties of any ship. As night fell, *Minden* covered the withdrawal of the fleet, while rocket boats and mortars bombarded the city.

The Dey accepted the British terms, which was fortunate as the fleet was effectively out of ammunition. There were 818 casualties, 128 fatal: a 16 per cent casualty rate which made this as bloody a battle as any in the age of sail. The Algerines soon disavowed the Treaty, and were only put out of business when France conquered the city in 1830.

The Greek War of Independence (1821–30) excited fears that Russia might secure an Aegean naval base. However, Greek piracy, allied to the barbarity of both sides, forced the European powers to intervene. The Egyptian commander, Ibrahim Pasha, sought an early end to the campaign by capturing the insurgent naval base at Hydra. Faced by piracy, intransigent belligerents, orders that included reference to settling the dispute by 'cannon shot', and in company with French and Russian forces Vice-Admiral Sir Edward Codrington might have been forgiven for doing nothing. In the event he elected to separate the warring parties, by force if necessary.

On 20 October 1827, Codrington, aboard his flagship the 84-gun *Asia* led the allied fleet to confront the Ottoman forces in Navarin Bay, a large natural anchorage in the south-western Peloponnese. Codrington hoped to persuade them to remain at anchor. In the event as the British ships moored opposite the great horseshoe arc of Muslim units, a defensive plan devised by a French officer in Egyptian service, it became clear that the day would not end peacefully. A boat from HMS *Dartmouth*, sent to warn off a fireship, was fired on and one crewman killed; by 14.25, the action was general. Neither the Turkish nor the Egyptian commander wanted to fight, the Egyptian held his fire for some minutes, but the close proximity of the fleets and the level of tension made battle inevitable.

The result was never in doubt, for although the Muslim fleets were more numerous, they had only four battleships and some frigates to face an allied fleet

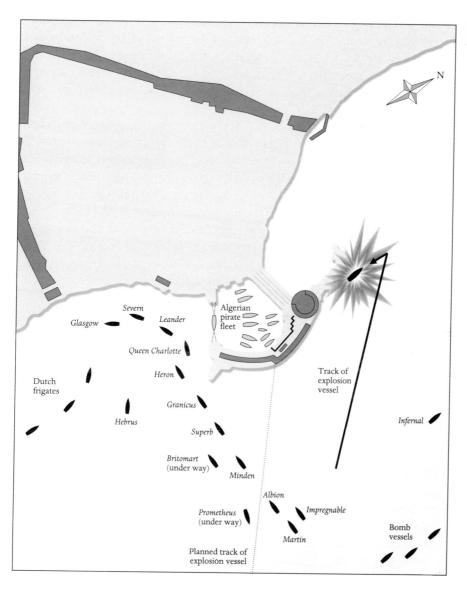

Battle of Algiers, 1815.

ANDREW LAMBERT

Massive gilt centre-piece in celebration of the Battle of Algiers, presented to Lord Exmouth by the officers of his fleet in 1817. It was modelled on the lighthouse battery, with the pairs of figures at each corner representing seamen, either freeing a slave, or striking a corsair.

with ten battleships. The *Asia*, a new and very heavily armed ship, quickly subdued both enemy flagships; one had her sides literally beaten in by concentrated broadsides of 32-pounder shot. Anchored by the bow and stern, with springs on her cables, the *Asia* hauled round to engage further targets. The French were only just entering the harbour when the battle began, while the Russians had yet to pass the castles. In consequence, the bulk of the fighting was borne by the three British battleships, *Asia*, *Albion*, and *Genoa*, although Admiral de Rigny's *Sirene* and *Breslau*, with Admiral Heiden's flagship *Azov*, were heavily engaged.

The battle ceased at nightfall. By the morning there was hardly a Muslim ship afloat; 60 had been sunk, burnt, or blown up. The allies lost 174 killed and 475 wounded, of which 75 and 197 were British. From the first shot the destruction of the Muslim fleets had been as inevitable as Greek independence. The victory,

168

while it cut the Gordian Knot for Britain—how to support Turkey yet show sympathy for the Greeks—was far from pleasing to the government. The duke of Wellington described it as an 'untoward event', and dismissed Codrington. In consequence it has a curious place in the history of the Royal Navy, being more highly regarded in Greece, Russia, and even France than in Britain. As a demonstration of the superiority of British ships, guns, officers, and men over those of Muslim fleets it was hardly necessary, while it contributed nothing to the main British interest, preserving Turkey and keeping Russia out of the Mediterranean.

By contrast, the next campaign in the Mediterranean, also against Egypt, but in favour of Turkey, was under strict control from London and secured Britain's political objects. In 1833, Mehemet Ali, pasha of Egypt, seized Syria from Turkey, and threatened Istanbul. This forced Turkey to accept Russian aid, for Britain, already committed against Holland and in Portugal, had no resources to spare.

The Battle of Navarino, 27 October 1827. This imaginary view manages to place Sir Edward Codrington's flagship, the *Asia*, between the Turkish and Egyptian flagships, which were, in fact, moored alongside one another. In addition, the action is rather too compressed, but the overall impression matches that provided by eye-witness accounts.

Russia secured the Treaty of Unkiar Skelessi, which the Foreign Secretary, Palmerston, considered inimical to British interests. In 1839, Turkey invaded Syria, but her army was defeated, the Sultan died, and the Turkish fleet deserted to Egypt. Mehemet Ali, supported by France, saw a chance to become independent. However, Egyptian rule in the Lebanon was unpopular, and the British fleet, led by Vice-Admiral Stopford and Commodore Napier, with a small Austrian squadron, landed arms to support the rebels, blockaded the coast, and drove the Egyptians from the coastal strongholds. Turkish troops secured positions ashore. Deprived of seaborne logistics the Egyptians had to evacuate Syria; their last foothold was the Crusader fortress of Acre, where Sidney Smith had denied Napoleon his destiny.

When Palmerston ordered an attack on Acre Stopford elected to bombard the two sea faces of the fortress, to create a breach on the south face for an assault by Turkish troops. The plan to tow the battleships in using steamships was abandoned, the fleet going into action on 3 November with a light sea breeze. The squadron attacking the western face, under Commodore Napier, *Powerful* (84 guns), ran in from the north before anchoring at 14.30. Some confusion was caused by later ships mooring astern instead of ahead of their leader before *Revenge* (74) filled the gap at the head of the line. By 17.00, the batteries had been silenced. The ships suffered little damage, the Egyptian guns being sighted on navigational buoys, which they took for moorings. As a result they fired too high. Before any correction could be made accurate British fire disabled the guns.

The southern squadron, three 74-gun battleships supported by British and Austrian frigates and corvettes, moored within 600 yards of the wall. Four

The Bombardment of Acre, 3 November 1840, viewed from off the south-western corner of the sea wall. The Egyptian magazine has just exploded. To the left are Admiral Stopford's temporary flagship, the steamer *Phoenix*, and beyond her the squadron led by Commodore Napier. To the right are the smaller ships engaged with the south wall, led by Admiral Walker.

steamships, including Stopford's temporary flagship *Phoenix*, fired while under way to occupy the guns at the angle of the fortress. At 16.20, a flash of light in the south-eastern sector of the city, followed by a catastrophic detonation, ended the battle. A shell, probably from the steamer *Gorgon*, struck the main magazine; 1,100 troops were killed and much of the town shattered. When the batteries resumed firing, some minutes later, the guns were quickly silenced, picked off by the trained gunners. At 17.50, Stopford ceased fire, although several ships had already run out of targets. Afloat only 18 men had been killed and 41 wounded: many more were temporarily deafened by the explosion. That night Acre was evacuated.

Within the month Napier signed an unauthorized but effective convention which ended the independence of Mehemet Ali and returned Syria to Turkish rule. French interest in the campaign collapsed when their naval mobilization failed to keep pace with Britain. In October, King Louis Philippe opted for peace and changed his government. France had been deterred by the superior reserves of the Royal Navy, providing Palmerston with one of his greatest triumphs. In addition he noted: 'Every country that has towns within cannon shot of deep water will remember the operations of the British Fleet on the Coast of Syria in . . . 1840 whenever such country has any differences with us.'

After Acre the Mediterranean fleet was run down; by 1844, it comprised only one battleship. However, that happy state did not last. From 1844 to 1849, a series of diplomatic incidents with France, many involving Admiral the Prince de Joinville, attacks on Morocco, and the Spanish marriages affair raised tension. The 1848 revolutions in Italy, the Refugees Crisis of 1849, when Russia once more threatened Turkey, and the blockade of Piraeus in 1851 ensured that the fleet rarely had the luxury of cruising in formation. Most ships were deployed in support of British diplomacy, moored singly in foreign harbours. Just as the period of crisis appeared to end, the affair of the Holy Places blew up.

Although British strategy in the nineteenth century was reactive and defensive, sustaining the 1815 settlement, once at war British policy-makers were prepared to exploit success to secure further advantages.

THE RUSSIAN WAR, 1854–1856: AN ATTEMPT TO CHANGE THE STRATEGIC BALANCE

The Russian War has been much misunderstood; most accounts ignore the strategic ambition of Palmerston as wartime premier. Although Palmerston did not start the war, he hoped to drive Russia from regions where she threatened British interests and prevent future advance. The theatres that attracted his interest, the Black Sea, the Baltic, and the Northern Ocean, had long been areas of tension. Palmerston established lasting limits on Russian expansion. Her position in the Baltic never recovered, her ambitions in the Northern Ocean were permanently halted, while the demilitarization of the Black Sea, even though abrogated in 1870, deprived her of effective naval power in the war of 1877–8, and preserved Turkey into the twentieth century.

The origins of the war lay in a squabble between Latin and Orthodox monks in the 'Holy Places' of Palestine. French intervention encouraged Russia to respond;

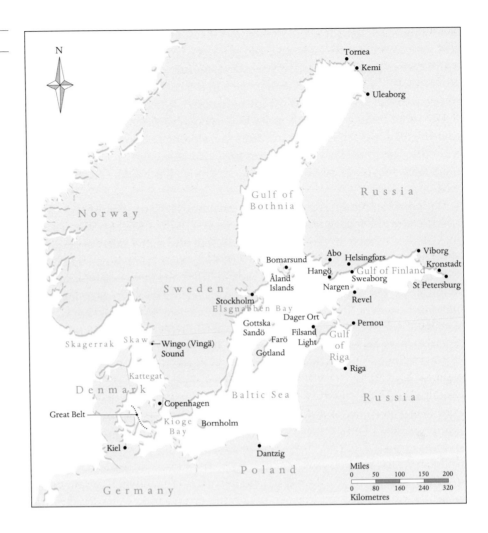

both used the threat of force to secure concessions from Turkey. News of the Russian mission in March 1853 prompted Sir James Graham, First Lord of the Admiralty, to begin planning for war. He did so reluctantly, believing a Russian war would distract attention from the real enemy, France. Given a free hand by the prime minister, and benefiting from prior experience of the post, Graham developed a grand strategy to win the war at the minimum cost. Russia could only be attacked effectively in the Black Sea and the Baltic. The plan was to destroy the Crimean naval base of Sevastopol by a grand raid. This would secure command of the Black Sea, open the Russian coast to attack, and prevent a Russian attack on Istanbul. For the Baltic Graham revived Nelson's plan to destroy the Russian battle squadron at Revel while those at Sweaborg and Kronstadt were imprisoned by ice. This would encourage Sweden to join the allies, providing troops and coastal craft for an offensive Baltic strategy. These plans were adopted by the allies.

In June 1853, the Mediterranean fleet linked up with the French at Besika Bay, committing the two powers to Turkey. However, this position was unsustainable. The fleets could not remain at Besika when the winter gales came; they could either enter the Dardanelles, in breach of the 1841 Straits Convention, or sail away, signalling a climb-down. The fleets reached Istanbul in October and the Turks declared war. On 30 November 1853 the Russians destroyed a Turkish frigate squadron at Sinope; this 'massacre' led the western powers to issue an ultimatum, demanding that Russia evacuate Turkish territory. In March 1854, war was declared.

During the winter Graham collected a Baltic fleet, under Vice-Admiral Sir Charles Napier. The ships were mobilized too late for training and few experienced seamen joined; they could earn more in the merchant service. When Napier entered the Baltic in March 1854 Revel was empty. Lacking a coastal flotilla and troops he could not carry the war to Russia, for the Russians waited behind stone walls. As a result he was restricted to a blockade, while he drilled his fleet and reconnoitred the forts, an area where he relied on Captain Sulivan, the Fleet Surveying Officer. Graham still hoped Sweden might join the allies in return for the Åland Islands, the archipelago dominating the Gulf of Bothnia. However, Sweden would only join the war if the allies were serious about defeating Russia; the islands would be little recompense for the undying enmity of the arbiter of northern Europe. Graham abandoned his efforts in the Baltic in favour of the Crimea, but the cabinet finally agreed to send 10,000 French troops to capture Åland, as much to 'do something' as for any sound strategic reason. In the mean time Napier had been joined by a badly manned French squadron, which further hampered operations.

In the Gulf of Bothnia British ships raided Finnish harbours, looking for gunboats, and burning shipbuilding materials. On 22 June, three steamers attacked Bomarsund, the fortress at the heart of the Åland archipelago, during which Mate Lucas of the *Valorous* earned the first Victoria Cross for throwing overboard a live shell. Napier reconnoitred Kronstadt in early August before the

Mate Lucas throwing a live shell from the deck of HMS *Hecla* during the attack on Bomarsund, 22 June 1854. For this act he was awarded the first Victoria Cross.

fleet went to Åland, to rendezvous with the French troops. The plan of operations was simple. The fleet isolated the islands, the troops landed and captured the fortress from the weaker landward face. The attack began on 8 August, Bomarsund surrendered on the 16th. The French troops then returned home, ending all offensive operations. Graham was heavily criticized for the limited results of the campaign, but rather than support Napier, who had called for gunboats, mortar vessels, and rockets, Graham sacrificed him to the public clamour, while building the equipment requested.

The Black Sea fleet, under Vice-Admiral Sir James Dundas, found itself tied to the movements of the allied army. On the outbreak of war Dundas sent a steamer to Odessa; her boat was fired upon while flying a flag of truce, which provided the excuse for the first naval action of the war. On 21 April nine steamships, supported by rocket boats and the sailing frigate *Arethusa*, the last major British warship to fight under sail, bombarded the military harbour. Considerable damage was done, while only three men were killed afloat. Rear-Admiral Lyons's steam squadron reconnoitred Sevastopol and cruised along the eastern coast of the Black Sea, capturing Redoubt Kaleh on 19 May. The Russians' only success came on 12 May, when the steam-frigate *Tiger* ran aground south of Odessa in a thick fog. Russian field-guns forced her to surrender.

Six days after the loss of the *Tiger*, the allied armies arrived at Varna and took control of the Black Sea theatre. As the French Commander, Marshal St-Arnaud, would not allow any significant detachment of warships to leave the army, the naval campaign stagnated for four months. The armies first watched the Russians evacuate the Danubian Principalities (Roumania) and then prepared to invade the Crimea. The clash between British and French strategies, the former dominated

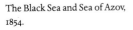

The Black Sea and Sea of Azov, 1854.

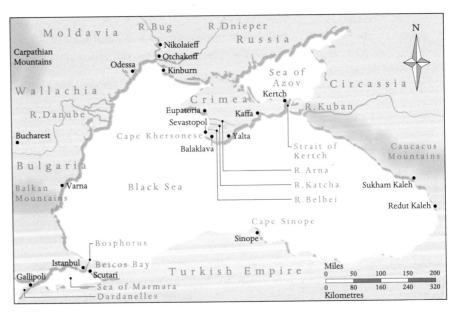

by naval concerns, the latter by military, hampered allied efforts throughout the war. At the end of June, the allied governments ordered a grand raid on Sevastopol, but the commanders were not ready until September.

This was the first major steam-powered invasion, demonstrating the strategic power of the British mercantile marine, and an increase in the value of amphibious power. The task of planning the movement of 50,000 men fell to Lyons, the British second-in-command. He delegated the work to his flag captain, William Mends. Mends provided steam-power for the British merchant ships, 27 steamers towing 52 sailing-ships, carrying troops and stores; the French and Turks were forced to crowd troops aboard their battleships. The fleet sailed from Varna on 7 September, and after a delay when St-Arnaud tried to change the plan, began landing on the 14th. Four days later the last horses, always a difficult item to handle, were ashore. Not one British soldier had been lost. The allies then advanced on Sevastopol, defeating the Russians at the Alma.

After marching round Sevastopol the British occupied Balaklava harbour and prepared for an assault. On 17 October the allied fleets engaged the sea defences of Sevastopol, as a diversion. The armies hoped to assault the city during the day. Dundas did not want to attack; as the harbour mouth had been blocked by scuttled warships he knew that nothing could be achieved. However, the new French Commander, General Canrobert, insisted. In consequence, French and Turkish ships bombarded the southern defences, the British those to the north. The battleships were led in by the steam-powered *Agamemnon*, the sailing-ships were towed by steamers lashed to the disengaged side. Lyons took his ships close to Fort Constantine, where they suffered considerable damage from red-hot shot and shells. Although a shell from the frigate *Tribune* wrecked part of Fort Constantine, Lyons's supporters were severely handled by the cliff-top Telegraph and Wasp batteries, and one by one they withdrew. Dundas kept the rest of the fleet further out, but even there ships were damaged. In all, 40 men were killed and 266

An artist's impression of the allied landing in the Crimea; curiously there are no merchant ships to be seen. By landing over 50,000 men in the Crimea the allies demonstrated that they had a working command of the sea, and the shipping to stage the largest amphibious operation to date.

wounded. The land attack had failed even before the ships sailed. Dundas was furious; in effect he had been ordered into an impracticable operation by a French General.

For the remaining months of 1854 the navy was restricted to logistic support for the army. Dundas, whose time in command was nearly up, attempted to run a naval campaign, but Lyons sensibly subordinated himself to the military. In addition, *The Times* attacked Dundas, accusing him of opposing the Crimean operation. Graham used Dundas and Napier as scapegoats for the 'failure' of the navy. In fact, both campaigns had been conducted in a thoroughly professional manner.

In late 1854, Lyons took command of the Black Sea fleet. He left most operational planning to the Fleet Surveying Officer, Captain Spratt. The struggle for Sevastopol settled into trench warfare, large armies dug in for a battle of attrition. The blockade of the Russian coast was reinforced, while the Sea of Azov, the key to Russian logistics, was the one area where naval forces could play a major role, if the French would release 6,000 troops to open the Straits of Kertch. After an abortive mission in early May, recalled when in sight of the target, Canrobert resigned. His successor, General Pelissier, sent a second expedition at the end of the month. This was a striking success. The Russians evacuated Kertch without a fight, allowing a squadron of gunboats, largely British, to roam the Azov destroying shipping, grain and fodder, stores, fisheries, and mills. Within a month the entire coastline was in a state of terror, not a boat could be seen, all coastal roads were empty. Nothing could reach the Russian army in the Crimea from the River Don, the principal source of food, fodder, ammunition, and weapons. Land transport could not compensate, from a lack of suitable animals and fodder. Within two months the Russians staged a forlorn assault on the allied position outside Sevastopol before preparing to abandon the city. The Russians were defeated when their logistics were cut, demonstrating that a maritime strategy could ruin the largest armies. Sevastopol fell to an assault on 9 September 1855, after a campaign of almost twelve months.

While the allied armies regrouped and settled for another winter in the Crimea, Lyons wanted more. Reluctantly, Marshal Pelissier agreed to capture Fort Kinburn, which guarded the confluence of the Bug and Dnieper rivers. The allied

The allied battlefleet steaming in to complete the bombardment of Kinburn, 17 October 1855. By this stage the fort was close to surrender, having been subjected to several hours of mortar, gun and rocket fire, in addition to the attentions of three French ironclad batteries. Here, in miniature, was a rehearsal of the tactics developed for the assault on Kronstadt in 1856.

fleets, including three new French armoured steam-batteries, were joined by 8,000 troops. The troops landed to isolate the fort while gunboats passed into Kinburn Bay. Then, again on 17 October, the fleets went into action. The mortar-boats and gunboats operated beyond the range of effective reply. Then the armoured batteries went into action at around 900 yards; although they were hit many times there were only three casualties. As the defensive fire began to slacken the battleships steamed up and poured in a series of shattering concentrated broadsides. The fort, literally shaken to pieces by a succession of blows to which it could make no effective reply, surrendered. This was a demonstration in minia-ture of the tactics developed for the capture of sea-forts; the implications were not lost on the Russians.

The Baltic campaign of 1855 demonstrated how well that of 1854 had been conducted. The new commander-in-chief, Rear-Admiral Sir Richard Dundas, relied on Sulivan to provide the campaign with logic and direction. Still without the resources to tackle Kronstadt, the most powerful sea-fort in the world, Suli-van persuaded Dundas to bombard Sweaborg. Sweaborg, a group of islands lying across the entrance to Helsingfors harbour, was defended by stone and earth batteries. On the reverse side lay the dockyard which supported both large warships and gunboats. This was the target of the attack, for without a base the gunboats could not operate on the northern shore of the Gulf of Finland, allow-ing the British to push their forces up to Kronstadt. Between 8 and 10 August, mortar-vessels, gunboats, and rocket-launches bombarded Sweaborg, while blockships and cruisers engaged flanking positions. Despite the premature failure of the British mortar-barrels the operation was a major success. While no allied servicemen were killed, the greater part of Sweaborg was destroyed, with Hels-ingfors left open to attack. This operation, with Kinburn, formed the basis for future planning. The remainder of the campaign involved clearing up Russian coastal positions, most of which were abandoned without a fight, and lying off Kronstadt making hydrographic surveys.

During the autumn and winter of 1855 it became clear that the French were losing what little enthusiasm they had ever possessed for the war. The capture of Sevastopol provided a sufficiency of *la gloire* for the army, while the emperor wished to cement his position with a peace conference. However, Palmerston, prime minister since January, hoped to inflict a major defeat on Russia, to push back her frontiers and stave off the day when her great resources could threaten Britain. Although more interested in the Black Sea, Palmerston was persuaded by the First Lord of the Admiralty, Sir Charles Wood, that the Baltic offered the great-est potential for a purely British strategic operation. Influenced by Sweaborg and Sulivan, Wood prepared a 'great armament' for 1856, to include 250 steam gunboats, 100 mortar vessels, nine armoured batteries, floating factories, flotilla depot ships, and support craft, in addition to the battlefleet and cruiser squadrons. The Russians knew the target would be Kronstadt and St Petersburg. The Tsar recalled General Totleben, the hero of Sevastopol, to prepare for the challenge. However, facing the inevitability of further defeats, with a crippled economy and

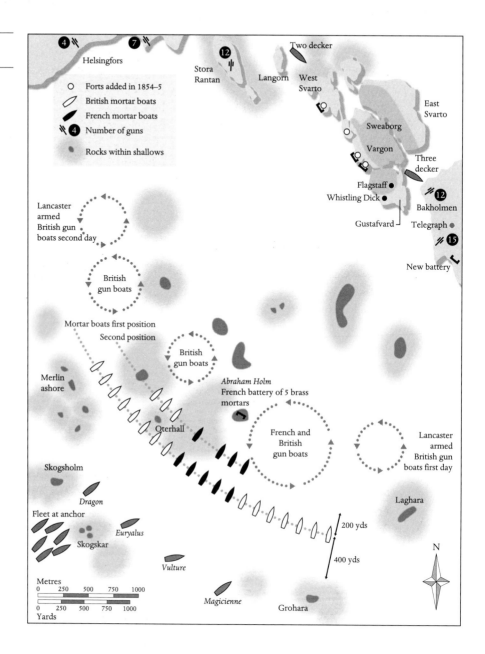

The Bombardment of Sweaborg, 1855.

defence industries unable to match the allies in steam, armour, rifled guns, or logistic support, the Tsar accepted allied terms. The Treaty of Paris destroyed Russia's influence in central Europe, and demilitarized the Black Sea and the Åland Islands, measures which crippled her power in both theatres. Less obvious was the impetus it gave to the rise of a unified Germany.

The other naval operations of the Russian war, in the Arctic and the Pacific, were of interest only for the unusual nature of the opposition at Archangel in 1854, where British warships exchanged fire with a Russian monastery, and at

The Bombardment of Sweaborg, 8–10 August 1855. To the left can be seen the silhouette of Helsingfors Observatory, in the centre a magazine on the island of Vargon explodes. In this major operation the Russian dockyard was destroyed without the loss of a single life in the allied fleet.

The first warships to be damaged by submarine mines on operational service were HM steamships *Merlin* and *Firefly*. While examining the defences of Kronstadt on 9 June 1855 the two ships were shaken by underwater explosions. The mines, using a primitive contact fuse, were too small to do any serious damage.

Petropavlovsk in the same year, when a nervous Rear-Admiral Price shot himself as the allied squadron went into action. Otherwise the enemy proved to be elusive and the targets of little value.

The peace was celebrated with a royal review of the 1856 Baltic fleet at Spithead on St George's Day 1856. The power of the Royal Navy had never been so obvious,

The Royal Yacht *Victoria and Albert* watches the Baltic fleet of 1856 pass round the pivot ships. The St George's Day (23 April 1856) Review was both a celebration of the end of the 'Crimean' war, and a conscious demonstration of British offensive seapower, intended for an audience on the other side of the Channel.

or so impressive. Britain could assault any naval base in the world. The navy's performance in the Russian War lacked only the drama of a great sea-battle to rank alongside those of earlier conflicts. It reached new heights of professionalism, particularly in the handling of logistics and the development of tactics for coastal operations. As *The Times* observed, a 'new system of naval warfare' had been created.

CHINA AND THE EAST: WAR IN SUPPORT OF TRADE

In 1834, the government abolished the East India Company's monopoly of trade with China. This business, conducted at Canton through established channels, was replaced by aggressive efforts to open China for western goods. However, as the Celestial Empire had no interest in western wares, the only alternative to paying for Chinese goods in silver was the illegal import of opium, specially grown in India. Opium turned the balance of trade in favour of the British. Facing internal collapse the Manchu Empire responded in March 1839 with new restrictions, seizing British representatives and merchants at Canton as hostages for the surrender of 20,000 chests of opium. Palmerston, an advocate of improved trade, sent naval and military forces to blockade Canton and the approaches to Peking, to settle British grievances, and secure trade. He hoped the Chinese would see reason: the reason of mid-nineteenth-century British liberalism.

Before any prompting from London could reach China, local difficulties resulted in the evacuation of Canton and a clash at sea. On 2 November the *Volage* (28), and the *Hyacinth* (18), engaged 29 war junks and put them to flight in half an hour. China was no match for the West at sea. The 'Opium Wars' had begun. Both London and Calcutta favoured the establishment of an offshore trading centre. In June 1840, a powerful fleet, including three battleships and several East India Company armed steamers, among them the pioneer iron warship *Nemesis*, with 20,000 British and Indian troops arrived. After seizing Chusan and entering the Yangtze, for unsuccessful negotiations, the fleet returned south to occupy Hong Kong on 26 January 1841. After the Bogue Forts had been crushed twice without the loss of a single man, the British forced the authorities at Canton to pay a massive indemnity and resume trade. The fleet moved north, under Admiral Sir William Parker, capturing Ningpo. On 9 August 1842, 73 ships arrived at Nanking, cutting the Grand Imperial Canal, the artery of Peking's food supply. China made peace, paying $6 million in silver, opening six major ports to trade, and ceding Hong Kong. Hong Kong joined Singapore as a great emporium for eastern trade, a coaling station and base for the East Indies Command, and from 1844, the China Station. However, as the treaty did not secure diplomatic representation at Peking, the extent of China's defeat was hidden from the regime.

Whatever the moral qualms of some at home at this form of commercial diplomacy, it is unlikely that China would have been left in peace for long by other rapacious nations. None could match Britain in projecting power in the east, few had bases, and none had the military power of India. This combination of bases and troops was used to annexe Burma, occupy Aden, and dominate the Persian Gulf. Seapower was critical to imperial development, the more so when armies could be moved by sea, and steamships opened up the great rivers of the world.

The Opium War: the attack on Shapoo, 18 May 1842. The difficulties of riverine operations, and the vital contribution of the steam ships are evident.

Trade with China, never amicable, was worsened by the Taiping Rebellion and the concomitant upsurge in piracy. By 1856, the authorities at Canton were harassing western merchant ships. Rear-Admiral Seymour, fresh from two seasons in the Baltic, seized the Bogue forts in October 1856, and occupied part of the city, but failed to secure the right of access. He then attacked the Imperial junks moored above Canton. On 1 July, Commodore Henry Keppel, with 20 gunboats, burnt 77 out of 80 junks in Fatshan Creek. British casualties, 84 killed and wounded, reflected the dangers of close-quarters fighting. Seymour noted that Chinese gunnery and discipline were much improved. Operations were suspended for the rest of the year, while troops and ships were recalled for the Indian Mutiny. Captain William Peel VC, hero of the Crimean Naval Brigade, formed another brigade. Heavy guns from HMS *Shannon* provided vital artillery support at Lucknow.

Finally, on 27–8 December 1857, 5,500 men, including 1,500 naval ratings and 1,000 French troops, captured Canton after a naval bombardment. The operation demonstrated the lessons of the Russian War, combining naval firepower with outflanking movement. When the Chinese ignored attempts to negotiate at Shanghai the fleet and army went north to the Peiho River, to threaten Peking. On 20 May, the Taku forts were abandoned after a naval attack. With the fleet at Tientsin, the jugular of Peking, the Chinese accepted commercial and diplomatic equality between nations; but they had no intention of ratifying the treaty.

In the interval the Chinese strengthened the Taku forts and studied the tactical combination of gunboats and troops. They wanted to demonstrate the insignificance of the western embassies by forcing them to travel to Peking overland and unarmed. The new Commander, Admiral Sir James Hope, lacking the experience of his predecessor, presumed that a repetition of the 1858 attack would succeed. However, his force was smaller, lacking French support and the troops used in the previous year. The Chinese had blocked the river, pinning the gunboats under the fire of the forts. Despite efforts to clear the barriers before the assault, on 25 June 1859, the flotilla was caught exactly where the Chinese had intended. At 14.00 40 heavy guns opened fire on Hope's flagship, the gunboat *Plover*, with devastating results; three-quarters of the crew were killed or wounded, the wounded including the Admiral. Hope transferred to another vessel, but an afternoon's bombardment produced little result. The earthworks absorbed British shells, while the occasional direct hit merely called up a fresh gun crew. As the tide fell the Chinese guns fired down on to the gunboats; four were sunk. Commodore Tattnall USN assisted, towing wounded men out of action and bringing in reinforcements from

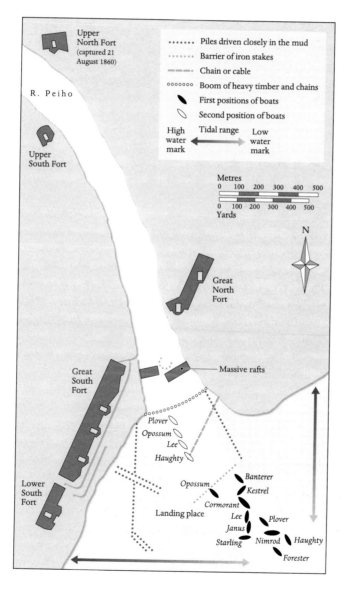

The Mouth of the Peiho, 1859.

the heavy ships; justifying his action with the famous remark 'blood is thicker than water'. Hope finally called off the attack after 93 had been killed and 111 wounded. As the light began to fade, marines, seamen, and sappers were landed on the mud-bank below the south fort in a vain attempt to recover the situation. Many men were simply swallowed up by the mud, too tired to carry on, having reached the walls and then been forced to retire. Over half were lost, or badly wounded. In all, 426 were killed, 345 wounded. This battle, the Royal Navy's only significant defeat in the nineteenth century, was John Fisher's baptism of fire.

This humiliating reverse had to be avenged, and avenged with interest. With 20,000 British and French troops, additional gunboats, and transports the allies had the power to force their way to Peking. The troops landed north of the forts on 1 August 1860, and approached from the landward side. After clearing away any resistance the north fort was bombarded with new Armstrong guns, while the flotilla distracted the south fort. At 11.00 on 13 August the north fort surrendered. The following day the south fort capitulated, and the river entrance was cleared. On 23 and 24 August the gunboats moved up-river to Tientsin, with the allied army marching alongside. After further demonstrations of Chinese bad faith, notably the murder of 20 negotiators, the allies captured Peking on 13 October and destroyed the Summer Palace. The subsequent Treaty of Peking restored the prestige of British arms, leaving China helpless. Trade was no longer troubled by central authority, and the gunboats took up the task of policing the great rivers, the arteries of commerce.

Across the China Sea, Japan, recently opened to western trade by the United States, also felt the weight of British arms. To enforce demands for justice after the murder of a British citizen, the clan fortress at Kagoshima was bombarded in 1863. The shore batteries proved more effective than had been anticipated, and although they were subdued, the ships did not come off unscathed. Shimonoseki was subjected to a more effective bombardment in the following year by Admiral Kuper's squadron, although the British ships found that the Armstrong breech-loading guns would not stand prolonged rapid fire. One of the ships engaged, the 90-gun *Conqueror* (ex *Waterloo* (120)) was the last British wooden battleship to fire her guns in anger.

After the Indian Mutiny, the Crown finally took control of India, and responsibility for naval defence in the Indian Ocean, hitherto the function of the Indian navy. Aside from the movement of troops and supplies to Abyssinia in 1867, there were no major seaborne operations in the eastern arc of empire after 1860. Any threat came from the land, not the sea. The Royal Navy was the undisputed master of eastern waters until 1895, deploying a small force of battleships and cruisers with gunboats. The station could be reinforced from Britain in wartime, while local French or Russian bases could be seized.

Between 1856 and 1914 the British battlefleet went into action but once, at Alexandria in 1882. However, the battlefleet made a vital contribution to national policy; the challenges of the period should not be ignored because they were resolved without conflict. The strength of the Royal Navy deterred war. Narrative histories of the nineteenth century, by concentrating on imperial conflicts, create the impression these were the *raison d'être* of the navy. Nothing could be further from the truth: the Royal Navy spent the bulk of its money on ships and weapons designed to meet first-class opposition and on training men to use these complex systems. On the imperial frontiers obsolescent warships, many, like the ubiquitous gunboat, originally built for a major war, were deployed on colonial service. The distant stations, gradually linked by the submarine telegraph, which Britain controlled, served as bases, depots for coal and ammunition and reservoirs of trained manpower.

This chapter will concentrate on the naval issues of the ironclad era; it will not discuss in detail the Royal Navy's contribution to the small wars of the Victorian empire, primarily a question of supporting the army. Naval brigades served ashore with the army in every war of the century, providing transport, artillery, machine guns, technical support, and the ability to solve problems which baffled the army.

The British Mediterranean Fleet, anchored at Malta in 1882. The ship nearest to the camera is HMS *Monarch*; behind her lie *Invincible*, *Alexandra*, and *Sultan*. The presence of the fleet at Malta reflected the central role of the Mediterranean in British strategy between 1815 and 1895; the age and obsolescence of the ships demonstrates that there was no serious threat of war.

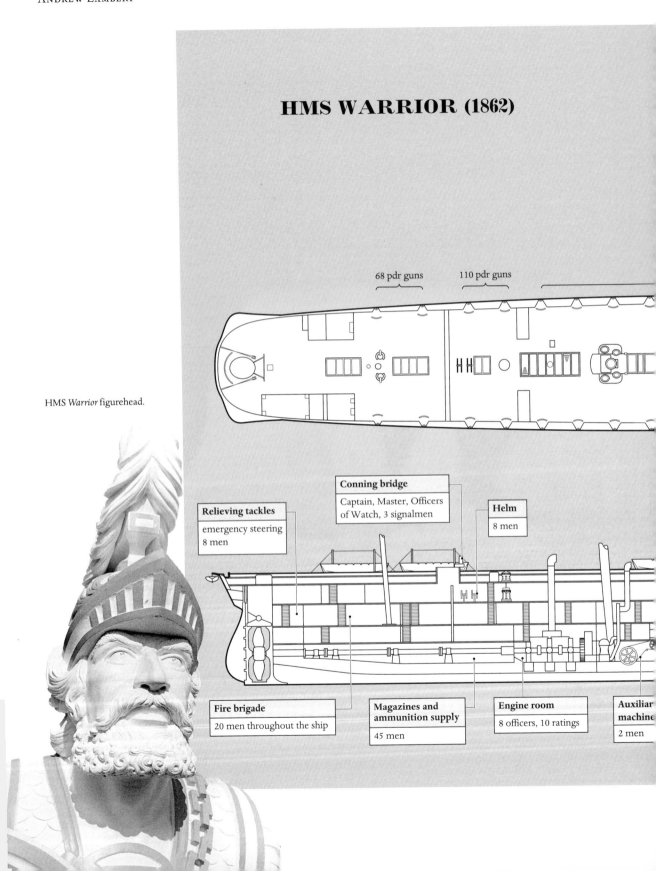

HMS WARRIOR (1862)

68 pdr guns

110 pdr guns

HMS *Warrior* figurehead.

Conning bridge

Captain, Master, Officers of Watch, 3 signalmen

Helm

8 men

Relieving tackles

emergency steering 8 men

Fire brigade

20 men throughout the ship

Magazines and ammunition supply

45 men

Engine room

8 officers, 10 ratings

Auxiliary machine

2 men

HMS *Warrior* at Portmouth, 1860s.

68 pdr guns

110 pdr guns

Main deck

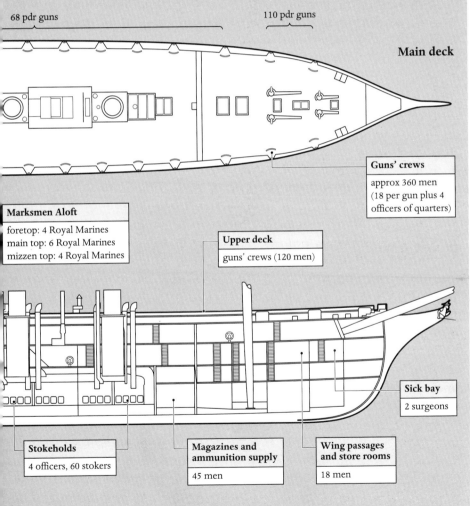

Guns' crews

approx 360 men
(18 per gun plus 4
officers of quarters)

Marksmen Aloft

foretop: 4 Royal Marines
main top: 6 Royal Marines
mizzen top: 4 Royal Marines

Upper deck

guns' crews (120 men)

Sick bay

2 surgeons

Stokeholds

4 officers, 60 stokers

**Magazines and
ammunition supply**

45 men

**Wing passages
and store rooms**

18 men

The navy moved the army to the theatre of operations, and finding nothing more to do, the logical solution was to detach men and guns for shore service. Yet, despite the glory and the career enhancement derived from conflicts with minor powers, the navy existed, first and foremost, to preserve for Britain the free use of the seas in peace, and to deny the sea to hostile powers in war. It could only guarantee those functions with a battlefleet superior to any possible combination of enemies. Between 1865 and 1885 there was no serious challenge to British naval mastery, leaving her with the luxury of undisputed, cheap, seapower. That situation could not last forever, but the adjustment to a new order of seapower would be painful.

The French Ironclad Challenge, 1858–1870 Following the Russian War the nature of seapower was radically revised, reflecting the experience of war and the concomitant rapid maturing of naval technologies. In France and Britain there were two major themes: the development of the seagoing ironclad and the new emphasis on coastal and amphibious warfare. After Kinburn, policy-makers in both countries recognized that seagoing ironclads were inevitable. Britain elected to wait on France before building any, although designs were prepared. The French laid down the first such ship, *Gloire*, in 1858. However, in seeking a temporary advantage, which was far less significant than most accounts would allow, French policy-makers revealed their weakness, and the inevitability of defeat. Britain had been alarmed by the progress of the French battlefleet during the war with Russia, and resorted to emergency programmes in 1858 to sustain her superiority in wooden steam-battleships. Now the new French ship would have to be matched; but, rather than building a similar ship British policy-makers reconsidered the issues. The result was the first modern warship, HMS *Warrior*. The French ship was built of wood, with iron plates screwed on to the hull: because the technology was largely pre-industrial such ships were easy to build, the only new infrastructure needed being rollers for 4½ inch wrought iron plate. *Warrior*, an all-iron ship, demanded a far larger investment in new technology, an investment already made by several private shipyards in Britain; by contrast, no private yards in France could build large iron ships. Furthermore, France did not produce enough wrought iron to build an iron fleet. In consequence, while France completed the first seagoing ironclad, and the first ironclad squadron, she was rapidly and comprehensively out-built by the superior financial commitment and industrial base of Britain. British statesmen would not allow another power to acquire even the most trifling of advantages at sea.

Unfortunately, the Prime Minister Lord Palmerston fell under the influence of military engineers, in particular General Burgoyne, who argued that in the face of steam-powered shipping the Royal Navy could no longer guarantee the security of the British Isles. He sanctioned 'Palmerston's Follies': the fortification of the dockyards, particularly Portsmouth, from both the seaward and landward sides, in addition to continuing work on wooden ships until 1861 and building an ironclad fleet. Between 1859 and 1865, eleven new iron-hulled ironclads were laid down, eight wooden-hulled ships begun or converted, and two powerful coast-assault ships completed. These ships, and in particular seven first-class ironclads, convinced France that she could not secure the desired political advantage, Britain's support for a programme to reshape Europe. The French programme, 16 ironclads laid down between 1858 and 1861, was crippled by lack of seasoned timber, weak artillery, and poor design. Only two had iron hulls. After 1865, the rise of Prussia diverted France from the sea, and in 1870–1, France was defeated. Twenty-five years would elapse before France again posed a threat at sea. In the interval Britain deployed a Channel fleet composed of large, obsolescent ironclads laid down before 1865. Modern units were built, in no particular haste, but in the absence of a serious threat the old ships proved adequate.

The second strand of new naval thinking concerned coastal warfare. Steam, armour, and long-range guns facilitated the capture of hitherto impregnable forts, the use of rivers, and opened the chance to destroy the enemy fleet in a defended harbour. The technical changes of the late 1850s reduced the list of major naval powers. Only Britain and France counted, Italy came in a distant third; other nations were more interested in coast defence than sea power. In this environment the major powers had to consider what to do in a situation where there was no opposition at sea. The experience of 1854–6 suggested they would face powerful fortifications, behind which sheltered any remaining warships. The campaigns of 1854 and much of 1855 had been crippled by the lack of specialist craft capable of solving this problem. France developed the 'Coastal Siege Train'; in Britain Captain Cowper Coles's armoured turret, based on his experience in the Sea of Azov, provided the final element for the ideal coast assault ship. The small, low freeboard turret-ships built in the 1860s and 1870s have been ridiculed as 'Coast Defence Ships'; but their design and the lack of any threat to the British coast suggest that their role lay elsewhere. Deep draught, low freeboard, and thick armour were only suited for offensive operations against, and into, fortified harbours. The technology of the 1860s facilitated the concepts and plans of the 1840s, the 'Cherbourg Strategy'. Where Nelson had opposed the use of ships against forts, the Russian War demonstrated that no fortress could withstand the weight of fire developed by gunboats, mortar-vessels, armoured-batteries, and blockships.

In the absence of a credible opponent *at sea*, the navy prepared for coastal warfare. The ships were mobilized for the Russian War scares of 1878 and 1885, but never fired a shot in anger. By a curiosity the reluctance of the British to indulge in theoretical speculation, allied to the absence of suitable targets, consigned this major area of nineteenth-century naval warfare to oblivion. Yet there were examples aplenty to show that coastal warfare worked in a war that threatened to involve Britain.

The American Civil War, 1861–1865 Anglo-American relations in the first half of the nineteenth century were punctuated by disagreements over the Canadian frontier, central America, and neutral rights, but there was little likelihood of war.

Palmerston's ambitious strategy to blockade the coast, raise a slave revolt, and employ West Indian troops required a long war. However, that did not diminish the validity of his analysis. He knew that the United States and Russia would become rivals for world empire; if they could be weakened before they matured Britain could preserve her unique position.

The federal blockade of the southern states, for whom most British politicians had some sympathy, or at least less active dislike, provided an incident to test the resolve of all the parties. On 8 November 1861 the USS *San Jacinto* removed two Confederate agents from the British mail steamer *Trent* on the open ocean. This clear violation of international law led Palmerston to make strong demands on the federal government. He hoped for French support, although Louis Napoleon was more concerned to establish an empire in Mexico. The intervention of the Prince Consort, his last in British politics, the reluctance of the French, and the return of the Confederate agents preserved peace. However, the civil war benefited Britain. Confederate cruisers destroyed the American merchant marine; Britain now dominated world shipping, while the United States concentrated on domestic issues until the 1890s. The US Navy, despite the much-vaunted *Monitor* and her type, was entirely coastal, inasmuch as it was modern; the only seagoing ships were obsolete wooden cruisers. The western hemisphere, quiet until the end of the century, witnessed a marked reduction in British naval presence. With the telegraph, powerful ships could be kept at home, where they were more easily maintained, to be sent when required. Only the most distant stations required a permanent force.

The Russian War Scares of 1878 and 1885 After 1856 Russia, defeated and humiliated, retired from European politics to rebuild and modernize her shattered economy, industries, and armies. She did not rejoin the major naval powers until the 1890s. In the interval, Russian naval policy, short of money and technologically backward, combined a measure of local defence with a handful of oceanic cruisers. Her security concerns were those of 1854–6: Kronstadt–St Petersburg and the southern coast. However, Russia did not renounce all forward policy, seeking conquest and glory in central Asia to distract attention from painful reforms. Furthermore, aggressive pan-Slav nationalism complicated the position of the government on issues involving Turkey. Britain had vital interests in both areas, interests that she would fight to protect.

Anglo-Russian relations in the early 1860s were poor; the Polish Rebellion of 1863 and the Schleswig-Holstein crisis of 1863–4 increased tension. However, neither threatened critical British interests. In 1877–8, Russia invaded Turkey, liberated Bulgaria, and approached Istanbul. Over the winter of 1877–8, Disraeli exploited public 'jingoism' and royal support to prepare for war to preserve Turkey in Europe. Indian troops were sent to Malta while Vice-Admiral Hornby's fleet went to Istanbul, but the most important signals came from Spithead, where a fleet of coast assault ships and ironclads was formed under Rear-Admiral Key with Fisher, the navy's expert on mines, as flag captain. Although the fleet never left home, it demonstrated British commitment, and threatened the heart of the Russian empire. The Russian climb-down and the Congress of Berlin were a triumph for Disraeli.

Thwarted in the Balkans, Russia continued her imperial drive across central Asia, subduing the muslim Khanates and threatening Afghanistan, a concern that dominated Indian strategy. War planning began in London in February 1884, after the occupation of Merv on the Afghan frontier. In March 1885, the reserves were called out, despite the large-scale commitment of troops to the Sudan. On the 30th, Russian forces overwhelmed Afghan positions at Penjdeh. Gladstone secured a vote of credit for £11 million, largely to prepare for war, persuaded the queen and the cabinet to evacuate the Sudan, and mobilized the fleet. The economics and strategy of imperial defence made it logical to send a fleet to threaten Russia in the Baltic, rather than an army to the Afghan frontier. Key, now First Naval Lord, appointed Hornby to the Baltic Fleet, with Fisher as flag captain. Russia backed down again, if grudgingly and incompletely. Hornby and Fisher

Admiral Hornby leads the Mediterranean Fleet into the Dardanelles on 13 February 1878. When the Russian army crossed the Balkan Mountains the British government used the fleet to warn the Russians not to occupy Constantinople. This gesture, combined with a powerful fleet assembling for Baltic service, persuaded the Russians to back down.

used the Particular Service squadron to experiment with harbour defences, mines, and torpedo nets. Had either crisis resulted in war the fleet would have moved to the head of the Gulf of Finland and considered an assault on Kronstadt. Russian fleet exercises in the early 1880s suggested there was little real opposition: two torpedo boats, obsolete monitors from 1863, small mines, and a handful of modern Krupp guns. Twice in eight years the offensive power of the Royal Navy had been mobilized to support British diplomacy, on both occasions Russia backed down. It is doubtful if naval coercion had ever been so effective.

The Bombardment of Alexandria, 1882 After the Russian War the Mediterranean fleet returned to the old routine, combining exercises with presence in a succession of wars and revolts that racked first Italy and then Spain. By the 1870s, the western basin was quiet. The Suez Canal, opened in 1869, made the Mediterranean the centre of the Empire. Disraeli's coup of 1875, purchasing the Khedive's shares, provided a measure of control over the company, and an excuse to intervene. In 1876, Egypt was bankrupt, leaving Britain and France to establish dual control. Egyptian nationalist officers, led by Colonel Arabi, mutinied in 1881, demanding an end to corruption, a constitution, and the removal of foreign control. Britain and France supported the Khedive, sending naval units to Alexandria.

At the end of May 1882 the two flagships were anchored in Alexandria Harbour. However, when this small naval presence had no effect, Vice-Admiral Seymour

The Bombardment of Alexandria, 11 July 1882, resulted in superficial damage to many of the ships involved; here the crew of HMS *Alexandra* pose with the evidence that she was one of very few British ironclads to fire her guns in anger.

192

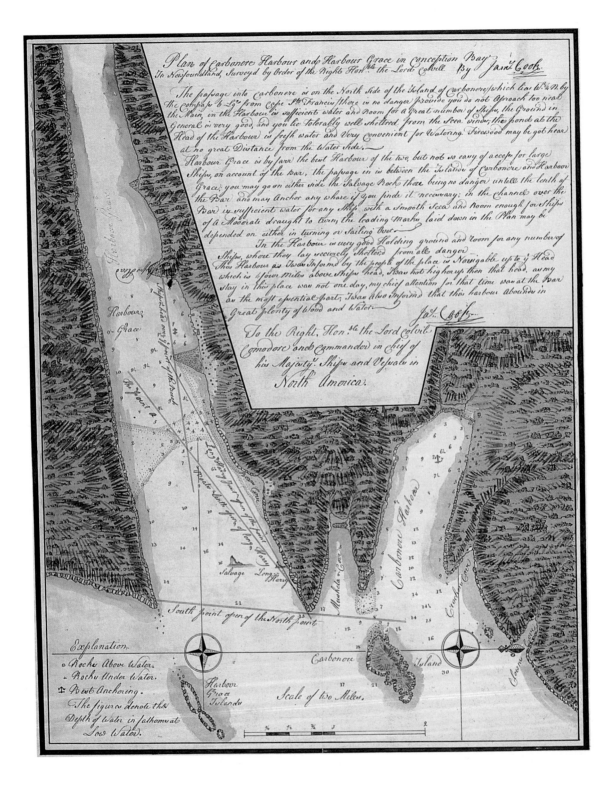

called up the Channel fleet to escort troops from Malta to the canal. Finally, on 11 June, riots broke out; 50 Europeans were killed and another 60 wounded. Ships arrived to take off the remaining European nationals, and on 3 July, the Admiralty gave Seymour permission to act on his own authority. His most pressing concern was the reinforcement of the forts. The French admiral withdrew. Meanwhile, the Admiralty covered the Red Sea, sent gunboats to occupy the canal, and reinforced Seymour. The Egyptians repeatedly denied that the works were being improved, although they were working within a mile of the British flagship. Seymour moved the fleet into the outer harbour, sending an ultimatum to Arabi. Unless the forts were temporarily disarmed they would be bombarded. The ultimatum expired at sunrise on the 11th.

When the battle opened the forts mounted 42 modern rifled guns and 250 old smooth bores, of which less than half were fired. The forts, and in particular the magazines, were primitive and vulnerable, while the two sections were too far apart for mutual support. Seymour's fleet, while powerful, was a mixed bag. The eight ironclads represented almost every type, save the one that would have been most useful: there were no mastless turret-ships of the 'Devastation' class. These were unpopular with the peacetime navy and remained in reserve. The fleet had a combined broadside of 43 heavy guns. Seymour did not concentrate his fire, electing to engage each battery.

At 7.07 the *Alexandra* fired a shot into Fort Ada, and the battle commenced. Fort Marabout at the western end of the bay, which Seymour had ignored, opened a harassing fire. Captain Lord Charles Beresford took the gun-vessel *Condor* under the fort, and when Seymour sent in four more gunboats the gunners were too busy to interfere with the battleships. The heavy ships were finding their work more difficult than anticipated. All dropped anchor, but the crews were unused to firing in a flat calm and troubled by smoke. In addition, not one ship in the fleet had been in battle before, so all were experiencing teething troubles and design deficiencies. Fisher, commanding the *Inflexible*, found that her 80-ton guns fired too slowly to deal with the smaller Egyptian pieces. However, the armour of the

The damage ashore was more substantial, although much of it was caused by landing parties after the forts had been abandoned.

Facing: Plan of Harbour Grace and Carbonear Bay in Newfoundland, surveyed and drawn by James Cook in his own hand when Master of the *Northumberland* in the summer of 1762. Note that north is to the right of the plan.

The Bombardment of Alexandria, 1882.

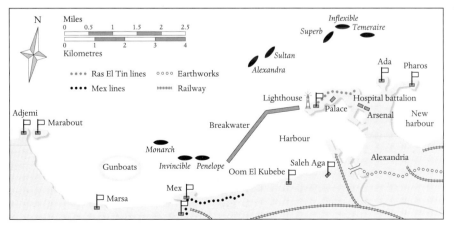

fleet demonstrated that the proving ground was no guide to battle performance. Not one plate was penetrated. British gunnery, notably that of *Inflexible* and *Temeraire*, was of a high standard. Individual guns were targeted, most were silenced by near misses. By 15.00 Egyptian fire began to slacken, allowing Seymour to concentrate his ships and gunboats on Fort Pharos and the Mex Lines. When Fort Mex ceased firing he sent a party ashore to spike the guns. By 17.00 Egyptian fire had ceased, and after driving the gun crews out of the temporary shelters Seymour followed suit at 17.30 The following day the Egyptians evacuated the city and most of the forts. Although only nine or ten modern guns had been disabled, and only two forts were completely out of action, partly as a result of inferior fuses, the Egyptians showed no inclination to test their mettle again. There were 53 casualties afloat. The subsequent campaign in Egypt involved Royal Navy and Royal Marine personnel in large numbers, manning armoured trains, Nile gunboats, and serving with the army.

A NEW ORDER OF SEAPOWER, 1885–1895

Although Queen Victoria died in 1901, and the traditional paint scheme of black, white, and buff was only abandoned the following year, the Victorian age of untroubled naval dominance ended in the 1890s. The geographer Sir Halford Mackinder considered the turn of the century marked the end of the Columbian era, in which seapowers were dominant. Henceforth transport developments would transfer dominance to land powers, in particular to the power controlling the eurasian heartland, Russia. This was merely a pseudo-scientific justification for Russophobia, based on fears for the security of India prompted by the completion of the Orenburg–Tashkent railway in 1905. Naval mastery and control of seabased transport remained critical. British naval mastery was ended not by any change in the geo-strategic situation, but by the dramatic growth of other navies. Before 1890, few powers built first-class battleships; Britain, France, and Italy were the only nations with battlefleets in the 1870s and 1880s. However, the stabilization of ship design, allied to the growth of imperialism, prompted an upsurge in construction. In the 1890s it became evident that the Royal Navy could not be strong everywhere. After 1900 Britain signed an alliance with Japan, and concentrated her resources to meet the main threat. That this would come from Germany was not evident until 1905; before that the danger came from France and Russia. However, moving resources to home waters was not without precedent, and reflected the underlying continuity of British strategy, not a radical shift of priorities.

The Naval Defence Act Aside from occasional alarms in the 1860s and 1870s, the public demonstrated little interest in the navy. However, in September 1884, W. T. Stead, editor of the *Pall Mall Gazette*, began a series of articles 'The Truth about the Navy', highlighting its supposed weakness, and the apparent equality of France in first-class battleships. These articles, based on conversations with Fisher, while exaggerated, demonstrated the value of publicity, an area where the 'silent service' had hitherto been noticeably reticent. The response was swift: in December, Lord Northbrook, First Lord of the Admiralty, secured £3.1 million for new construction and £2.4 million for guns and coaling bases. However, the alarm, based on a 'creative' manipulation of the numbers of first-class ships, was premature. The ships of the Northbrook programme, the last ironclads, including the ill-fated *Victoria*, quickly became obsolete.

After a period of calm, and reduced estimates, public anxiety was renewed by fears of a Franco-Russian alliance, symbolized by a combined fleet in the Mediterranean. Eventually, the prime minister, Lord Salisbury, with becoming reluctance, introduced the Naval Defence Act of 1889. The two-power standard was formally restored by spending £21.5 million on ten battleships, 38 cruisers, and other craft. Similarities with Castlereagh's memorandum of 1817 are remarkable, and yet hardly surprising. Once again France and Russia were the powers whose imperial interests were in conflict with Britain; notably in Persia, Afghanistan, China, Turkey, and Africa. However, their real reason for their alliance was the military might of the German empire.

Mahan, Public Opinion, and Seapower Despite the alarm generated by Stead's articles and Beresford's speeches in parliament, the single greatest contribution towards popularizing the navy came from an American naval officer with a profound aversion to the sea. In 1890, Captain Alfred T. Mahan published *The Influence of Seapower upon History: 1660–1783*, based on his lectures at the US Naval War College. Before the end of the century it would run through 15 editions. Two years later he drove home his message in *The Influence of Sea Power upon the French Revolution and Empire*. Mahan argued that seapower, and in particular battlefleet-based sea-control, had been critical to Britain's rise to empire. If the supporting text was only well-presented naval history, the introductory analysis provided an intellectual framework. Mahan became an instant celebrity. He was not working in a vacuum: the field had been prepared by John Laughton, Admiral Colomb, and others; however, no one could match the impact of Mahan's weighty prose.

While Mahan's purpose was to sell oceanic seapower to his countrymen, his work confirmed the wisdom of the Royal Navy in persevering with a battlefleet strategy. By 1895 few politicians would have been unaware of seapower. Where five years before, the very expression would have caused most to pause, now even Gladstone admitted the influence of the seapower volumes. Mahan helped foster the new navalism of the 1890s, his work being translated into all major languages and 'devoured' by Kaiser Wilhelm II. In consequence, if Mahan opened the eyes of the British to their strength, he also persuaded other nations that seapower was the golden road to world empire. The Sino-Japanese War of 1894, and the Spanish–American conflict in 1898 emphasized the relevance of Mahan's ideas.

The Spencer Programme and the End of the Victorian Age When the Naval Defence Act orders were complete, further agitation was inevitable; industrial seapower had created a military–industrial complex. Salisbury's government planned a new programme for 1892, but Gladstone's return to power delayed the order, creating a crisis in the shipbuilding industry. Further support for new construction was provided by the Navy League, founded in 1893 to promote naval reform, but quickly taken over by the City of London, which was more interested in the continued prosperity of commercial shipyards in peacetime than the defence of trade in war. More immediately the resolution of Admiral Sir Frederick Richards, the First Naval Lord, and the support of Earl Spencer, the First Lord, secured cabinet support for the Admiralty programme. When the cabinet met on 9 January 1894, only two members held out for less. Gladstone, speaking passionately against militarism, made it an issue of confidence. His colleagues let him resign. The Spencer programme—7 battleships, 6 cruisers, and 36 destroyers—reflected Britain's commitment to naval mastery. It proved timely; the secret Franco-Russian alliance had been signed on 4 January.

The programmes of 1889 and 1894 established the style of the navy for two decades. A new standard battleship, the 'pre-"Dreadnought"', allowed Britain to accumulate ships; by 1905 the old dominance had been largely re-established. Both France and Russia found their programmes hard to sustain, producing inferior

Admiral Sir George Tryon (1832–93). Although he is now remembered only for the manner of his death in the collision between the *Victoria* and the *Camperdown*, Tryon was the leading figure of the naval revival of the 1880s, and poised to take the highest office. Eton-educated and a late entrant, Tryon secured rapid promotion by his abilities displayed in the Crimean Naval Brigade, as Commander of HMS *Warrior*, in organizing the Abyssinian expedition of 1867, at the Admiralty, and as a commander-in-chief, first in Australia, and later in the Mediterranean. Until the day of his death Tryon stood head and shoulders above his contemporaries, as a fleet commander and naval administrator.

ships, usually badly overdue and far more expensive than British products. The speed and efficiency of British shipyards was the reward for sustained order books.

If the Franco-Russian challenge was never as serious as the alarmists argued, it was the only threat that could justify increased estimates, improved war readiness and a new professional emphasis. As designs settled, the problems of naval tactics were considered with more intelligence than had been evident in the decades of ram mania and torpedo scares. Two schools of thought developed to deal with the control of fleet actions. Traditionally, the admiral exercised direct control, leaving no initiative to subordinates. The alternative adopted by Vice-Admiral Sir George Tryon, one of the great men of the Victorian navy, was for the fleet to act on simple guidelines without excessive direction from above. Junior admirals and captains were expected to show initiative.

On 22 June 1893, a clear and calm day, the Mediterranean fleet, commanded by Tryon, was about to anchor on the Lebanese coast. Tryon signalled for an impossible manœuvre: the two columns were to turn inward simultaneously when there was clearly too little space. This was either a simple error on his part, or a test for his subordinates. Unfortunately no one on board his flagship, the *Victoria*, or Rear-Admiral Markham's *Camperdown*, had the courage to question the signal, for Tryon, a man of outsize physique and character, did not trouble to hide his contempt for inadequate subordinates. The two ships collided, the *Victoria* sinking with heavy loss of life. Tryon was never seen again. Many blamed Markham for failing to question, or disobey, the order, yet the court martial laid the blame on Tryon. Fleet tactics in the years up to the First World War fell into the hands of taciturn centralizers, epitomized by Arthur Wilson, who witnessed the disaster as captain of the *Sans Pareil*. While technical improvements in gunnery and fire control allowed heavy ships to hit the target, sterile tactical systems deprived large fleets of the flexibility to exploit fleeting tactical opportunities. John Jellicoe, a disciple of Wilson, narrowly escaped drowning in the *Victoria*.

Between 1815 and 1895, the Royal Navy changed its ships and guns, its method of recruiting and retaining seamen, and many things besides. However, the basic issues of naval policy did not change; the strategic problems remained those of a maritime world empire. The two-power standard remained in force, with France and Russia the most likely enemies. Battleships were the measure of power, but the true indication of strength lay in financial commitment, dockyards, and factories, and the political support that underpinned the fleet. Despite the low estimates of the 1830s, the 1870s, and early 1880s, the Royal Navy had been equal to any task that had arisen. The politicians and officers who administered the service were generally men of sound common sense, who recognized the importance of their task. Serious errors of policy were the exception. If the British empire was never *Pax Britannica*, neither was the Royal Navy a colonial police force run by reactionary fools.

CONCLUSION

THE ADMINISTRATION OF THE NAVY, 1815–1895

During the nineteenth century the administration of the Royal Navy was subjected to a series of reforms that had the cumulative effect of streamlining and improving the day-to-day running of the machine, while depriving it of any capacity for long-term policy formulation. This process reflected the growing importance of finance and domestic politics in the period after the first, or great, Reform Act of 1832, a trend reinforced and emphasized by the second Reform Act of 1867. The roles of the naval lords and the politicians, who once formed a solid block of ministerial voting strength on the Treasury benches, were gradually separated until, by the 1890s, naval lords no longer sat in the House of Commons. This separation of function emphasized the professional role of the naval lords, and their apolitical position, something that can be traced back to the administration of Lord Auckland (1846–9). This had the advantage of removing from the House of Commons a body of men who were noted both for their lack of interest in the questions of the day, and their ability to empty the benches during technical debates.

In 1815 the administration of the navy stood largely unaffected by the long war. Indeed the post-war Board, under the second Earl Melville, in combination with the Controller of the Navy, Admiral Sir Thomas Byam Martin at the Navy Board, was among the most effective policy-making administrations. However, the bitter political faction fights begun by Earl St Vincent in 1802–3 had left a deep legacy. The Navy Board had

been identified as a target for administrative and economic reform by the Whig opposition. In 1822, they forced the government to reduce the number of lords of admiralty by one, but it was not until they took office in 1830 that any major change was carried out. In the interval, the duke of Clarence, later William IV, spent a year, 1826–7, as Lord High Admiral, a post for which his career and interests made him well suited, before his indiscretion and impulsive nature forced the duke of Wellington to dismiss him.

In the Reform Ministry of 1830, Earl Grey appointed Sir James Graham, a noted critic of defence spending, First Lord. Graham had a mandate to make radical changes, both to have a long-delayed political revenge on the Navy Board, and to provide the savings that would be necessary to secure support for the Reform Bill. Graham adopted Bentham's principle of individual responsibility, abolishing the Navy Board in 1832; in its place the lords of admiralty became administrative department heads with specific duties concerned with the day-to-day running of the navy. Overnight the Admiralty had been converted from a small political decision-making body into a vast administrative machine. Inevitably, day-to-day administration dominated their deliberations, because cost-cutting was the main concern of succeeding ministries.

By the early 1850s, the blatant use of the navy as an electioneering machine had been ended, a fact that was impressed on the Tory party by the Stafford Inquiry of 1853. During the Crimean War the Admiralty performed far better than the chaotic army administration, but it was grossly overstretched and had no role in strategic decision-making. Graham, once again First Lord, consulted a variety of officers from both services, including Captain Sir Baldwin Walker, the Surveyor of the Navy, but took the key decisions alone. His naval lords lacked the stature, experience, and ability to contribute; they were left to run the machine, a task that they carried out with notable success. The post-war situation, rapid retrenchment followed by invasion scares and the introduction of the ironclad, left little time for reflection. The key decisions were taken by the cabinet, often by the prime minister alone. Only rarely did the First Lord possess the political weight to fight his corner in cabinet. In consequence the Palmerston administration produced large estimates, while the subsequent Disraeli and Gladstone ministries competed to put forward the lowest figures.

Gladstone's first ministry (1868–74) saw two reforms and a long-overdue recognition of the stature of the First Lord. Hitherto, the Admiralty had received the directions of the cabinet, although the First Lord was a member, through the Secretary of State for War. Lord Cardwell, the incoming Secretary of State, stressed that this was unrealistic. The First Lord, Henry Childers, a man whose political existence was committed to Gladstone, introduced a radical reform by Order in Council in 1869. Childers, who had served as Junior Civil Lord in 1864–5, ended the meetings of the board. He became the executive head of the navy, consulting the lords of admiralty on an individual basis. The object was to complete the subordination of naval administration to Treasury–political control.

Childers, already a sick man, suffered a major breakdown in the aftermath of the loss of the *Captain*, and left office in early 1871. His replacement, George Goschen, reintroduced board meetings, but did nothing to upgrade the board's decision-making role. In the years after 1830, the pace of technical change had increased dramatically, and as new ships and guns were the most costly items on the budget technical questions, and the *matériel* departments that understood them, soon came to dominate the board's deliberations. With a small board trying to oversee a vast machine, it was inevitable that many issues would be ignored. In the aftermath of the Russian War scare of 1885, the Naval Intelligence Department was established, building on foundations laid in 1879. This, for the first time, provided a body, albeit small, with a dedicated task to consider the development of foreign fleets, and the needs of war. The Committee of Imperial Defence, established in 1902, provided a forum for the key political and professional defence planners to meet. However, the navy reached the end of the century without a true naval staff. Naval administration was economical and well run, but there was little time or demand for wider considerations. War-planning was left to the First Lord and the First Naval Lord, both of whom spent the bulk of their time arguing the naval case in cabinet.

7

WOOD, SAIL AND CANNON-BALLS TO STEEL STEAM, AND SHELLS 1815–1895

DAVID K. BROWN
RCNC

$\mathcal{B}$etween the end of the Napoleonic War in 1815 and the introduction of the *Dreadnought* in 1905, the Royal Navy moved from the wooden sailing-ship to steel, turbine-driven ships firing shells to some 20 times the maximum range of Nelson's guns. This chapter shows that the Admiralty managed these changes wisely, using the talents of their own staff and the skills of industry in a usually happy partnership. The power of the Royal Navy was an emotional issue, debated with great heat and not a little ignorance at the time but, with hindsight, one may be impressed with the Admiralty's judgement.

The last generation of wooden, sailing battleships were some 50 per cent longer than the *Victory* and fired twice the weight of broadside, an advance due very largely to Sir Robert Seppings (Surveyor, 1813–32). He realized that owing to the opposing forces of weight and buoyancy the loading of a ship in waves would cause the hull to flex with the planks in the side sliding over each other, an action which he likened to the behaviour of a five-bar gate without the diagonal. He proposed a new structural arrangement incorporating diagonal frames, and trials showed that such a structure was far more rigid. In consequence, water was less likely to get into the seams between planks and such ships were hence much less susceptible to rot.

Seppings also introduced the round bow and stern which increased the strength of the ends, giving some protection against raking fire and permitting heavier bow and stern batteries. British designers of the period are often accused of neglecting 'French Science', but much of their hydrodynamics was quite wrong whilst Seppings's approach to structural design was truly scientific in the way he worked from loading to disposition of the timbers and in the subsequent measured trial of his ideas.

Seppings's much stronger structure enabled a heavier armament to be carried. At that time all major navies were moving towards a uniform armament of 32-pounders in place of the mixture of 18-, 24-, and 32-pounders (so designated by the weight of the shot the guns fired) mounted during the wars. Shortage of funds meant that this rearming could take place only slowly, but from about 1830, ships in commission generally carried the uniform armament. Many of the bigger guns needed were obtained by boring out the older, smaller guns—a procedure made

Facing: Gorgon of 1835 was a big and powerfully armed paddle sloop and was seen as a great success, 28 more, generally similar ships being built. Gun ports on the main deck can be seen though guns were not carried on that deck.

Diagonal stiffening designed by Robert Seppings to prevent the structure racking, seen very clearly in this photograph. Less obviously, the bottom timbers have been filled in solid, also adding rigidity and helping to reduce pockets in which water could lie, causing rot.

PADDLE WARSHIPS

possible only as a result of the improved casting technology introduced by Thomas Blomefield at Woolwich at the end of the previous century.

The Admiralty also took a number of steps aimed at improving the design of ships, as it was believed, incorrectly, that French designs were faster than British ships. In 1811, a School of Naval Architecture was set up at Portsmouth with a seven-year course blending theory and practical skill. Initially, the graduates met with great hostility, but by 1860 they held most senior technical posts in the Admiralty, and many outside.

The Admiralty also introduced 'experimental sailing', in which ships by different designers were raced against each other. These races continued into the 1840s, but the skill of the captain in sailing and in trimming the ship obscured any small differences due to hull form. These races did at least show the determination of the Board of Admiralty to obtain the best possible ships from any source.

Captain William Symonds had designed some ships which had been successful in the early trials, largely owing to his excellent seamanship, and in 1832 a new Board appointed him Surveyor in place of Seppings. It is likely that the Board intended that Symonds should set the requirements for new ships with the design being carried out by his very able assistant Surveyor, John Edye. On these broader issues Symonds had some success, breaking away from the old 'Establishment Rules' and producing spacious ships with gun decks well clear of the water which were well liked though very expensive. Unfortunately, Symonds believed that his hull form, with a broad beam and high rise of floor, had unique virtues, and though such ships did quite well in later races, the excessive stability which he gave his designs made them roll rapidly, and hence they were poor gun platforms. Symonds was also prejudiced against scientific design and new technology such as propellers and iron hulls.

Even during the wars, the Admiralty had begun to introduce new technology into the dockyards under Samuel Bentham and Marc Brunel. The latter's block mills in Portsmouth were the first mass-production factory in the country and the roofs of the covered building slips were the largest unsupported spans. From about 1840 onwards, Symonds lost the confidence of the Board and increasingly his designs were subject to review by his opponents.

Early experiments with steamers in 1793 and 1816 were unsuccessful and a trial in 1819 using a steamship to tow sailing-ships out of harbour showed that they were barely powerful enough. Despite this, the Admiralty built its first steamship, the *Comet*, in 1821. She was 115 feet long with a two-cylinder engine and was primarily intended as a tug for use on the Thames. Two similar ships were built in 1823, one of which, *Lightning*, accompanied the expedition to Algiers in 1824, the first operational deployment of a British steamship. Further favourable experience with steamships was obtained by Royal Navy officers employed by the East India Company, by Greek insurgents, and others, notably with the *Karteria*.

A number of generally similar ships were built, and by 1830 there were 11 in service. They were still used mainly as tugs or mail packets though from 1828 they appeared in the Navy List. Some slightly bigger ships were built in 1832, including the *Rhadamanthus*, the first British steamship to cross the Atlantic, though her limited coal capacity forced her to use sail much of the time. The increased size of these ships meant that for the first time an effective armament could be carried, typically three 32-pounders. The engines were all of the side lever type, a modification of the beam engine in which the beam was cut in half longitudinally and each portion arranged either side of the cylinders, lowering the centre of gravity of the engine. The boilers were rectangular iron boxes, full of sea water, heated by flues through which the furnace gases passed. Hot, salt water is very corrosive and the life of a boiler was about three years. Such machinery was very bulky and heavy; that of *Medea* (1832) occupied one-third of the length of the ship, and weighed 245 tons, with 320 tons of coal which was burnt at the rate of one ton an hour at the full speed of 8 knots. In 1836 a steam factory was opened at Woolwich which not only overhauled the growing number of engines but also trained both seagoing engineers and designers.

Progress was steady, and by 1837 the Royal Navy had 29 steamships as well as 37 mail packets. The *Gorgon* of 1835 marked a major advance in design; she was bigger at 1,610 tons and her new direct-acting engines provided much more power from the same weight and space, giving a speed of about 9 knots. She also mounted a heavier armament of two 68-pounders and four 42-pounders, all on the upper deck. The armament of paddle-steamers always suffered from the problem that the paddles, some 27 feet in diameter, and their sponsons, occupied nearly one-

Rhadamanthus of 1832 was typical of early paddle warships. She was the first British steamship to cross the Atlantic though she had to use sail much of the time to save coal. Under sail, the lower paddle boards would be removed to reduce drag; an unpleasant task for which the stokers were rewarded with extra rum. Note the 32-pounder swivel guns forward and aft.

third of the side, restricting broadside mounting. In consequence, most paddlers mounted a few, very large guns on the upper deck. The *Gorgon* was seen as a great success, 29 generally similar ships being built. As Edye gained in authority he moved away from Symonds's extreme form, adopting a more conventional shape, better suited to the installation of machinery.

From 1842, a small number of bigger, first-class paddle frigates were built, with guns on the main as well as the upper deck. The largest of these was the *Terrible*, with a displacement of 3,189 tons, a length of 226 feet, and an armament of eight 68-pounders and eight 56-pounders, a truly formidable battery.

Paddle warships fought no major action but they were frequently in action with shore batteries during various rebellions, colonial wars, and during the Crimean War. These demonstrated the tactical value of the steamship, with its independence of the direction and strength of the wind, and, strategically, its ability to arrive at its destination at a time planned in advance; and there were many letters from admirals asking for more steamers. At a time when much of the sailing fleet was laid up in reserve, all the steamships were either in commission or refitting. The vulnerability of paddle wheels to gunfire had been thought of as a serious disadvantage, but, on the only two occasions on which a wheel was smashed, the ship was able to proceed at only slightly diminished speed on the remaining wheel.

IRON HULLS

Iron had been used to build canal barges from about 1787, and in 1821 the *Aaron Manby* was built for commercial service between London and Paris. A number of other iron ships followed, but none could go out of sight of land as there was no way to use a magnetic compass in an iron hull. This problem was solved in principle in 1838 by Sir George Airy, working for the Admiralty, and a number of sea-going iron ships were started.

Brunel changed the design of the *Great Britain* to iron, the Admiralty took a cautious step with the packet *Dover*, whilst the East India Company ordered the first iron warship, *Nemesis*. She was a paddle gunboat, 184 feet long, and was given enthusiastic reports when serving alongside the Royal Navy in the China War (1841–3). The iron hull was about 20 per cent lighter than one of wood which enabled her to float at lesser draught, whilst damage from grounding and from gun shot was not serious and was easy to repair. The smaller iron frames also gave about 20 per cent more internal space. There were similar, encouraging reports from the British-built, Mexican frigate *Guadeloupe* in the war with Texas (1843).

As a result of these reports, the Board ordered a considerable number of iron ships from 1843 onwards, beginning with the sloop *Trident*. She was followed by the paddle frigate *Birkenhead* and by a number of smaller ships. In 1845, four very large, iron, screw frigates were ordered, of which *Simoom* was the biggest with a displacement of 2,920 tons and a length of 246 feet, not much smaller than the *Great Britain*.

The great French naval architect, Dupuy de Lôme, visited British shipyards and his report of 1844 gives a valuable insight into the building of iron ships. He wrote

of iron ships that there were '*Beaucoup de préjugés, de doutes raisonnables, et de difficultés réelles*'. The Admiralty's iron ship programme did indeed suffer from prejudice, reasonable doubts, and real difficulties. It became a political issue with the Whigs opposed to Tory plans but, by 1845, a number of people realized that the effect of shot and shell on iron hulls had not been adequately examined.

A number of very careful trials were carried out at Woolwich Arsenal extending over some months from August 1845 using iron from several manufacturers and exploring the effect of varying the plate thickness, striking velocity, and type of lining. It was found that shot hitting at high velocity would make a clean hole but those hitting at lower velocity would make a jagged puncture, hard to plug. Thick plates would break the shot into a lethal shower of splinters. While these tests were not conclusive against the use of iron, they weakened the defence against political attack. Iron hulls were also liable to rapid fouling which could reduce speed by a knot or two very quickly.

Fears and doubts increased, and in July 1846, the Board ordered another trial in which an iron harbour launch, the *Ruby*, was blown apart by heavy guns. Even at the time, it was realized that this trial was meaningless and its importance has

Birkenhead was the first iron frigate and the only one with paddle propulsion. As a result of well-founded doubts concerning the behaviour of iron under the impact of gunshot, she was converted to a troopship and lost after hitting a rock near Simonstown on 26 February 1852, with the loss of 455 lives.

been much exaggerated by some modern writers. Initial reports of damage to iron gunboats under fire in the Parana river in 1847 caused further alarm and the Whig government ordered the iron frigates to be converted to troopships, though the smaller ships were retained as warships.

A parliamentary committee of 1850 realized that the evidence for or against iron hulls was unconvincing and further test firings were carried out at Portsmouth against sections representing the frigate *Simoom* together with variations on it. Captain Chads's report concluded that 'iron is not a suitable material of which to build ships of war', and most contemporary naval architects agreed with him.

For a century the difference between the results of these tests and the behaviour of *Nemesis* and *Guadeloupe* in action remained unexplained and Chads was derided as a reactionary. Finally, tests on samples taken from the *Warrior* during her restoration showed that the resistance of iron to sudden impact depends on temperature, the material becoming brittle at quite modest temperatures. The battles were in warm water, the crucial tests in an English winter; Chads was justified at last. It is hard to fault the Admiralty for any of the decisions made during the iron ship programme, as the crucial facts would not be available for another century.

SCREW SHIPS

Though there were many proposals for screw propellers in the early nineteenth century, some of which worked quite well, none was developed fully. Two men working quite independently in London in 1836 produced successful schemes for screw propulsion, bringing together the design of hull, machinery, and propeller. John Ericsson completed the *Francis B. Ogden* in 1837 and demonstrated it to the Board. To his amazement his proposal was rejected, apparently because Symonds believed that it would not be possible to steer a ship propelled at the stern.

Petit Smith, a farmer, was advised by Sir John Barrow to approach the Steam Department who were favourably impressed by his trial boat, *Francis Smith*, and encouraged him to build a ship, *Archimedes*, at his own expense, which completed in 1839. After a few teething-troubles had been rectified, she was raced against the fastest paddle packets of the day. The Admiralty trials officers' report was most enthusiastic and the Board decided to build a screw warship. Isambard Brunel followed the Admiralty's lead and changed the design of the *Great Britain* to screw. Both Brunel and Smith were engaged as consultants by the Admiralty and, initially, there was some friction both between these great men and with the equally brilliant Thomas Lloyd, the Admiralty's chief engineer, but enthusiasm for the project soon led to lasting friendship between them.

The new screw ship, *Rattler*, was a near sister of the paddler, *Alecto*, but with finer stern lines to improve the flow into the propeller. She completed late in 1843 and the main trials programme began the following year with 28 propellers from various designers, varying diameter, pitch, and number of blades. Smith's original propeller, now displayed at the Royal Naval Museum, Portsmouth, was marginally the best and by the end of the year the Admiralty had ordered a considerable number of screw ships, including the big iron frigates mentioned above.

In March 1845, *Rattler* was matched against her paddle half-sister, *Alecto*, in a series of races, under steam alone, under sail, and with both sail and steam, and also a number in which one ship towed the other. At first sight these trials show a considerable advantage to the screw but *Rattler* had more power, and on equal terms there was little difference. In one of the later trials the two ships were lashed stern to stern and in a tug of war tried to pull each other backwards, with an easy victory to the screw. It is often said that this trial convinced the Admiralty but the numerous orders the previous year show that the Board was already convinced.

Since there was no way of designing a propeller until the Froudes established a sound method, based on model tests, in the 1870s, there were many trials, seeking experience. The critical aspects of screw propulsion were shown in one such trial in which the screw-ship's hull was 95 tons lighter and a further 45 tons were saved in the machinery. The screw machinery was entirely below the waterline giving protection, and the ship was more easily prepared for sailing.

The next step followed fast. At the end of 1844 it had been decided to fit engines to four old 74-gun ships to serve as mobile coast defence batteries. By the end of 1845, plans had changed and these 'blockships' completed as seagoing battleships. *Ajax* completed on 23 August 1845 as the world's first steam battleship and reached 7 knots on trial. The space taken up by the machinery left them very cramped so that the number of guns was reduced to 60, though these were all of large calibre: most suitable for their original role of coastal defence and, later, for attack of enemy coasts. These simple conversions were favourably reported on, both during trials and in later exercises; indeed, in 1850, it was noted that it was unlikely that the navy would ever fight again under sail. A true steam battleship was designed by Edye in 1848 and was ordered in July 1849 as the *Agamemnon*.

By then, the French had taken a brief lead; their first screw battleship, *Napoleon*, designed by Dupuy de Lôme, had been laid down in 1848. Isaac Watts, the Chief Constructor, and Lloyd, now Engineer-in-Chief, visited France in 1851 and reported on the rapid progress of *Napoleon*. As a result, work on the *Agamemnon* was speeded up and she completed in October 1852, only three months after her French rival. The two ships were of much the same size (see Table 7.1).

STEAM BATTLESHIPS

Table 7.1 English and French screw battleships

Ship	Displacement (tons)	Length (feet/inches)
Agamemnon	5,080	230′ 0″
Napoleon	5,120	234′5″

Agamemnon reached 11.4 knots on trial with 2,500 indicated horse power (ihp: the power potentially available in the steam entering the engine; some 25 per cent was used in overcoming internal friction, working auxiliaries, and in transmission losses before reaching the propeller). There are no measured mile data for *Napoleon*, but after her first, unreliable engines were replaced it would seem that her speed was of the same order. The navy thought highly of *Agamemnon* and ten generally similar ships were built together with a number of slightly stretched derivatives, reflecting great credit on Edye's original design.

Symonds had been replaced as Surveyor by Captain Baldwin Walker, an enthusiast for the screw fleet who won the respect of all and healed past feuds. Walker set the requirements for new ships leaving the design to his assistant, first Edye, soon replaced by Watts and later restyled as Chief Constructor.

· THE CHIEF DESIGNERS ·

In the period under discussion, the first chief designer was Sir Robert Seppings, a former Master Shipwright, who from 1813 to 1832 was fully responsible for the design and building of His Majesty's Ships. In 1832 there was a reorganization: the Navy Office was abolished, and Sir William Symonds was appointed surveyor. He was not a technical man and though he chose the overall dimensions and form, the design was carried out by John Edye. Symonds was replaced in 1848 by Sir Baldwin Walker whose duties were defined as administration, leaving design first to Edye and then to Isaac Watts. The titles of the chief designer continued to change but there was no doubt who was in charge.

Table 7.2

Name	Dates	Title
Robert Seppings	1813–32	Surveyor
John Edye	1832–48	Assistant Surveyor
Isaac Watts	1848–59	Assistant Surveyor
	1859–63	Chief Constructor
Edward J. Reed	1863-70	Chief Constructor
Nathanial Barnaby	1870–72	President of the Council of Construction
	1872–5	Chief Constructor
	1875–85	Director of Naval Construction
William H. White	1886–1902	Director of Naval Construction
Phillip Watts	1902–12	Director of Naval Construction

Walker embarked on a rapid expansion of the screw fleet; three sisters of *Agamemnon* were laid down together with a slightly larger ship. All these were two-deckers. The first three were 91-gun ships, with 34 8-inch on the gun deck, 34 32-pounders on the main, together with a 68- and 22 32-pounders on the upper deck. The *Saint Jean D'Acre* carried 101 similar guns and was probably Watts's first major design. Resources of all kinds, timber and shipwrights, as well as money, were inadequate and to increase rapidly the number of screw-ships some sailing-ships, still on the building slip, including the 131-gun *Duke of Wellington*, were lengthened and given engines. These major conversions were almost the equivalent of new construction and by the outbreak of the Crimean War the navy had eight screw battleships, France had nine, and no other country had any.

The screw frigate story is complicated, as the iron frigates were not put into service and the frigate 'blockship' conversions were unsuccessful. The first moderately successful frigate was *Amphion*, converted while building, and completed in 1846. It was thought that most frigate tasks would still be carried out under sail and the early ships were given low-powered engines for use in a calm. In consequence, they were found to be under-powered in service, and at the outbreak of war there were only six screw frigates. The need for large numbers of such ships for trade protection was not appreciated; by 1849 there were 1,100 steam merchant ships, totalling 255,000 tons.

A number of changes, some apparently minor, in training and equipment, had much improved the fleet's gunnery. An old battleship, *Excellent*, had become a gunnery training and experimental establishment designing and testing new equipment such as sights, and carrying out trials on the effectiveness of gun-fire.

Saint Jean D'Acre, a wooden, screw battleship carrying 101 guns on two decks. She was designed by Isaac Watts and was quite similar to John Edye's *Agamemnon*, a slightly smaller ship.

In one such trial, in 1835, both solid shot and shells were fired into the old wooden battleship *Prince George*, exploring the penetration which could be achieved in both direct and ricochet fire. Of 80 shells fired, 38 failed to explode, but two which were exploded inside the ship caused severe damage. Other trials showed the inaccuracy of shell-fire, but these findings were not well known and both the unreliability of shells and the damage which a successful burst would cause came as a surprise in the Crimean War.

That war is covered in Chapter 6, but it must be noted that there were many technical changes; rifled guns and shells were used in some numbers, while the Russians laid the first contact mines. These mines caused some damage but the navy quickly improvised sweeping gear which was effective. A very large number of technically advanced small craft were built for coastal warfare; the 'floating batteries', discussed later, were the first armoured warships.

The 156 gunboats were remarkable for innovation in production technology. Twenty yards built the hulls which were taken to a dockyard for completion; a special yard for the purpose being designed by Isambard Brunel at Haslar, opposite Portsmouth. All the machinery, using high-pressure steam, was ordered from Penn or Maudslay, who introduced the first large-scale use of mass production with a large number of sub-contractors making interchangeable parts. Orders for the ships were placed at a fixed price, but, since wage-rates trebled during the war, the builders suffered heavy losses, a major factor in the collapse of shipbuilding on the Thames. The gunboats had a very shallow draft and proved most successful both in their design role and later as peacetime policemen.

During the war, considerable development took place in the Royal Dockyards, with new basins and factories to refit the increasing numbers of engines, but the small craft programme took most of the resources and few battleships were completed. After the war, Britain and France resumed their traditional hostility, which led to a building race in wooden screw battleships. The threat was exaggerated and, in 1858, when the Admiralty believed that France had achieved equality, the true figures were 34 British and 28 French.

The new battleships were either new construction, based on *Agamemnon*, conversions from ships which were still on the slip and which were lengthened, or older ships into which engines were fitted without lengthening. There was little to choose between the first two categories but the unlengthened ships were slower, about 9 knots, and were overcrowded. Though less effective, they could become available more quickly, used fewer resources, and were at least as effective as the older French ships.

Steam frigates were given less priority, and by 1858 there were still only 12 available. But by 1866, numbers had risen to 36 including some of the longest wooden ships ever built, designed by Isaac Watts. In addition, there were about 70 wooden steam corvettes and sloops together with over 50 gun vessels, all mounting a heavier armament than could be mounted in a merchant ship converted to a commerce raider.

During the Crimean War, Napoleon III suggested the use of armoured, floating HMS *WARRIOR* batteries, with small steam-engines, to attack coastal fortifications. The designer was Dupuy de Lôme and, following a suggestion from Thomas Lloyd, they were protected by rolled plates, about four inches thick. Watts designed similar ships for the Royal Navy, some with iron hulls, but they were delayed by the First Lord who insisted on repeating the armour trials carried out in France. The French ships completed in time for the bombardment of Kinburn in October 1855, where their armour successfully resisted enemy shot, as would be expected since these were only of medium size, fired at long range.

In 1857 Dupuy de Lôme began the design of the *Gloire*, the first armoured seagoing warship; though the French were aware of the advantages of an iron hull for armoured ships, *Gloire* was built in wood, as French industry was not capable of building such a large ship in iron. It is strange that the French chose to move the competition in battleships to an area in which Britain's industrial superiority gave her such an advantage.

The *Gloire* was laid down in March 1858 and completed in 1860. The Table below compares her main features with the British response, *Warrior*.

Table 7.3 *Warrior* and *Gloire*: comparative details

	Warrior	*Gloire*
Displacement (tons)	9,180	5,630
Length (feet)	420.0	255.5
Horse power (ihp)	5,270	2,500
Speed (knots)	14.0	12.5
Guns (design)—all smooth bore, muzzle loaders	40 68-pounders	36 6.4-inch

The order for *Gloire* caused alarm in Britain, and in March 1858, Walker asked Watts to design a 26-gun armoured ship. Walker also wrote to the First Lord outlining the case for such a ship in words which must be quoted at length since they so well describe the Admiralty's conservative but rational philosophy for technical innovation in the previous half century.

It is not in the interest of Britain—possessing as she does so large a navy—to adopt any important change in ships of war which might have the effect of rendering necessary the introduction of a new class of very costly vessels, until such a course is forced upon her by the adoption by foreign powers of formidable ships of novel character requiring similar ships to cope with them, yet it then becomes a matter not only of expediency but of absolute necessity . . . This time has arrived.

Funds were allocated in July 1858 for two armoured ships, and in November the Controller was directed to proceed with a wooden steamship having 4.5-inch armour. It is clear that Watts, aware of structural problems with his longest wooden frigates, never even contemplated a wood hull for *Warrior* and his design for an iron ship was submitted in January 1859. This design was circulated to shipbuilders, including the Royal Dockyards, inviting alternative designs with the same armament and speed. Fifteen proposals were received and were forwarded to the Board, together with his reasons for selecting his own design, listing the failings of the others. The Board approved Watts's design in April 1859.

Warrior's most novel feature was her armour and a number of full-scale tests were carried out on different arrangements. It seemed that thick wooden backing was essential but the true problem was later recognized as shock waves in the exposed bolts, causing them to fail. The final design was tested in October 1861 when a section 20 feet by 10 feet successfully resisted the impact of 29 shot weighing up to 200 lb. The structure used both longitudinal and transverse framing and a recent study shows that it was well matched to the loading and the properties of wrought iron.

Warrior was a big ship, second only in size to the enormous but uneconomic merchant vessel *Great Eastern*. The length was set by the number of guns, their separation of 15 feet, together with the fine ends thought necessary for speed. The ends were unarmoured as it was believed that heavy weights at the ends would cause severe pitching. In consequence, only 13 guns each side were protected. It was calculated that if the unprotected ends were entirely flooded she would still be safe and would only lose a knot or so of speed. Ten of the 68-pounder guns were replaced by 110-pound breech-loaders during building but these new guns proved disappointing in service.

Warrior had a Penn trunk engine with a fuel consumption of about 3.5–5 lb. per horse power per hour giving her an endurance of 2,100 miles at 11 knots; the ten boilers each contained 17 tons of sea water. She sailed well, though her length made her slow in tacking. Her accommodation was spacious and she was a popular ship. She was laid down at the Thames Iron Works in May 1859 and completed in August 1861 at a cost of £190,255 plus £74,409 for the machinery; a sister and a number of generally similar vessels followed, including four smaller, cheaper ships which proved of little value.

Warrior drew together separate developments in steam and screw propulsion, in iron hulls, and in armour. She only showed small advances in each separate aspect as her design team, Watts, his deputy Large, and Lloyd, the engineer, adopted the best, proven technology of the day; a wise approach when the whole concept is novel. As the table above shows, she was bigger and faster than *Gloire*, initiating a competition between gun and armour which lasted until the battleship died in the Second World War. Her iron hull proved durable—she is still afloat in 1993—and was a triumph both for British industry and for the School of Naval Architecture where her designers were educated.

CENTRE BATTERY SHIPS

Gun-makers soon built guns which could penetrate *Warrior*'s armour at fighting range. The response was thicker armour, but iron plate is heavy; each square foot weighs 40 pounds for every inch of thickness and there are many square feet on the side of a battleship. The new Chief Constructor, Edward Reed, introduced the centre battery ship in which a few big guns were mounted in a short, well-protected battery amidships above a narrow armour belt which protected the full length of the waterline. Reed was a graduate of the Admiralty's second school of naval architecture, but soon became frustrated and left to edit a technical magazine which gave him an unusual ability to write clearly on warship design. Some freelance proposals won favour and he was given the job of Chief Constructor when Watts retired in 1863.

Reed's first major design was the *Bellerophon*, which mounted ten 9-inch muzzle-loading rifles (MLR) in a short battery. The hull was shorter and this, together with a more efficient structure, saved 1,000 tons of weight, enabling the armour thickness to be increased to six inches. *Bellerophon's* structure used the 'bracket frame' system in which the double bottom extended above the turn of bilge with five deep longitudinal girders either side of the keel joining the inner and outer bottoms. The transverse frames were widely spaced and, within the double bottom, were fitted as plate 'brackets' between the longitudinals.

It was, and still is, customary to credit major advances to the chief designer who will certainly be blamed for any failure. However, in many cases, the idea comes from a younger man, and, in this case, the bracket frame system seems to have been due to Barnaby, who also contributed much to the design of *Bellerophon*. There are many other cases in which the chief received credit for the work of his staff but, as with Barnaby, reward sometimes came in later promotion to the senior post.

Until about 1871 the structure was designed subjectively by careful comparison with a previous ship but, in that year, Reed read a paper which set out the logical basis for calculating the scantlings. The initiative was Reed's but the work was by William White, who would also rise to the top. Reed had been prepared to pay a penalty in power for the short hull of *Bellerophon* and was pleasantly surprised to find the penalty was not too high.

Reed developed the centre battery style which culminated in Barnaby's *Alexandra*, which mounted her 12 heavy guns in a two-tier battery with armour 12 inches thick. The later such ships had the sides cut away fore and aft of the battery to give some degree of end-on fire, though blast damage limited the usable angle very considerably (about 30° off the bow) whilst the cut-aways threw up vast clouds of spray. Reed also designed for overseas service a number of smaller centre-battery ships which he claimed had used William Froude's theoretical work to reduce rolling.

It had long been recognized that high-pressure steam, expanding in two stages, would be more economical and lead to reductions in machinery weight. There were many problems to be overcome, most apparently trivial, such as the choice of packing material for sealing joints in steam pipes, but by 1875 the *Alexandra* had engines working at 60 lb. per square inch with a fuel consumption of about two-thirds that of *Warrior*.

TURRET SHIPS

Captain Cowper Coles devised a scheme based on his wartime experience for mounting big guns on turntables, protected by armour. The Coles turret consisted of a cylindrical box which passed through the upper deck and rested on rollers on the deck below and which could be turned by steam or hand power. The Admiralty had a prototype made which was tested in 1861 on the *Trusty*. The turret-mounted gun could fire nearly twice as fast as the traditional mounting whilst a later trial showed that the turret could still train after 29 hits from heavy shot.

The coastal defence ship *Royal Albert*, designed by Watts, was ordered in February 1862 with four single turrets mounting 9-inch guns. In April 1862, the wooden three-decker *Royal Sovereign* was converted to a similar design with one twin and three single turrets. Later, in 1866, *Bellerophon* fired three shots at 200 yards range against one of these turrets without causing significant damage. The advantages of the turret were also confirmed by USS *Monitor*'s action during the American Civil War with CSS *Virginia* (ex *Merrimac*) in which the former mounted an Ericsson turret.

All these turret ships were for coastal use, without sails. Fuel consumption was still too high for ocean passages without sail but the upper deck of a fully rigged ship was a most unsuitable site for a turret. After various impractical suggestions from Coles, a committee of naval officers set out the requirements for a rigged turret ship and Reed designed the *Monarch* with two twin 12-inch turrets and a 7-inch belt. She displaced 8,300 tons, could steam at 15 knots, and had a long, successful life.

Reed was still not convinced that a rigged turret ship was either effective or safe. Coles's objections to *Monarch* were stronger and, backed by popular opinion, the First Lord allowed him to set the style of a new ship, to be designed by Lairds. They submitted their proposals in July 1866 with a low freeboard of 8 feet though still with a forecastle and poop, and proposed to abolish most standing rigging by using tripod masts and to operate the running rigging from a flying deck over the turrets. Reed and the Controller objected strongly to the low freeboard and, after a few days' consideration, Reed pointed out that the centre of gravity would be much higher than Lairds assumed—for they had made no attempt to calculate this

Captain showing the low freeboard which was to lead to her tragic capsize in 1870. This photograph shows how Coles had tried to reduce the amount of rigging by fitting tripod masts whilst the remaining running rigging was worked from a flying deck over the turrets. Even so, the arcs of fire were very restricted, supporting the view that turrets and sails were incompatible.

vital parameter! Despite these objections, the First Lord placed an order for *Captain*, though without defining the division of responsibility between Lairds and the Admiralty.

Lairds' weight estimates proved grossly in error and *Captain* completed 735 tons over-weight, reducing her freeboard to 6 feet 7 inches. The importance of free-board in helping ships to recover from large angles of heel had not been appreciated as, though the basic mathematics had been published by Attwood, the equations were too difficult to solve until another of Reed's assistants, Barnes, developed a solution while *Captain* was building.

Captain was highly praised on her first two voyages, as she sailed well, was steady in a seaway, and, though the turret bases were often awash, they could still be operated. Following an inclining experiment to determine the true height of the centre of gravity, Barnes began the laborious task of calculating the righting moment at large angles of heel. He showed that the maximum righting moment was at 21° and if heeled beyond this angle, recovery was unlikely.

Captain was already at sea again and on 6 September 1870, she sailed all day with her deck edge heeled under; at midnight a gust blew her over and she capsized with the loss of 473 lives including Coles. Reed had already resigned, tired of the quarrelling, and the Controller, Spencer Robinson, who had warned of the danger, was forced to resign. The inquiry blamed Lairds for their poor estimates, but none of the politicians involved was censured for disregarding the advice of their staff. Coles was not directly responsible for the design but much of the acrimony was surely his fault.

Reed had his own ideas for safe, effective turret ships. The new engines, with lower fuel consumption, made it possible to design steamships without the sails whose heeling moment had been the direct cause of the loss of *Captain*. He was influenced by Ericsson's *Monitor* but recognized her fatal weakness with openings close to the waterline. His plans were first shown in the coast defence ship *Cerberus* for the State of Victoria, which had a low hull protected by 8-inch armour. Above the hull there was a breastwork also of 8-inch armour, which carried two twin 10-inch turrets and also a light superstructure which carried the access and ventilation at a safe height. She was the prototype of all modern battleships and her remains may still be seen at Melbourne (1993).

Reed designed the *Devastation* in 1869 as an enlarged *Cerberus* of 9,330 tons. She was the first battleship without sails and it was intended that she should carry enough coal to cross the Atlantic. She had two twin turrets with 35-ton guns and armour up to 14 inches thick. *Devastation* and her sister could steam at 14 knots; a third ship was planned at the time when *Captain* was lost.

As has been described, this tragedy was due to the combination of low free-board exacerbated by bad estimating and full sailing rig, but this was understood by few. The Admiralty set up a committee of distinguished engineers, scientists, and naval officers to examine the safety of current ships and new designs. *Devastation* and her sister were thought to be safe, though some light structure was

Devastation, the first major battleship without sails. Without their overturning moment the freeboard could be low with safety. Her designer, Edward Reed, brought the openings for access and ventilation into a fairly high armoured breastwork which carried the turrets giving them excellent arcs of fire.

added which improved stability and provided valuable living space. The third ship was redesigned by Barnaby and White with the armoured breastwork brought out to the side and renamed *Dreadnought*. One member of the committee was William Froude who had recently published a new theory of rolling. This was followed up by the committee and large-scale model tests of the *Devastation* showed the value of bilge keels in reducing roll.

The evidence given to the committee showed that extremely large guns would be built in the near future and that it would not be feasible to provide complete protection against such weapons, a problem which was to dominate the design of the *Inflexible* in 1873. She was originally intended to be armed with four 38-ton guns but completed with 81-ton weapons mounted on a heavily armoured citadel, just long enough to contain the machinery and the two turrets, which was designed to float, without capsizing, even if both ends were flooded. These end spaces were given some protection by a three-inch deck and close subdivision and there was a light superstructure above to provide living-space.

Inflexible's armour belt consisted of 24 inches of wrought iron with 36 inches of teak backing, weighing 1,100 pounds per square foot. The two turrets mounted 16-inch Rifled Muzzle Loaders (RML)—guns loaded with shells through the muzzle which had studs to engage in grooves in the gun—and had 16 inches of compound armour in which a hard steel face was welded to the tough iron back. It was equivalent to iron of 1.25 times the thickness. The guns were loaded by hydraulic machinery below deck. She was full of novel features: electric light, anti-rolling tanks, torpedo launchers, and Reed was to refer to her as a 'Steam Being', with her 39 auxiliary engines for steering, ventilation, and other services.

While she was completing, Reed claimed that she was unsafe, a charge investigated by yet another committee. Thanks to tests by Froude using a very large model, both in still water and in waves, it was shown that *Inflexible* was safe with the ends flooded. Whilst *Inflexible* was a very clever design, she would probably not have been very effective in battle as her guns could only fire every 3–4 minutes. However, the style was copied in several other navies and four smaller versions were built for the Royal Navy.

Up to this date, gunpowder was used as the propellant in the guns. As it burnt fast, only a short barrel was needed to impart momentum to the shell, and muzzle loaders proved to be cheaper, more reliable, and at least as effective as the early breech-loaders. Slow-burning propellants were now introduced which needed longer barrels, making breech-loaders essential, and these were introduced in the 'Edinburgh' class of 1882.

The problems with iron had never been overcome and it was regarded as unsuitable for unarmoured ships. Barnaby had long advocated the use of steel instead of iron for the hull, with a potential weight-saving of about 20 per cent, but Bessemer steel, which had been available since about 1860, was inconsistent in its properties and the French took a lead in the use of the more reliable open-hearth steel. By 1875 the Landore Works in Wales was producing good steel, used in the cruiser *Iris*, and once some problems were overcome it was introduced in the battleship *Edinburgh* (1879).

To sink a ship it is necessary to let water in, difficult to achieve by gun-fire, and most navies took an interest in ramming, particularly after the battle of Lissa in 1866. Most large ships were designed with rams, but in 1863 the French designed a

**UNDERWATER ATTACK–
RAMS AND TORPEDOES**

The *Inflexible* of 1873 carried the thickest armour ever fitted to a British ship, with two thicknesses of 12-inch iron and 36 inches of teak backing. Her muzzle-loading guns and machinery were concentrated in a short armoured citadel amidships. She was full of novelties: electric light, torpedoes, and the use of auxiliary steam engines for many tasks.

ship specifically for ramming. The navy followed with *Hotspur* in 1868, and later with *Rupert*. Though these were failures, two larger rams were built in 1879 and 1884, with a twin 12-inch turret forward. Despite many attempts to ram in battle, successes were rare, and in many cases the attacker suffered more than the victim.

The Royal Navy purchased its first torpedoes from Whitehead in 1870 and, as well as adapting existing ships and boats, developed three types of specialist torpedo vessel. The *Vesuvius* of 1873 and the *Polyphemus* of 1878, a much larger vessel of 2,640 tons, with a cigar-shaped hull, largely submerged, and with the exposed part covered in three-inch armour, were 'stealth' craft. The *Lightning* of 1876 was a launch of 32 tons with a trial speed of 19 knots. She was seen as the right style for torpedo boats and similar, but bigger boats were built in large numbers by most navies despite their lack of seaworthiness.

From the early 1870s, battleships were fitted with machine-guns to counter the threat from torpedo boats. These were found to have insufficient hitting-power and from about 1880, 3- and 6-pounder Hotchkiss quick-firing guns were fitted. Armstrong developed the principle and by 1887 a 40-pounder (4.7 inch) fired 10 rounds in 47 seconds compared with 5 minutes 7 seconds for the same number of rounds from a conventional, breech-loading 5-inch gun.

THE DARK AGES

The period from about 1870 to 1890 was a difficult one, with no obvious enemy, no clear naval policy, low budgets, and yet with a wide range of technical options. Barnaby received little guidance from the Board and the widely differing styles of ship reflect this confusion.

However, during the so-called Dark Ages the science of naval architecture was put on a firm basis by Reed, Barnaby, Froude, Barnes, *et al.* The last problem was that of estimating the power needed for a given speed and selecting the right hull form. In the mid-1860s, Froude showed that model tests, correctly interpreted, could give reliable answers. Reed was convinced and enabled Froude to obtain £2,000 from the Admiralty to set up the world's first ship-model test tank at Torquay. These advances meant that naval architecture was recognized as a professional discipline and that ill-informed views such as those which led to the loss of *Captain* had no place.

THE 'ADMIRAL' CLASS

The muddled nature of requirements is well shown in the genesis of the 'Admiral' class. Barnaby was asked to design a ship of 10,000 tons which would be superior to the bigger French *Formidable* of 11,720 tons. Not surprisingly, this proved impossible and it was decided to match the smaller *Caiman* but with some resemblance to the *Dreadnought*. *Collingwood* emerged in 1887 as a very successful ship of 9,500 tons with an 18-inch compound belt extending over 140 feet, with a barbette at each end mounting a pair of 12-inch guns. The barbettes were devised by Rendel and took the form of a fixed pear-shaped structure with armour 11.5 inches thick inside which the guns rotated on a turntable. Though there was no overhead protection to the guns, the crew were covered in the loading position.

As built, *Collingwood* had a secondary battery of 6-inch breech-loaders, but

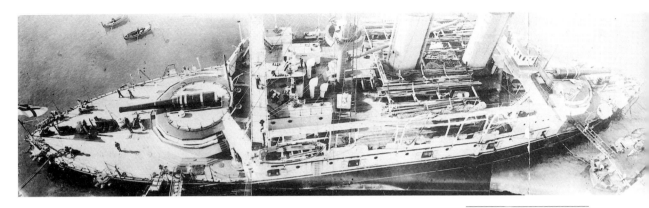

these were replaced in 1896 with quick-firing guns. She was the first battleship to use forced draught but, though this gave her a speed of nearly 17 knots on trial, the overloaded machinery was not very reliable and her low freeboard limited speed in other than a calm. She was criticized for her unarmoured ends, but a trial showed that her speed was reduced only by a quarter-knot when these were flooded.

The Board were well satisfied with *Collingwood* and five generally similar ships were laid down in 1882–3 giving, for the first time for many years, a homogeneous squadron. The first four had 13.5-inch guns, but completion was delayed by late delivery of the guns from Woolwich. The last ship, *Benbow*, received two 16.5-inch Armstrong guns which were not very successful as the barrels drooped and its rate of fire was very low, about one round every 3½ minutes.

Since the 13.5 was still slow in production, the 16.5 was used in the next class, with two guns in a single turret forward. *Victoria* was based on the unsatisfactory ram, *Hero*, but with a belt similar to the 'Admiral' 's and 17 inches of armour on the turret. She had an unprotected 10-inch gun aft and a secondary battery of twelve 6-inch. She was the first battleship with triple expansion engines, working at 135 lb. per square inch, giving a much reduced fuel consumption of about 2.5lb. ihp per hour. The triple expansion engine, whose introduction was largely due to Dr Alexander Kirk, was reliable, relatively light and compact, and would remain in service until well after the Second World War.

Victoria was sunk in collision with *Camperdown* on 22 June 1893 and White's technical enquiry led to a number of inconspicuous but vital improvements in later ships. Bulkhead doors and valves will leak after damage and bulkheads should be unpierced, a lesson since relearnt in every war.

It was not fully appreciated that even a single commerce raider will force the opposing navy to deploy many ships for trade protection. In consequence, there was a tendency to build single ships as a counter to individual foreign vessels. A clear example of this approach came after the American Civil War when the US navy built some fast cruisers of the 'Wampanoag' class which, though very impressive on paper, were failures in service. The Admiralty proposed to counter

Benbow carried two single 16.5-inch breech-loading guns, which were not very successful. She was the last of the 'Admiral' class—the first homogeneous group of ships for many years.

CRUISERS

219

these with a mix of frigates and corvettes. Reed's *Inconstant* had a speed in service of over 16 knots, could steam 3,000 miles, and carried ten 9-inch RML. The iron hull was sheathed in teak, partly to reduce the splinter problem and partly so that she could be given a copper sheath against fouling. *Inconstant* was very expensive and only two similar ships were built. One of these, the *Shah*, fought the Peruvian turret ship, *Huascar* (now preserved), during which she fired 280 rounds, scoring about 30 hits, none of which penetrated armour. *Huascar*'s slow firing guns scored no hits.

The big corvettes, such as *Volage*, were of about 3,000 tons, could steam at 15 knots, and carried six 7-inch and four 64-pounder guns. They, too, were expensive and many smaller corvettes were built for work in distant waters, where coaling stations were far apart, and sails were essential, making them inferior to French ships designed for European waters. Barnaby's *Iris* of 1875 is often seen as the first modern light cruiser. Once Froude had matched the propeller design and form, she was very fast at over 17 knots due also, in part, to her lighter steel hull.

Armstrong's 'Flat Iron' gun-boats, developed from a gun-testing barge, also date from this period. They carried a 9- or 10-inch RML forward on a small, shallow draught hull, and were intended to protect the entrance to ports and, possibly, to attack enemy dockyards. They seemed attractive because of their low cost, but the problems of operating a large gun from a small craft in a sea-way made them virtually useless.

NAVAL DEFENCE ACT

As described in Chapter 6, the mid-1880s were marked in parliament and press by a new concern about the vulnerability of British maritime interests and the ability of the Royal Navy to protect them. Barnaby retired in 1885, handing over to White, and they wrote a joint memorandum expressing their own concern which was followed in 1887 by a detailed study from White, listing 72 ships which should be replaced. The report of the 1888 manœuvres, presented to parliament in early 1889, revealed deficiencies in all aspects of the navy and was instrumental in the passing of the 1889 Naval Defence Act, which authorized a five-year programme, costing £21 million, to build 70 ships—10 battleships, 42 cruisers, and 18 torpedo gunboats. It marked the beginning of a virtually continuous rise in building programmes and an increasing standardization of ship classes.

In the meantime, White had completed a major reorganization of the Royal Dockyards, where most of the ships would be built. He had also initiated a series of trials on different systems of protection against both shells and torpedoes, using the old battleship, *Resistance*.

THE 'ROYAL SOVEREIGN' CLASS

There was a special meeting of the Board with White in 1888 to consider a number of alternatives for the new battleships. It was decided that seven of them should have a high freeboard with two widely spaced twin barbettes each with 13.5-inch guns together with a secondary armament of ten 6-inch quick-firers, widely spaced on the main and upper decks. As a sop to the First Sea Lord, it was decided that the eighth ship should be a low freeboard turret ship. All would be bigger and

The guns of the *Royal Sovereign* were mounted in barbettes rather than turrets. The barbette was a fixed ring of armour inside which the guns were carried on a turntable. To load, the breeches ran back under cover and the shells and charges were loaded by hydraulic machinery.

more expensive than previous ships at 16,000 tons. The design was then developed by White's assistant, Whiting.

The barbettes extended down to the citadel to protect them from shells bursting underneath. The secondary armament weighed 500 tons, much greater than in previous ships, mainly because of the amount of ammunition needed by quick-firers. The widely spaced, well-armoured, casemates were intended to protect the guns, but those on the main deck were virtually useless in a sea-way as in many other ships of the period.

The main belt was of 18-inch compound armour and there was a three-inch protective deck over. There was also a four-inch nickel steel upper belt which the *Resistance* trials had shown to be necessary to protect against high explosive shells bursting high in the ship. The barbette ships could steam at 15.5 knots under natural draught and, thanks to their freeboard of 18 feet, were excellent sea-boats for their day.

Battleship design was still a matter for heated and often ill-informed debate. To counter such opposition, White read a very detailed paper on the design to the Institution of Naval Architects in April 1889 and in discussion routed his critics, notably Reed, with factual evidence demonstrating the very professional approach of the Admiralty.

Between 1889 and 1902, White was responsible for the design of 50 battleships and countless smaller ships. The battleships are usually grouped together as 'Pre-"Dreadnoughts"', and this grouping, together with their similarity of style, conceals the very real advances which were made.

THE 'PRE-"DREAD-NOUGHTS"'

Table 7.4 Main features of the Pre-'Dreadnoughts'

Class	No. of ships	Armour Belt Type	Thickness (inches)	Speed (knots)	Displacement (tons)	Notes
'Royal Sovereign'	7	Compound	18	16.5	14,150	—
'Hood'	1	Compound	18	16.5	14,150	Turret ship
'Centurion'	2	Compound	12	18.5	10,500	2nd class
'Renown'	1	Harvey	8	18.0	12,350	2nd class
'Majestic'	9	Harvey	9	17.0	14,560	—
'Canopus'	6	Krupp	6	18.0	13,150	—
'Formidable'	3	Krupp	9	18.0	14,500	—
'London'	5	Krupp	9	18.0	14,500	—
'Duncan'	6	Krupp	7	19.0	13,270	—
'King Edward VII'	8	Krupp	9	18.5	15,585	—
'Lord Nelson'	2	Krupp	12	18.0	16,090	—

The second-class battleship *Renown* introduced two important changes in armour. The sides of the protective deck, which was level with the top of the belt, were sloped down to meet the bottom of the belt, both supporting it and providing a second defence against a shell piercing the belt. Her belt was of Harvey armour, with a hardened face, providing twice the resistance of wrought iron. The 'Majestic's also had a Harvey belt, but from *Canopus* onwards, the armour used was made under licence from Krupp with 2–2.5 times the resistance of iron.

Table 7.5 Resistance to penetration

Material	Resistance relative to iron
Wrought iron	1.00
Compound	1.25
Harvey	2.00
Krupp	2.00–2.50

Considerable care was taken to position the coal bunkers for additional protection with tests showing two feet of coal equivalent to one inch of iron. The tests of torpedo protection on *Resistance* showed the need for numerous bulkheads without penetration. Unfortunately, longitudinal bulkheads were fitted in the machinery spaces of the bigger ships and the asymmetric flooding which these caused after wartime damage caused several to capsize.

All the first-class battleships, from *Majestic* on, mounted two twin 12-inch guns which, in later ships, could be loaded at any angle of training. The turrets bore little resemblance to Coles's pioneering design but may be seen as *Royal Sovereign*'s barbettes with an armoured shield over the gun breeches. Most had a secondary armament of ten or twelve 6-inch, but four 9.2s were added in *King Edward*, while the *Lord Nelson* had ten 9.2s and no 6-inch. The secondary armament was to attack the unarmoured portions of enemy battleships and there was a tertiary armament of 12-pounders against torpedo boats.

Great attention was paid to weight-saving; the weight of the hull only increased by 70 tons between the *Majestic* of 14,900 tons and the *Lord Nelson* of 16,500 tons. All these ships had triple expansion engines, but from 1892, watertube boilers were introduced working at about 260 lb. per square inch. These were smaller, lighter, and more efficient, but the higher pressures led to a number of problems for which the boiler was somewhat unfairly blamed. Building times were steadily reduced with the Royal Dockyards taking about a year less than the best commercial yards, and Portsmouth by far the fastest yard, mainly due to its early introduction of electric lighting inside ships under construction. The dockyards had more skilled management at both senior and junior levels, better labour relations, and more investment in equipment.

The battles of the Russo-Japanese War vindicated White's design style, as the Japanese ships, based on his designs, proved superior to the Russian ships of French style. White's reputation has suffered since his ships lasted long enough to fight in the First World War when they were obsolescent and easy prey to more modern weapons.

White had briefly worked for Armstrongs, who had a high reputation for cruisers designed for foreign countries, and much was expected from him when he returned to the Admiralty. He was responsible, with his assistants, Whiting and Deadman, for designing altogether 128 cruisers for the navy which were often attacked by his critics for inadequate speed, armament, and protection in relation to their size. White defended his designs vigorously, pointing out the need for

Sir William White, the greatest warship designer of all time. As a young man he was responsible for several major technical advances; he was the founder of the Royal Corps of Naval Constructors and, as Director of Naval Construction, he was responsible for 50 battleships, 128 cruisers, and many smaller ships.

CRUISERS

Bacchante was a typical cruiser, launched in 1901. She had a 6-inch belt of cemented armour and an armament of two 9.2-inch guns in turrets and twelve 6-inch in casemates, the lower tier of which were useless in a seaway.

long endurance, so costly in weight, in ships of the Royal Navy; that speed was often quoted in unrealistic conditions; and that the thick armour listed often extended over a very small area.

Prior to the introduction of cemented armour (steel alloy armour with a very hard, high carbon face and a tough, low carbon back), about 1897, it was not possible to give cruisers a useful belt, and protection was confined to a thick deck, close to the waterline. The 'Cressy' class had a 6-inch cemented belt which would keep out 6-inch armour-piercing shells and high explosive shells of most sizes. The introduction of satisfactory anti-fouling paints around 1890 obviated the need for copper sheathing to prevent loss of speed though annual dockings were still needed to renew the paint.

There were 21 armoured cruisers, costing nearly as much as a battleship, sacrificing armour and armament for 3–4 knots in speed. The smaller cruisers were for trade protection and to serve as colonial policemen. One of these, *Amethyst*, introduced the turbine to big ships in 1905.

DESTROYERS

The threat from early torpedo boats was almost certainly overrated, but by 1885 it was decided that a counter was required and the torpedo gunboat *Rattlesnake* of 550 tons was laid down with a speed of 19 knots. Such ships were thought to be too slow and the torpedo boat destroyer was conceived as an overgrown torpedo boat with a heavier armament, generally one 12-pounder, five 6-pounders, and two torpedo tubes. Some 110 destroyers of this style were built with a low freeboard,

Contest was a '27-knot' destroyer, built by Lairds in 1894–5. She was broken up in 1911. These early destroyers were appalling sea-boats having only about half the freeboard which a modern designer would select. The turtle deck forecastle did little to keep green seas off the bridge and their sea speed in good weather was more like 23–4 knots.

a turtle deck forecastle, and an unrealistic trial speed of 27–30 knots in calm water, several knots slower in service.

They were all designed by their builders to a loose specification with a bonus for exceeding the specified speed and a heavy penalty for failing to achieve it. This forced the pace of machinery development, lightweight hull construction, and propeller design but it also encouraged builders to employ unrealistically large numbers of skilled stokers and hand-selected coal to squeeze the last fraction of a knot. Almost all these destroyers had watertube boilers and most had triple expansion engines. The Engineers-in-Chief, John Durston and Henry Oram, kept closely in touch with Charles Parsons's development of steam turbines and attended the early trials in his experimental vessel, *Turbinia*. In 1898, the Admiralty ordered a destroyer, *Viper*, with turbines. Many early destroyers had difficulty in putting their power into the water through their propellers, and often several sets of propellers would have to be tried. This problem was much worse in the turbine ships with very high shaft rotational speed, and Parsons identified the cause as the water 'boiling' in the very low pressure round the screw which he called cavitation. Palliatives, mainly larger blade area, were tried, but the problem was only solved just before the First World War when geared drive was introduced, permitting slow-turning propellers.

Armstrongs built a turbine destroyer for 'stock' and despite a report criticizing her strength she was purchased for the navy as *Cobra*. She broke in half, in moderate weather, during her delivery voyage and the Admiralty ordered a far-reaching investigation into destroyer design. Known loads were applied to the destroyer *Wolf* in dry-dock and the strains measured. These were then compared with the strains measured in a sea-way, the results justifying the Admiralty design method, though some refinements were introduced. There was also a very detailed questionnaire on sea-keeping to be completed by all captains showing that Palmers's ships were generally preferred.

The last destroyers built under White's direction, the 'River' class, were of very different style, with trial speeds of 25 knots which could be maintained in a sea-way thanks to their high forecastle.

In their time, both the wooden sailing battleship and the *Dreadnought* were the largest and most complicated man-made artifacts of their day and they were separated by less than a century. This brief account has only mentioned the major developments, but many such changes were possible only after apparently minor problems had been overcome, often in materials technology.   REVIEW

The Admiralty managed their industrial revolution with great skill and were the leaders in most aspects either using their own staff or by encouraging and sponsoring the work of others. This superiority rested on British industry's lead in engineering and the gradual erosion of this lead from about 1860 is reflected in occasional advances from other navies such as that of the French use of steel hulls.

The Admiralty's lead in naval architecture is clear through the work of Reed, Barnes, Froude, and others. The Controller's department was highly professional, resting on the teaching of successive schools of naval architecture and, later, marine engineering, which provided technical education of a standard which only later was copied by universities.

This professionalism was put on a firmer basis in 1883 with the formation of the Royal Corps of Naval Constructors, suggested by White, which brought together education, training, and career development. Naval architecture embraces a number of disciplines but these are brought together in creative design. It is never easy to say which are good designs but the Admiralty designs were widely copied and performed well in their intended tasks, reflecting great credit on the designers from Seppings to White and their staff.

$\mathcal{B}$y the middle of the eighteenth century the general shape and size of the world was known, but there were many details to be filled in. Did the Great Southern Continent, shown on most maps, really exist? Was there a navigable passage round the north of the American continent? What island groups lay scattered across the Pacific Ocean?

Traders and adventurers with the ear of kings and ministers in all major European countries wanted to find the answers to these questions and to exploit any discoveries, whether lands, peoples, or natural resources. The eighteenth century was the Age of Reason, when knowledge was esteemed for its own sake. Scientists and men of letters wanted to fill the gaps on the charts and to find out about the nature of undiscovered lands, and also petitioned their kings and powerful patrons. Those same kings and ministers saw advantage in being the first to find and claim the new lands, and involved royal forces in the search. So their Admiralties sent ships of their navies out with instructions to explore and survey the unknown.

INTRODUCTION

An inquiring mind, with the ability to observe natural phenomena and deduce facts from them, was expected of any educated man, and naval officers were no exception. During their voyages captains and masters of king's ships were expected to maintain 'remark books' to record details of navigational or scientific significance, and many survive today in the archives of the Hydrographic Office at Taunton.

From the days of Edmond Halley and the *Paramore* pink (1698–1701), Royal Naval vessels were sent with orders to concentrate on scientific discovery. Halley, a scientist, not a professional naval officer, had trouble with his officers, culminating in their court martial at which the principal charges were dismissed. After this unfortunate episode, the command of king's ships was always entrusted to king's officers, even on scientific voyages.

Of all the naval officers given command of scientific voyages, the most deservedly renowned is James Cook. He conducted his first published survey, of Gaspé Bay in Nova Scotia, in 1758. From then on he was more or less continuously employed on surveying and exploration work until his death in Hawaii.

COOK

The sciences of navigation and astronomy made great advances in the eighteenth century, and with them the accurate recording of maritime exploration. Nevil Maskelyne, appointed Astronomer Royal in 1765, published the first *Nautical Almanac* in 1767 in a format which has remained little changed to this day. John Harrison built the first precise, transportable timepieces. Jesse Ramsden invented a dividing engine for mechanically subdividing the scales of instruments, so that those instruments could be made smaller and more cheaply without loss of accuracy.

By the third quarter of the century the navigator, explorer, or surveyor could fix his position anywhere in the unknown, carrying his longitude by chronometer

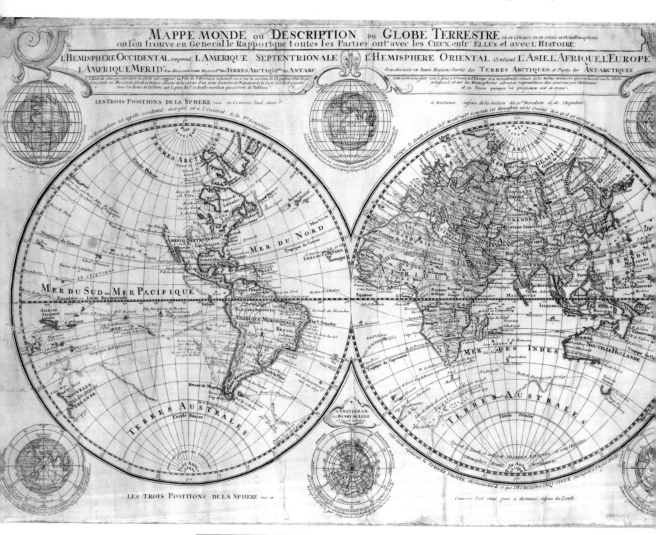

World Map of 1740 published in Amsterdam by Henry de Leth. Though Europe, Asia, and Africa are accurately shown, the north and west coasts of North America and the east coast of Australia are blank. New Zealand is shown as the west coast of an arm of the Great Southern Continent, which occupies much of the South Pacific.

and landing to check it by a series of astronomical observations, which he could make accurately with light, portable instruments.

In July 1758 Cook was the master of the *Pembroke*, 64 guns, in the fleet besieging Louisburg at the mouth of the St Lawrence. On shore the day after the fortress surrendered he met Samuel Holland, an engineer making a plan of the captured fortress using a plane table. Cook was so interested that he returned the next day to be instructed in the use of the instrument. Encouraged by his captain, John Simcoe, Cook invited Holland on board, and during that summer Holland conducted what was in effect the first field surveying academy in the Great Cabin of the *Pembroke* and on the shores of Cape Breton Island.

First under Simcoe, and later under another influential captain, Lord Colville, in the *Northumberland*, Cook carried out surveys of ports and anchorages and wrote a 'Description of the Sea Coast of Nova Scotia, Cape Breton Island and Newfoundland'. His work, carried out in the intervals of his normal duties as master and navigator of the *Northumberland*, was less than complete. He knew it, and said so in his report.

The governor of Newfoundland, Captain Thomas Graves, agreed and had Cook appointed to his staff to make a complete survey, with a small schooner, the *Grenville*, from which to operate. Nothing in Cook's later work, despite its sweep across unknown oceans delineating new islands and continents, has the meticulous detail of the charts of the southern and western coasts of Newfoundland which he made between 1763 and 1767.

Scientists wished to know the dimensions of the solar system, and particularly the distance between the earth and the sun. This could be measured by timing the transit of the planet Venus across the face of the sun from places as far apart as possible. This transit occurs on two separate occasions, separated by roughly eight years, every hundred years. War, weather, and sheer misfortune had allowed only an imperfect series of observations in 1761. The Royal Society determined that the opportunity in 1769 should not be botched. As well as observations in Europe, an expedition should be sent to the Pacific, where both the required distance from Europe and the weather to ensure good observations should be obtained. The Royal Navy agreed to provide ships. Two former Whitby colliers were purchased, renamed *Endeavour* and *Resolution*, and fitted out. Cook was appointed to command the expedition.

As well as the astronomical purpose of the voyage, Cook's instructions directed him to search for the continent believed to lie in the South Pacific to the south of the tracks of previous navigators. He was to explore and chart any land he discovered in detail—no longer would their Lordships be content with brief, ill-supported reports of land glimpsed from a distance.

Cook and his team, which included the astronomer Charles Green, the naturalist Joseph Banks, and the biologist Daniel Carl Solander, with clerks, draughtsmen, and servants, successfully observed the transit of Venus at Tahiti on Saturday, 3 June 1769. They went on to sweep across the South Pacific, showing it clear of land, delineated the whole coast of New Zealand, proving it to be two

Captain James Cook. Born of humble parents in Yorkshire in 1728, raised in the Whitby collier trade, Cook volunteered into the Royal Navy in 1755 and qualified as Master two years later. He was promoted captain after his second voyage, in 1775. The portrait was painted at this time by William Hodges, artist on the voyage.

islands not joined to any greater land to the south, and charted the east coast of New South Wales and Queensland, not without excitement and grounding.

On the return of the *Endeavour* to England in the summer of 1771 both Cook and Banks were acclaimed for their discoveries; Cook was promoted commander, and was received by King George III, to whom he explained his charts and voyage.

Two more Whitby colliers were fitted out for another voyage into the Pacific in 1772, named *Resolution* and *Adventure*. Joseph Banks was to have gone again with Cook, but when the additional cabins he demanded for himself and his people made the *Resolution* so crank that they had to be removed Banks declined to sail. He and Cook remained friends, and he was one of the signatories of Cook's nomination as a Fellow of the Royal Society in 1775.

The voyage was much like the first, but now they were filling in the detail on a

Cook's second voyage. *Resolution* and *Discovery* lying at anchor off Point Venus in Matavai Bay, Tahiti, with native craft in the foreground. One of the series of topographic paintings by William Hodges now in the National Maritime Museum, Greenwich.

chart on which the outlines were already inked in. They explored the deeply indented coast of south-west New Zealand, which bad weather had prevented them from charting fully before. They also made sweeps into the Antarctic seas which conclusively proved that no Great Southern Continent extended out of the ice into temperate regions. This second voyage lasted from April 1772 to July 1775.

One serious gap remained in the knowledge of the Pacific—the north-west coast of the American continent. There was still hope among both the military and the merchants that the fabled North-west Passage—a navigable route round the north of the continent—might exist. A third Pacific expedition was mounted, and Cook, now promoted post-captain, given the command.

Leaving the Nore at the end of June 1776, Cook sailed eastabout to North America, discovering the islands of Hawaii, which he named the Sandwich Islands after the First Lord of the Admiralty. He was received in the islands as a demi-god—an honour which was to have fateful consequences.

The *Resolution* reached the Canadian coast in March 1778 and worked her way north throughout the summer. Though Cook took her through the Bering Strait and eastward along the Alaskan coast until he was stopped by impenetrable ice, it was clear that no great gulf pierced the continent. He returned to winter in the Hawaiian Islands, where again he received divine honours.

When the ships sailed for North America in January 1779 they were soon driven back by bad weather. The *Resolution* needed a new foremast, and one was made from spare spars and stepped while the ships lay anchored in Keleakekaua Bay. Gods, though, should not be subject to the ills and disasters which are the lot of mortal men, and faith in Cook's divinity was shaken. Irritated by the natives' thieving, he went ashore to remonstrate, and in a spontaneous riot was killed on the shore.

A man of the Sandwich Islands. Engraving after a drawing by John Webber of a Hawaiian noble in ceremonial costume, made on the third voyage. Cook and his team of scientists and artists made extensive notes of customs, language, and art. These records are a valuable source of information on Polynesian life before the advent of Europeans changed it.

George Vancouver had been a midshipman on Cook's second and third voyages, and afterwards had risen slowly but surely up the naval tree, surveying only as opportunity offered, until in 1790 he was promoted captain and given command of two ships for an expedition to the north-west coast of America.

VANCOUVER

The ships were the *Discovery*, not Cook's ship but a new vessel launched the previous year, and the brig *Chatham*. The purpose of the expedition was twofold. First it was to resolve the dispute over the settlements at Nootka Bay, on what is now Vancouver Island, which had brought England and Spain to the brink of war. Secondly, it was to complete Cook's search for the North-west Passage by charting the previously unexamined fiords and inlets of the British Columbian coast.

Though Cook had proved that no great gulf existed, there was still a hope that some inlet or river might communicate with the Great Lakes, and so provide a convenient passage north of Spanish territory. Vancouver and his people took three years to chart the tortuous coast, beset then as now with fogs, gales, and swift and irregular tidal streams. Much of the work was done by boats under oars, going away from the ship for days at a time. In these unknown, rocky waters disaster was never far away, and the *Discovery* grounded more than once.

Vancouver's *Discovery* aground on a rock pinnacle in Queen Charlotte Sound. She fell over on her side as the tide fell, but after twelve hours of anxious hard work she was lightened and brought off. *Chatham* stands by, and canoes of interested natives watch.

Some of the names Vancouver gave to new features help us to see the area with his eyes. Desolation Sound and Starve-Gut Cove speak for themselves, but Salmon Cove conceals another disappointment, as he notes that 'the salmon, though incredibly plentiful, were of poor quality'.

Winter weather on the coast was too inclement to allow surveying to continue, so the ships followed Cook's practice of withdrawing to Hawaii. Vancouver was dismayed by the increase in violence in the islands since his earlier visits. He determined to try to stop it and to this end he established good relations with the chiefs, specially with Kamehameha, the paramount chief of the largest island. He negotiated the cession to the British Crown of Kamehameha's lands. Since, in the following years, Kamehameha conquered the remaining islands to become the first king of Hawaii, his gift eventually included the whole group.

The cession was entirely unofficial. It was, though, recognized by the British government for years after the event. It was allowed to lapse by default later in the century, and the islands were annexed by the United States in 1900.

FLINDERS

Even after the work of Cook, Vancouver, and the earlier, mainly Dutch, explorers, it was still uncertain whether New South Wales and New Holland were separated by an extensive sea, or whether they were joined to form one large Australian continent.

By 1800 the governor of New South Wales felt that too little had been done to explore the continent since the British first settled in Botany Bay, and petitioned the home government to do something about it. The Admiralty agreed, purchased a merchant ship of 334 tons, named her the *Investigator*, fitted her out for foreign service, and appointed Matthew Flinders to command her.

Flinders had learnt surveying under William Bligh, master of the *Resolution* on Cook's last voyage. He nearly scuppered his chances of success at the outset of the voyage when, on passage from Deptford to Portsmouth, the *Investigator* grounded off Dymchurch. It appeared that the captain was paying more attention to his new wife, who was on board without Admiralty authority, than to the navigation of his command.

Repairs having been completed, and Mrs Flinders landed, at Portsmouth, the *Investigator* sailed from Spithead on 18 July 1801. Their Lordships had given Commander Flinders voluminous instructions, set out in order of priority, telling him to fill in the gaps in existing charting of the coast of the Australian continent (incidentally it was Flinders who was responsible for the adoption of the name 'Australia').

New Holland was sighted bearing north-east, distant 10 leagues, at 19.00 on 6 December 1801, near the present Cape Leeuwin. During a stop in King George Sound to set up the rigging after the long passage, Flinders took the opportunity to improve on Vancouver's sketch survey. He then worked eastward along the shore of the Great Australian Bight, constantly in danger from the rocks and islets on a lee shore.

Investigating the mouth of Spencer Gulf, the first promising indentation in the coastline, disaster struck. Mr Thistle, the master, was sent inshore in the cutter to find a safe entrance for the ship. Returning at dusk in a blustery wind, the boat disappeared while still three miles from the ship. Darkness fell, and when a search was made next day only the broken hull of the boat was found on rocks inshore, with no sign of the master or his men. Cape Catastrophe commemorates the incident.

Flinders reported to the governor at Port Jackson, and then continued his careful and accurate surveys. His work in the Gulf of Carpentaria was not superseded until the Royal Australian Navy's resurvey started in 1962. His work here and in the Great Australian Bight conclusively proved that Australia was one large conti-

Kangaroo Island. Flinders too carried artists to record topography and wildlife. This picture by William Westall, painted in 1811 after a sketch he made on the spot in March 1802, shows a bay on the north side of Kangaroo Island, in the Great Australian Bight, with some of the animals after which the island was named.

nent. But the *Investigator's* hull, weakened by alterations during her conversion for naval service, was leaking more and more. When she returned to Port Jackson in June 1803 Governor King ordered a complete survey of the ship, and she was condemned as beyond economical repair.

Various schemes were considered for continuing his surveys, but it was eventually decided that Flinders should go home to report, with the hope that he would be given another ship with which to complete his work.

Taking passage home, his ship, the *Porpoise*, was wrecked on a shoal in the Great Barrier Reef. Flinders sailed to Sydney in the *Porpoise's* cutter, organized the rescue of his people, and then restarted the voyage to England in the schooner *Cumberland*.

Calling at Mauritius, then in French hands, Flinders was told that his passport, guaranteeing him safety from actions of war, was not for him personally but for the *Investigator*. Unfortunately a feud developed between Flinders and the governor, General Decaen. As a result, he was detained on the island, albeit in fairly comfortable circumstances, for six and a half years. He used the time to redraw many of his charts, and to reconstruct those lost, noting the fact in exquisitely written notes on the charts which remain in the archives of the Hydrographic Office at Taunton to this day.

Before 1795 the captains and masters of king's ships had to provide themselves with such charts as they could find and afford. Similarly, even those captains sent out by the Admiralty on voyages of survey and exploration had to arrange for the publication of their discoveries at their own expense. When it was realized that more ships were being lost by grounding than by enemy action, something had to be done. By Order in Council in August 1795 the office of Hydrographer of the Navy was created, and Alexander Dalrymple appointed to it.

Dalrymple made slow progress in bringing order into the miscellaneous collection of charts and surveys deposited in an attic in the Admiralty building in Whitehall, and in publishing charts for the use of the fleet. He was replaced in 1808 by Captain Thomas Hurd, the first in an unbroken line of naval officers in the post to this day.

Hurd speeded the production of charts, and started issuing them to ships in boxes, forerunners of the modern folio system. Following the end of the Napoleonic War there were ships to spare, and a series of brigs and bomb vessels were commissioned for surveying at home and abroad. Trade was expanding; where traders went the navy followed, often led by surveying vessels, and the process was started whereby, during the next century, the Royal Navy surveyed and charted the world.

In 1817, the Navy Board was instructed to keep a separate list of surveying ships, and a special rate of pay was established for officers employed in surveying. Thus the Surveying Service was born.

Under Hurd's successor, W. E. Parry (Hydrographer 1823–9), the first Sailing Directions, written supplements to the charts, were published, as were the first

THE HYDROGRAPHERS

A

CATALOGUE

OF

CHARTS, PLANS, & VIEWS,

PRINTED AT

The Admiralty Office,

FOR THE USE OF

HIS MAJESTY'S NAVY;

AND NOW SOLD,

(WHOLESALE AND RETAIL)

FOR THE USE OF NAVIGATORS IN GENERAL,

By the undermentioned Agents :

J. WYLD, Geographer to His Majesty, *Charing Cross* ;
A. ARROWSMITH, Hydrographer to His Majesty, *Soho Square* ;
Messrs. NORIE & Co. 157, *Leadenhall Street* ;
KINGSBURY, PARBURY, and ALLEN, *Leadenhall Street*;
R. H. LAURIE, 53, *Fleet Street* ;

AND BY

Messrs. J. & A. WALKER, *Pool Lane,* ⎰
T. JONES, 5, *Harrington Street,* ⎱ *Liverpool.*

PRINTED BY

G. HAYDEN, LITTLE COLLEGE STREET, WHITMINSTER, PRINTER TO
THE ADMIRALTY-OFFICE.

Chart Catalogue. Permission was granted to sell Admiralty charts to the general public in 1819, but the first receipts, £72 for the year, were not recorded until 1823. The first catalogue of charts was published in 1825. This is the title-page.

The headland of Selinde, with the ruins identified by Beaufort as the ancient Trajanopolis. An engraving from an original sketch by Beaufort himself, probably in 1812. The sketches and descriptions of naval surveyors in the early nineteenth century show many classical ruins which have long since disappeared.

Light Lists and catalogues of charts. Francis Beaufort, the longest serving and greatest of all the Hydrographers (1829–55), introduced Tide Tables and the promulgation of amendments to published charts by Notices to Mariners, and the essentials of the full range of services provided today were in place.

Beaufort was the first Hydrographer to have been involved in practical surveying before taking office. The son of a parson, he was a scholar as well as a seaman. Scholarship was then synonymous with the classics, and in Beaufort surveying and classical scholarship combined to produce an unrivalled series of charts and views incorporating both navigational information and a record of the classical antiquities of the areas he was surveying.

ARCHAEOLOGY

He took command of the *Fredericksteen* in December 1810, with orders to survey the south coast of Asia Minor. In the next eighteen months he produced a series of charts, views, sketches, and descriptions which give a wonderful picture of the remains of the Greek and Roman cities of that coast. He identified no fewer than 60 classical sites, only seven of which have not stood the test of modern scholarship. He waxed ecstatic over Side, queen city of Pamphylia, charted and drew its walls, agora, and harbour, and made a calculation of the seating capacity of its theatre (15,250) which is still the best estimate. His work was cut short when he was shot and seriously wounded by a fanatical Turk at Ayas in June 1812.

As Hydrographer, too, he made sure that the tradition which he had started was continued. Thomas Graves, first in the *Blossom*, then in the *Beacon*, spent ten years in the eastern Mediterranean, and produced almost 100 charts of the Aegean archipelago, all with archaeological detail. In 1841 he was instructed to take Sir Charles Fellows to Xanthos to bring back the marbles he had discovered there. Graves disagreed with his passenger about the method Fellows proposed for dismantling the monuments, and months of argument ensued before the stones were removed and brought to England. The Harpy Tomb and the Nereid Monument stand now in the British Museum, a monument to the surveyors who brought them out of the parlous condition in which they then languished.

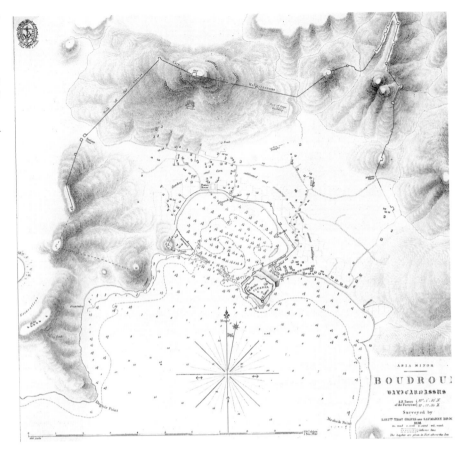

Chart of the harbour of Bodrum, the ancient Halicarnassus, surveyed by Thomas Graves in the *Blossom* in 1838, as published by the Hydrographic Office in 1844. Again this shows ancient ruins, many later robbed or built over.

OWEN AND AFRICA

As a young lieutenant, William Fitzwilliam Owen had been a prisoner on Mauritius with Matthew Flinders. Whether he learnt any surveying from the older man is not known, but after the end of the Napoleonic Wars Owen was sent to Canada in charge of a party surveying the Great Lakes. His greatest claim to fame, though, lies in his Great African Survey, which lasted from 1821 to 1826.

He was given two ships, the sloop *Leven* and the brig *Barracouta*, commanded by William Cutfield. Among his officers were several who became eminent among nineteenth-century surveyors, including Vidal, Mudge, Boteler, and Richard Owen, William Fitzwilliam's nephew.

The task set him was the survey of the whole east coast, from the Cape of Good Hope to the Horn of Africa. Starting at the Cape, the ships leapfrogged their way up the coast, one ship going on ahead and working back until she met the one coming north. From the start fever was the scourge of the surveyors. It was mostly malaria, then believed to be caused by unhealthy vapours arising from the mudflats and mangrove swamps of the shoreline. When the tender *Cockburn*, under Richard Owen, was left in Delagoa Bay to survey English River, her entire crew of 20 officers and men were stricken, and only five survived. It was not until the ships met a Frenchman in Madagascar in 1824 who recommended quinine to them as a control if not a cure, that mortality was brought within bounds. An early casualty from fever was Captain Cutfield of the *Barracouta*. To replace him Alexander Vidal was promoted.

From the start Owen nurtured an implacable hatred of the slave trade, and little scruple about what he did to encompass its destruction. Early on he made a treaty with a local chief on the border of what is now Mozambique, by which the 'Kingdom' came under British protection. This protection endured until 1872, when international arbitration gave the land to the Portuguese.

His most adventurous political exploit came at Mombasa, where the local rulers, the Mazrui, were in revolt against their suzerain, the Sultan of Oman. In exchange for a pledge to abolish slavery Owen granted Mombasa the protection of the British flag and secured the withdrawal of the besieging Omani fleet.

By the time the *Leven* left the Cape homeward bound in October 1825, 5,000 miles of coastline had been surveyed. When the Kenyan coast was resurveyed in 1962, Owen's work was found to be still extraordinarily accurate.

But the little squadron's task was not finished. At the Cape orders were received to delineate the coast from the mouth of the Congo to the Gambia on their way home. This they duly did, throwing in for good measure a running survey of the coast from the Cape to the Congo, since that was on their way!

The two ships arrived home at the end of August 1825. Their work comprised in all some 300 sheets of fair charts, of which a later Hydrographer, remembering the mortality from fever during the surveys, said that they had been drawn and coloured with drops of blood.

When the former Spanish colonies in South America gained their independence the continent was opened to British trade, and where trade went the Hydrographer's ships followed. One area in dire need of accurate charts was the tortuous archipelago at the southern tip of the continent. Two ships were commissioned in 1826, the *Adventure*, under Captain Phillip Parker King, with the sloop *Beagle*, Commander Pringle Stokes, under him.

BEAGLE

King was the son of the Governor of New South Wales, and often told of his memory, as a small boy, of Flinders coming sunburnt and bedraggled into Government House after his epic voyage from Wreck Reef. His task now was to survey the coast of South America from the River Plate through the Magellan Straits north to the Island of Chiloe on the west coast.

King determined to tackle the most arduous part of the work first, and sent the *Beagle* to the western end of the Magellan Straits, starting himself from the east. The area was as hostile an environment as any, with narrow inlets, pinnacle rocks, violent and irregular squalls, and treacherous currents. It broke Pringle Stokes's spirit, and after two years of unremitting struggle he shot himself.

When the two ships withdrew to Rio de Janeiro to rest and refit they met the Commander-in-Chief of the South America station, Sir Robert Otway. Instead of confirming *Beagle*'s first lieutenant, Skyring, in command he appointed his own flag lieutenant, one Robert Fitzroy.

A watercolour sketch by Conrad Martens of the *Beagle* in the Beagle Channel, among the islands south of Tierra del Fuego, painted on the spot in 1833.

King and Fitzroy continued their work until August 1830, by which time the Straits and the archipelago out to Cape Horn had been charted. There was, though, a gap in the surveys, from the Plate to Port Desire, some hundred miles north of the eastern mouth of the Straits. In 1831, the *Beagle* was recommissioned, with Fitzroy again in command, to complete the surveys, and to carry an accurate chronometer traverse round the world, checking old-established stations and making new ones.

Fitzroy, considering that it would be useful to take a naturalist on the circumnavigation, asked Beaufort if he could recommend a young gentleman to share his cabin and record the natural history of the seas and lands they would visit. Beaufort found Charles Darwin. The two men, Fitzroy himself only 26, Darwin 23, soon became friends.

After two years completing the South American survey, the *Beagle* went on to visit the Galapagos, Tahiti, New Zealand, and Australia. Everywhere the young Darwin found sights and phenomena to engage his attention. He returned to England full of the spirit of scientific enquiry, was lost to the church for which he had been destined, and set out on the career which led to *The Origin of Species* and the theory of evolution.

THE ARCTIC

Vancouver's work proved that there was no passage round the north of the American continent south of the Arctic regions. Earlier explorers and fur-trappers had found that an ice-bound archipelago extended north of the Canadian mainland. Throughout the nineteenth century the Royal Navy led the exploration of this archipelago, no longer searching for a practicable route for trade, but simply adding to the sum of human knowledge. The Second Secretary of the Admiralty from 1804 to 1845 was John Barrow, a geographer and traveller, who supported the use of Naval vessels and crews for this work.

In 1817 it was reported that the Greenland Sea was unusually free of ice, and in 1818 two ships under Commander John Ross were sent to search westward of Baffin Bay for the North-west Passage. Ross sailed into Lancaster Sound, but a bank of cloud led him to believe that a great range of mountains barred the way further west, and he turned for home.

It became clear that, for any serious exploration, expeditions must spend the winter in the ice. Accordingly, two bomb vessels, *Hecla* and *Griper*, strongly built and fully provisioned, were sent out under Lieutenant Edward Parry. On 1 August 1819 they entered Lancaster Sound and sailed through Ross's mountains. They penetrated to longitude 110° West, almost through the archipelago into the Polar Ocean, but increasing ice and the onset of autumn made them retreat to a secure berth in Winter Harbour on Melville Island.

Here they snugged down for the long winter night, roofing the ships in, producing plays and a local newspaper, running classes for the illiterates among the ships' companies, keeping fit and alert for ten immobile months and setting a pattern for all subsequent successful polar expeditions.

The ships returned to England in October 1820. The following year Parry tried again, this time through Hudson's Strait and Foxe's Basin, along the coast of the mainland. Again they overwintered. A friendly Inuit woman, Iligluik, drew them a map of the land to the north and west. When the ice released them they confirmed the accuracy of Iligluik's map, but the strait she had depicted was blocked by old fast ice.

Meanwhile Lieutenant John Franklin was exploring the rivers, lakes, and coasts of the northern edge of the mainland by canoe and on foot, with the aid of

The search for the North–west Passage.

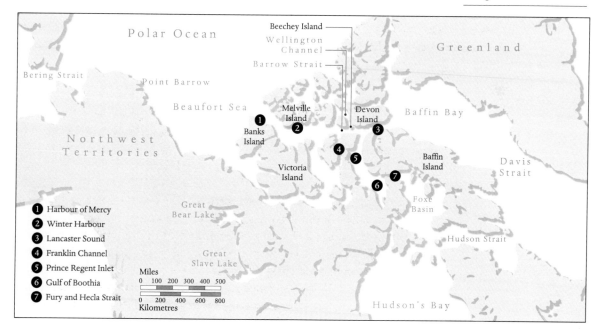

Polar Ocean

Beechey Island
Wellington Channel
Barrow Strait

Greenland

Bering Strait

Point Barrow

Beaufort Sea

Melville Island

Devon Island

Baffin Bay

1 Banks Island

2

3

Northwest Territories

4

5

Baffin Island

Davis Strait

Victoria Island

7

6

Foxe Basin

Great Bear Lake

1 Harbour of Mercy
2 Winter Harbour
3 Lancaster Sound
4 Franklin Channel
5 Prince Regent Inlet
6 Gulf of Boothia
7 Fury and Hecla Strait

Great Slave Lake

Hudson Strait

Miles
0 100 200 300 400 500

0   200   400   600   800
Kilometres

Hudson's Bay

trappers and Indians. Conditions were arduous, their guides not always reliable, and several of Franklin's companions died before he returned to England in 1822, having charted 550 miles of the shore of the Polar Ocean.

Parry, by now Hydrographer of the Navy, planned a campaign of exploration to solve the question of the North-west Passage once and for all. He would himself take charge of the *Hecla* and *Fury* to attack the passage through Lancaster Sound and Prince Regent Inlet. Franklin would return overland to the north coast of the mainland. A third party, under Captain Beechey in the *Blossom*, would sail into the Pacific round the Horn, north through the Bering Strait, and approach the passage from the west. In the event Parry was unable to penetrate Prince Regent Inlet, being blocked by ice and losing the *Fury*, though none of her people.

'The perilous situation of HM Ships *Hecla* and *Fury* on 1 August 1825'. This contemporary engraving shows the two ships trapped in the ice in Prince Regent Inlet. On this occasion they escaped, though the *Fury* later had to be abandoned and only the *Hecla* returned to England, carrying the crews of both ships.

Painting by Stephen Pearce showing the Arctic Council meeting in 1851 to discuss the search for Franklin. Beaufort sits in the centre before a chart of the Arctic. Parry is second from the left, Sir James Clark Ross behind Beaufort with John Barrow in the background next to him, and Beechey sits on the right.

Beechey and Franklin missed joining up by only 160 miles, but both had to retreat before the winter ice.

For the next few years, though some naval officers sailed on private expeditions, official interest shifted to the Antarctic, where James Clark Ross, with the *Erebus* and *Terror*, discovered the Ross Sea and charted much of the coast of the Antarctic continent.

Then in 1845 it was determined to make another push in the north, with the two ships who had served so well in the south. James Ross, knighted for his Antarctic exploits and newly married, declined the command, and it was given to Sir John Franklin, now 60 years old. They failed to break north from Lancaster Sound through Wellington Channel, and after overwintering sailed in 1846 southwest into Franklin Channel. Here they were beset and spent their second winter. Next year the pack never released the ships, and Franklin died in June 1847. During the third winter their supplies ran short, and the crews weakened. It was decided to abandon the ships and to make for the mainland. The party straggled and all died without reaching safety. Their fate was only discovered many years later.

By 1848 the authorities in London were becoming worried, and the first of a succession of expeditions was sent to find Franklin and his ships. Beaufort, the Hydrographer, took the advice of the most eminent Arctic explorers as to the best methods and places to search. Much of the later detailed delineation of the Canadian Archipelago stems from the expeditions sent, either by the Admiralty or by the increasingly desperate Lady Franklin, to search for Sir John and his people.

In April 1853 Lieutenant Pim of the *Resolute*, sledging west from Dealy Island in Barrow Strait, came across Captain McClure's *Investigator*, working east from the open sea and beset in the Harbour of Mercy on Banks Island. The two ships were separated by some 150 miles of pack ice. This was the closest anyone came in the nineteenth century to making the North-west Passage.

241

On 25 March 1807 the act abolishing the slave trade received King George III's Royal Assent, and in 1811 another act made slave-trading a felony, punishable by transportation. But it was not to be expected that so profitable a trade as the round voyage carrying piece goods to Africa, black slaves to America and the West Indies, and sugar, tobacco, and rum back to Europe, would stop overnight.

Before the acts, British ships had carried a large proportion of all the slaves taken across the notorious Middle Passage, and this may have stirred the national conscience of Britain to prosecute the long and thankless war against the slavers, on both the west and east coasts of Africa.

In the west the Slave Coast ran from Cape Verde, with its ancient slave town of Gorée, to Benguela and Loanda in Portuguese Angola. Only Sierra Leone and Liberia, founded respectively by British and US philanthropists as homes for freed black slaves, broke the network of creeks, rivers, and inlets through which the blood of the Dark Continent haemorrhaged across the Atlantic into slavery in the early years of the nineteenth century.

The coast was divided into three: the northern division, the Bights, and the southern division, and as we shall see the stamping out of the trade proceeded generally from north to south.

The insuperable problem was that only ships of countries which supported British efforts to suppress the trade could be boarded and detained. And even countries which did support our efforts were reluctant to allow the maritime superpower of the age to interfere with their shipping.

Slowly, by dint of incessant diplomatic pressure, concessions were won. Lord Palmerston, often criticized for heavy-handed 'gunboat diplomacy', here exerted his persuasion in a purely humanitarian cause. Until 1822 it was necessary to find slaves on board a vessel before she could be detained. Thereafter, only 'clear and undeniable proof that slaves had been on board' was needed to condemn the vessel, but what constituted 'clear and undeniable proof' was often disputed in the courts.

As in its own continent, the United States was divided on the slave trade. Laws were enacted to suppress the trade, but were honoured as much in the breach as in the observance. US ships were sent to the Slave Coast from time to time, and sometimes co-operated with the British ships there. When HMS *Wolverine* and USS *Grampus* were together in 1840, their captains agreed that either ship might detain ships of any nation, but that US ships would be handed over to the *Grampus* while ships of nations in treaty with Britain would be handed over to the *Wolverine*. When the news of this agreement reached Washington, though, it was repudiated by Congress.

In the first 40 years of the attempt to suppress the trade, roughly one in four of the slavers were captured, but only about one in eight of the slaves exported were liberated. The trade was virtually stamped out in the northern division, some infamous spots lingered in the Bight, but in the south, at the mouth of the Congo and in Angola, it still flourished.

Arguments were now raised that, far from easing the lot of the slaves, the

African Squadron was increasing their suffering. Treatment of the slaves on board slavers had not improved, while they were often thrown overboard if a king's ship was sighted and gave chase. It was even suggested that free trade and moral suasion would do more to stamp out the trade than armed policing—a suggestion greeted with horror by the president of Liberia. The cost of the African Squadron, both in men (the station was notoriously unhealthy) and money was cited by its critics, but though it included a large number of ships (over 20 in its heyday) they were small and often old, while once the use of quinine became general deaths from fever diminished.

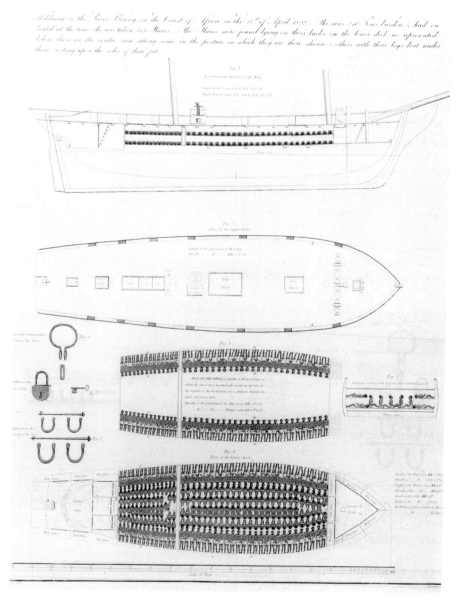

Plan of the slaver brig *Vigilante*, captured in the Bonny River in 1822, showing how the slaves were stowed, some lying on their backs, some sitting with bent legs, all tightly packed in rows. Conditions for the slaves were considerably worse than for farm animals of the time, and the crews of slavers were brutal, harsh, and inhuman.

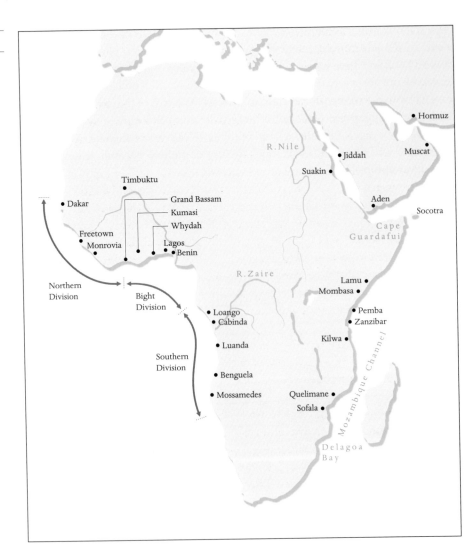

The trade was finally killed on the west coast by 1850. The last outpost was the import of slaves into the southern United States through Cuba from the Congo and Angola. It was the victory of the abolitionist north in the American Civil War which stopped this branch of the trade.

In East Africa the decay of the Portuguese empire left the coast north of Mozambique a hotch-potch of petty sultanates and sheikhdoms, all under the loosely exercised suzerainty of the Sultans of Oman, and all exporting slaves to Arabia, the Gulf, and India.

Owen's abolitionist protectorate over Mombasa lasted only a short time after his departure from the African coast. Without a base nearer than the Cape or Bombay, a squadron nevertheless struggled to suppress a trade just as noxious as that across the Middle Passage. Indeed, the short voyage from Kilwa to Zanzibar could be far worse than the Atlantic crossing. The Kilwa passage took, with favourable winds, about 24 hours. A dhow's cargo of slaves would be fed and watered before embarking to save carrying food or water. If the wind failed, or turned contrary, there was every expectation that the cargo would either starve or suffocate.

The *Lyra* captured a dhow shortly after it put out from Kilwa, and found on board 100 young girls bound for the harems of Southern Arabia. The stench when they boarded was indescribable, and the girls were crowded so closely that none of them would have survived the passage.

On the east coast there were few diplomatic inhibitions. The Squadron's problems here were simply the length of the coast to patrol, and the absence of any strong local authority on whom to bring pressure to stamp out the trade. The slaving dhows could creep along the shore, and were often faster than the sailing vessels of the squadron. Only with the advent of steam did serious inroads begin to be made in the volume of trade, and only with the establishment of a British protectorate over Zanzibar (which included the coast from Kilwa to the Somali border) in 1890 was the trade virtually stamped out. A trickle of smuggled slaves still reached Arabia and Iran until quite recently.

While drama and tragedy stalked the stage with the explorers of Arctic Canada, and pure humanity ruled the actions of the anti-slavery patrols off the coasts of Africa, the charting and detailed exploration of the Pacific proceeded less eventfully. THE PACIFIC

From 1817 to 1823 Phillip Parker King continued Flinders's work in delineating the coasts of Australia. As a result of his surveys Port Essington, the first settlement in the north-west, was founded. Later, after her epoch-making round-the-world voyage, the *Beagle*, under John Wickham, extended King's surveys both in the north-west and on the coast of what is now Victoria.

The Torres Strait, between the northern tip of Australia and New Guinea, is the only practicable route west from New South Wales to India and Europe. It is strewn with reefs and has strong currents; accurate charts are essential for safe navigation. Surveying this, and the passage inside the Great Barrier Reef up the east coast to it, was the task, from 1842 to 1846, of Francis Blackwood with a small flotilla, his own ship the *Fly*, the smaller *Bramble*, *Prince George*, and *Midge*. Blackwood carried with him a zoologist, a geologist, and an artist, following in the tradition of scientific inquiry dating back to Cook and beyond. They provided the first systematic descriptions of the flora and fauna of New Guinea, exploring much of the south side of that great island, and mapping the course of the Fly River, named after the flagship of the flotilla.

Brief mention must be made of Owen Stanley's work in the *Rattlesnake*. He was already an experienced surveyor who had worked with King in Australia and with Back in the Arctic. Not only did he continue Blackwood's work in New Guinea. He also carried with him as assistant surgeon one T. H. Huxley, whose commission in the *Rattlesnake* was responsible for turning his career from medicine to science.

With the Australian gold rush in 1851 traffic between America and Australia increased, and better charts of the eastern approaches to the continent were demanded. The *Herald* was sent out under Captain Henry Denham, with a broad remit to survey practically the whole of the western Pacific. In nine years he charted the reef-strewn area from Tonga and the Fiji Islands to the New Hebrides (now Vanuatu) and the Solomon Islands.

The rise of the Queensland sugar industry led to a demand for cheap labour which was met by shipping in islanders from the Pacific islands. The activities of the 'Blackbirders', as the procurers of cheap labour were called, were little less horrific than those of the African slavers. The islanders naturally became suspicious of all white men, and the captains of queen's ships had to contend with hostile natives as well as to watch out for and to apprehend the blackbirders, while still surveying to their usual accuracy.

By the 1860s the governments of the Australian colonies agreed with the Hydrographer to provide boats and crews for use by Royal Naval surveying officers charting their coasts. From these beginnings the surveying service of the Royal Australian Navy was to grow.

SURVEYING IN WAR

So far we have seen the navy surveying in peacetime. It might be thought that in war surveying ceased until peace returned and allowed the surveyor to go about his work undisturbed by alarums and excursions. However, it is a characteristic of war that its operations rarely occur in the areas planned beforehand, and for those operations the planners require charts.

So it was in the nineteenth century, in the opium wars with China, and most particularly in the Russian War, both in the Crimea and in the Baltic. Belcher in the approaches to Canton, Spratt in the Black Sea, and Sulivan in the Baltic all earned the praise of their senior officers for their invaluable contributions to the success of the operations their charts helped to make possible.

The fruits of an early reconnaissance operation. A sketch of the forts and the Russian fleet at Kronstadt made by Captain B. J. Sulivan of HM Surveying Sloop *Lightning*, 26 June 1854, in preparation for the planned assault.

Though there had been some scientific interest in both the depth of the oceans and the chemical composition of their water in the eighteenth century, it was the advent of the submarine cable as a means of communication which gave rise to the need for data on the profile of the sea-bed along the route of the cables, and of the currents and density of the water above them.

As always, the navy assisted in work which was of general benefit to mankind. To lay the first Atlantic cable a large vessel was needed, both to carry the enormous quantity of wire and to be as steady as possible in all weathers. HMS *Agamemnon* and USS *Niagara* collaborated in the great work, which was brought to a successful conclusion in 1858.

We have already met Thomas Graves, surveying and becoming involved in classical archaeology in the Aegean in the *Blossom* and the *Beacon*. With him in the *Beacon* was the naturalist Edward Forbes, who carried out dredging of the sea floor to examine the life inhabiting it. He noticed that different fauna inhabited different depths, and then propounded the Azoic Theory, which stated that there was no life at depths greater than about 300 fathoms.

Dredging carried out at almost the same time by Joseph Hooker from the *Erebus* in the Southern Ocean found living organisms at 400 fathoms, but it was not until the voyage of the *Lightning* in 1868 with Charles Wyville Thomson and William Carpenter, the first oceanographic cruise specifically ordered as such by the Admiralty, that the Azoic Theory was definitely demolished.

After his voyage in the *Lightning*, Carpenter, a vice-president of the Royal Society, persuaded that body to draw up a plan for a major expedition of marine research. The Hydrographer, Captain George Richards (1864–74), was on the committee drawing up the recommendations. Admiralty approval was given for the expedition in April 1872, and the wooden steam corvette *Challenger* fitted out at Sheerness with laboratories, accommodation for the scientists, and special winches and overside arrangements for dredging and deep sounding.

She sailed from Portsmouth in December 1872, under the command of Captain G. S. Nares, with Wyville Thomson as senior scientist. She was away from home for three-and-a-half years, and with her voyage the science of oceanography can truly be said to have begun.

All the observations now commonplace in oceanographic work were taken from the *Challenger*. The temperature of the water was measured at all depths, using the newly devised reversing thermometers which, for the first time, allowed the temperature at a selected depth to be measured instead of, as hitherto, the lowest temperature reached in the course of the cast. Samples of the water at various depths were taken for chemical analysis. Sediment samples were taken from the sea-bed, and biological samples were taken both from the sea-bed and at various levels in mid-water.

After the *Challenger* returned Wyville Thomson set up the Challenger Office in Edinburgh, from which 50 volumes of scientific reports were published between 1880 and 1895.

The co-operation between the navy and scientists who we can now begin to call oceanographers continued after the *Challenger* expedition. Staff Commander T. H. Tizard, who had been a lieutenant in the *Challenger*, took command of the brand-new *Triton* in 1882. His first work was to examine the bottom topography and the water chemistry in the waters between Scotland and the Faeroe Islands to investigate the anomalies found by Wyville Thomson and Carpenter in the *Lightning* 15 years earlier.

A particularly dramatic investigation was that carried out by the *Magpie*, Commander Foley Vereker, in the Sunda Strait after the eruption of Krakatoa in 1883. Among other things he found that the small crater from which the explosion had begun, then over 500 feet above sea level, now had a depth of more than 100 fathoms over it.

PIRACY IN THE FAR EAST   With the advent of the steamship, with the opening of Japan to outside influence, and with the imposition of the Unequal Treaties on China, trade in Far Eastern waters expanded enormously. Where trade went the navy, and in particular the hydrographic surveyors, followed. In the second half of the century the coast of China was surveyed and charted, as was the coast of Malaya and much of the Indonesian archipelago.

An early anti-piracy patrol. For work among the islands and reefs of the Indonesian archipelago the navy took local craft into service under the White Ensign and armed them. This lithograph shows HM armed prau *Jolly Bachelor* beating off the attack of two pirate praus in Darvel Bay, Borneo, in 1843.

In the dying years of the Manchu dynasty authority had little control over the outlying provinces of the empire, and piracy was rife on the Chinese coast. The converted gunboats of the surveyors often had to go to the aid of merchant vessels beset by these sea wolves, or aid the Chinese Customs officials, themselves mostly westerners, in attacking pirate strongholds.

Further south, in Malaya and the Indonesian islands, there was no pretence of a central authority away from the few European settlements. Sultans and rajas ruled anything from large provinces to a single town and its immediate vicinity, and between this mosaic of states roved pirates of half a dozen races and as many different faiths.

The surveyors, before the days of the telegraph or wireless, had to make such accommodation as they could with local authority to let them carry out their task of charting. As representatives of the premier maritime nation some of them also considered it their duty to suppress piracy, real or imagined, wherever they met it.

Edward Belcher was one such. A brilliant surveyor, he was a harsh superior with a persecution complex, who made life in his ships a hell for his officers and seemed perpetually to have half of them under arrest. When in command of the *Samarang* on the Borneo coast in the late 1840s he attacked and routed a fleet of supposed Illanon pirates.

He called on a number of the minor sultans on the Borneo coast, and negotiated treaties of alliance between Great Britain and three of them. The Hydrographer at home was not best pleased. 'Your last letter is really all Hebrew to me. Ransoms and dollars, queens, treaties and negotiations? What have I to do with these awful things; they far transcend my limited chartmaking facilities', he wrote in reply to one of Belcher's reports.

With ships virtually all over the world, it was inevitable that naval vessels were often first on the scene after a disaster, natural or man-made. In 1847, six small steam-vessels were allocated to famine relief in Ireland, causing the Hydrographer, Beaufort, to revert to the use of hired vessels for home surveys.

DISASTER RELIEF

The story of the *Calliope* steaming out to sea in the teeth of a cyclone from Apia is well known. Less emphasized is that when she returned to find the rest of the international squadron wrecked and the town in ruins she set to assisting both the ships' companies and the townsfolk to recover from the disaster.

Earthquakes, storms, or famine, victims of all were helped by ships' companies of naval vessels on the spot soon after disaster struck. Perhaps the last major occasion before the First World War was the deployment of the Mediterranean fleet to Messina after the great earthquake of 1907.

This chapter has ranged over a wide variety of subjects and locations. In its very diversity it shows the multifarious tasks and very wide influence of the navy on humanitarian and scientific affairs, far beyond the business of sea-fighting, in the century and a half of its heyday.

9

LIFE AND EDUCATION IN A TECHNICALLY EVOLVING NAVY 1815–1925

JOHN WINTON

'*My* examination [on board *Victory*] to enter the Navy was very simple, but it was adequate,' wrote Admiral Fisher. 'I wrote out the Lord's prayer and the doctor made me jump over a chair naked, and I was given a glass of sherry on being in the Navy.'

In fact, Fisher had previously passed an examination. He wrote from dictation 20 or 30 lines of English and about the same number of sentences from a morning paper, and answered some simple questions in common arithmetic. He had also undergone a strict medical examination and provided a certificate that he was 'free from defect of speech, defect of vision, rupture, or any other physical disability'.

When Fisher joined the navy in June 1854, aged 13, most naval cadets were nominated by the Admiralty, but admirals on hoisting their flags and captains on their appointments in command also had a limited number of nominations in their gift. Fisher himself was nominated by Admiral Sir William Parker, who hoisted his flag as Commander-in-Chief Plymouth in 1854. Sir William was a neighbour of Fisher's godmother, Lady Wilmot Horton, in Derbyshire, who asked him to take her godson to sea.

The system of nomination owed much to continuity and tradition. As was described in Chapter 5, officers in the eighteenth-century navy generally entered the service at an early age through the 'interest' of family or friends. No qualifying examination was required of 'captain's servants', as the youngest entrants were called, and numerous instances exist of boys well under 10 years of age, borne on a ship's books, with the captain drawing their pay, until they were old enough, normally at 12 or 13, to go to sea.

An alternative method of entry, the Royal Naval Academy, established in 1733, was generally less popular; in 1801, Lord St Vincent wrote, 'The Royal Academy at Portsmouth, which is a sink of vice and abomination, should be abolished . . . '. It was not abolished but it was reconstituted and renamed the Royal Naval College in 1806. The headmaster was the Revd James Inman (of the Nautical Tables). Inman established the first School of Naval Architecture at Portsmouth in 1811. It amalgamated with the college in 1816 and lasted, with some 24 students, until 1834.

Henry Keppel, a future Admiral of the Fleet, joined the college in February 1822. He had been coached for the college examination, although 'How I got

Facing: Kit Muster in the boys' training ship HMS *Ganges*. Built at Bombay in 1821, *Ganges* was the last sailing-ship to serve as a seagoing flagship. She was moored in Falmouth Bay from 1866 until 1899 when she transferred to Harwich. *Ganges*, the shore training establishment at Shotley, commissioned in 1905.

through,' he said, 'I forget.' He met Professor Inman—'a tall man in black with an austere countenance; but there was that in him that I liked.'

'Our uniform', Keppel said, 'was a blue tail-coat, stand-up collar, plain raised gilt buttons, round hat, gold-lace loop with cockade, and shoes.' The course lasted two years, and the syllabus was much as it had been for the last 80: 'We learned to pull as well as to steer under sail. We had, in addition to school, French, drawing, and dancing masters, also fencing.'

College success carried extra seniority. Astley Cooper Key, a future admiral, who joined in 1833, won the college silver medal and was granted a year's sea time, which meant that he could offer himself to pass the qualifying examination for the rank of lieutenant after five years at sea, instead of the normal six. Key went to sea as a college volunteer of the first class in 1835.

However, many boys continued to join the navy straight from school. William Peel, son of prime minister Sir Robert, a future captain and KCB, who won the Victoria Cross in the Crimea and led HMS *Shannon*'s Naval Brigade in the Indian Mutiny, went to Harrow before joining the line of battleship *Princess Charlotte*, flagship of the Mediterranean Fleet, as a Volunteer First Class in April 1838.

Princess Charlotte's captain, Arthur Fanshawe, had had to recruit and enter his own ship's company, even in 1837, in a way which had not changed much for centuries. The marines came from their barracks and the seamen boys from the port flagship, but everybody else had to be recruited by the captain himself, but without the aid of the press gang, which was still legal but had become socially unac-

The Midshipmen's Berth. A vivid picture of life in the 'cockpit' in the 1840s, with plenty of evidence of the young gentlemen's main leisure interests—bottles, cigars, cards, and dice, but with cutlass and a sextant hanging on the bulkhead.

ceptable and was almost never used after the end of the Napoleonic wars. The only men who could still be pressed into the navy were convicted smugglers, who were often sentenced to serve up to five years in a man-of-war.

A captain could bring a nucleus of men with him from his previous ship, but the rest he had to find—by bribery, persuasion, his own reputation, and advertising by word of mouth, by posting placards, and by setting up recruiting rendezvous in the commissioning port, in London, and possibly also in Bristol and Liverpool. The rendezvous was manned by officers and petty officers from the ship who attracted attention to themselves by posting the name of the ship and the captain prominently, posting up notices extolling the advantages of service in the navy, and draping bunting, flags, and Union Jacks to catch the eye.

If few ships had commissioned lately and the captain or the destined station were popular, a ship's company was quickly assembled. But a ship with an unpopular captain, or commissioning for an unpopular station (that is, unpopular for deserting, such as West Africa) could take months to complete her crew.

The sailor of the post-Napoleonic War period was still the sailor of Trafalgar, Copenhagen, the Nile, and a hundred lesser fights. He went into action bare-footed and bare to the waist, with a scarf twisted round his forehead to keep the sweat out of his eyes. He often wore a pigtail and chewed tobacco which was prepared in long rolls, the same colour, appearance, and almost the same consistency as hemp rope. He was paid 25s. 6d. a month, with 8s. a month more as able seaman, and his pay was always months and sometimes years in arrears.

The standard rations provided by the navy were based on hard ship's biscuits, salt beef or pork which was sometimes many years old, and for breakfast a mixture of oatmeal and molasses known as 'burgoo'. Supplements of fresh produce might be found if in contact with the shore, but this was a matter for the initiative of the command and the purser. The sailor's drinking-water was stored in wooden casks and went fetid after a very few days at sea. He was entitled to half a pint of spirits, or a pint of wine, or a gallon of beer, every day. Not surprisingly, he was often drunk, ashore and on board. Drunkenness and consequent leave-breaking and insubordination were by far the most frequent punishable offences.

The sailor of 1816 was still subject to the lash. He could still be flogged round the fleet, so many lashes inflicted at the gangway of each ship. For very serious offences, such as mutiny, he could be sentenced to death and hanged on board. Awaiting court martial, he could be gagged, and bound, hand and foot, in irons, for indefinite periods. Lesser punishments were 'toeing the line', that is, standing on the same spot on the quarterdeck for hours on end, or banishment to the rigging for half a day.

He had very little formal training, but picked up his craft from his fellows or from his own experience. Shore leave was a matter to be decided by the command; some captains never allowed their crews ashore during a commission, for fear they would desert. On discharge from a ship a sailor was free to go where he liked, and to join another ship, or not, just as he pleased. There were no arrangements, except by a few far-sighted captains, to improve his professional competence, health, living conditions, or morals. He had no uniform; he wore what was convenient, or what he could afford, though the makings of a standard range of clothing were available from 'pusser's slops'.

A commission might last five, six, or seven years, but the sailor had no formal terms of service or sickness benefits. If he was discharged to shore with wounds, neither he nor his wife and dependants were any official concern of service or state, though help might be available from the Chatham Chest. There was little prospect of becoming an officer except by some act of extreme gallantry in the presence of the enemy. Occupational diseases were many and severe: tuberculosis, from the damp, generally insanitary living conditions, and overcrowding on the messdecks; typhoid, cholera, dysentery, and tropical diseases from service abroad; ruptures, from the heavy manual labour; scurvy, from lack of antiscorbutic acid in his diet over long periods; and syphilis from contacts ashore.

The warships of 1816 were wooden-hulled and powered entirely by sail. Nearly 1,000 men were needed to man a line-of-battleship; the great majority of them would be required to work aloft at some time or another. Every sailing warship suffered a steady wastage of men lost overboard and drowned, or killed or maimed after falling from aloft. Throughout the nineteenth century, these 'operational' losses far outnumbered casualties due to enemy action.

The main weapon was still the broadside, in which muzzle-loaded guns of various calibres fired solid iron roundshot simultaneously—or as near to simultaneously as the state of gunnery training on board that individual ship would allow. Other weapons were the Congreve rocket, fired from the deck, or more usually from a ship's boat; or simply boarding an enemy with cutlass and pistol.

Although there were enormous improvements in every aspect of life in the navy during the nineteenth century, in many ways a sailor's social status and the way he regarded his position in society hardly changed until the last quarter of the century. The bluejacket considered himself—just as society did—as someone apart, from another sphere. Sailors on a run ashore were creatures from another existence.

Jack always reacted to officers in the same ways, ready to work and fight his heart out for good officers, sullen, unwilling, and very occasionally murderous with bad. He hated ship's thieves and ship's police equally. He was always suspicious of those who issued his pay, his rum, and his food. Slumps, depressions, and declines in local industries affected recruiting, but it was still remarkable how many boys joined the navy because they came from seafaring families and had been brought up to the sea, or simply had a sudden, young man's urge to run away and see something of the world for themselves.

Jack was also, like the navy itself, very conservative. Although the radical spirit grew and flourished in the nineteenth century, none of the navy's necessary reforms was brought about by the sailors themselves. Very few were even pioneered by the Admiralty, indeed many reforms were bitterly opposed by the majority of naval officers. The navy was often forced to change its ways almost entirely because of outside pressure, from parliament and public opinion.

The British tar might be lauded on the early nineteenth-century stage and acclaimed in songs as all that was brave and honest and true, but within the navy there were growing misgivings about his professional skill, and in particular his gunnery. In Nelson's day the simplest, and generally preferred, tactic for a British man-of-war was simply to close her opponent and fire broadsides into her as hot and fast 'as she could suck it', until the enemy hauled down his flag to three ringing British cheers.

The rapidity of the British fire, their superior seamanship, and their sheer dogged courage had been too much for their French and Spanish opponents. But they were no longer enough as the nineteenth century progressed. In the war of 1812, the larger, better-armed and better-fought American frigates had proved too much for their Royal Navy counterparts, with the famous exception of the *Shannon*. The British tars at Navarino had expected the Turks to strike speedily, and often asked whether they had 'doused the moon and stars yet', but no Turkish ship struck her colours all day.

In the Royal Navy of 1815 there was no common accepted gunnery doctrine, nor any official means of passing round knowledge of improvements. Every captain was free to train his gun crews as often and in such a manner as he himself thought fit, and to exchange information with his fellow captains, if he so wished, by word of mouth or by letter.

A NEW EMPHASIS ON TRAINING

HMS *Excellent* and HMS *Calcutta* about 1865. *Excellent* was commissioned as a gunnery training school at Portsmouth in 1830. *Calcutta* was moored ahead of her as an accommodation ship in 1865.

Individual captains, such as Exmouth in the Mediterranean and Philip Broke in the *Shannon*, might pay special attention to gunnery, carrying out intensive training with constant drills and repeated firing practice at various targets, but in general the handling of the guns was a matter for the gun captains. These, though brave and worthy seamen, took their own time and fired at what they could see through the smoke, almost invariably aiming along the barrels of their guns, always assuming that shot rose upwards, and hardly ever making allowances for the tapering of the barrel, heeling of the ship, over-shooting, or different strengths of charges—in fact, almost totally ignorant of the science of gunnery.

Criticisms of the navy's gunnery began during the French wars, but nothing was done until 1829, when Commander George Smith sent the Admiralty his 'prospectus of a plan for the improvement of naval gunnery, without any additional expense'. Smith suggested that an establishment be set up for the training of gun crews and for the testing and evaluation of new gunnery equipment. In June 1830, a Board Minute authorized such a gunnery school at Portsmouth, with Smith as the first captain. The ship used was the *Excellent*, which was moored in a position in Portsmouth harbour from which guns could be fired across mudflats without endangering anybody.

The Admiralty addressed a new prospectus to the Commander-in-Chief, Portsmouth. It was a seminal document and its first paragraph, though nobody realized it at the time, was the dawn of a new era for the British bluejacket:

Their Lordships having had under their consideration the propriety and expediency of establishing a permanent corps of seamen to act as Captains of Guns, as well as a Depot for the instruction of the officers and seamen of His Majesty's Navy in the theory and practice of Naval Gunnery, at which a uniform system shall be observed and communicated throughout the Navy, have directed, with a view to the formation of such establishment, that a proportion of intelligent, young and active seamen shall be engaged for five or seven years, renewable at their expiration, with an increase of pay attached to each consecutive reengagement, from which the important situation of Master Gunner, Gunner's Mates, and Yeoman of the Powder Room shall hereafter be selected to instruct the officers and seamen aboard such ships as they may be appointed to in the various duties at the guns, in consideration of which they will be allowed 2 shillings per month, in addition to any other rating they may be deemed qualified to fill, and will be advanced according to merit and the degree of attention paid to their duty, which if zealously performed, will entitle them to the important situations before mentioned, as well as that of Boatswain.

A more revolutionary paragraph had never before been published in the Royal Navy. It was studded with key words and phrases—*permanent* corps, *uniform system*, engaged for five or seven years, *renewable*, increase of pay at each *re-engagement*, advanced *according to merit*—which were the first seeds of a permanent, professional career for a seaman in the navy.

Excellent was established as a sixth-rate, with a complement of 200, including Royal Marine Artillery instructors. The captain, Thomas Hastings, who relieved Smith in 1832 (much to Smith's disgust), was given instructions as to training which were themselves quite unlike anything seen in the navy before. The officers and seamen who came to *Excellent* were to be taught:

the names of the different parts of a gun and carriage, the dispart in terms of lineal magnitude and in degrees how taken, what constitutes point blank and what line of metal range, windage—the errors and the loss of force attending to it, the importance of preserving shot from rust, the theory of the most material effects of different charges of powder applied to practice with a single shot, also with a plurality of balls, showing how these affect accuracy, penetration and splinters, to judge the condition of gunpowder by inspection, to ascertain its quality by the ordinary tests and trials, as well as by actual proof.

They were to practise as teams, firing all sorts of guns, loaded with all sorts of shot, on a range laid out from *Excellent*.

The old-time sailor needed agility in the rigging and manual dexterity, but little else. The *Excellent*'s crew needed these qualities, and very much more. They had to be intelligent and they were encouraged to learn how to read, write, and cypher. They soon realized that they could not get the most benefit from their training without these skills, and they were eager to learn. It was quite common for literate sailors to teach their mates.

The sailors liked *Excellent*. One wrote to his father in 1838:

I have had to buy several things already; a set of cards of instruction I have also bought, 3/6 the set, and a few clothes as we must dress all alike, when we are mustered which is every morning and must not dress in white in the winter. I pay 2/- a month for washing, subscribe twopence per month to an excellent library and 1/- entrance money. I am very comfortable and happy here and quite well; what time I have to spare is fully taken up in learning my cards of instruction. I must now conclude for they are piping for Hammocks to be slung.

Hastings's training methods were a surprise to his students. He did not simply give orders to a sailor, but showed which method was best. One visitor to *Excellent* noticed how Hastings handled a sailor who had got into the habit of using the handspike lever to train a gun in a most awkward and inefficient way. 'Now', said Hastings, 'take notice how *I* use the handspike, and you shall try the difference yourself.' The sailor watched with great attention, then resumed the handspike and found, to his surprise, that the gun now moved about with as much ease as if the 32-pounder had been changed to a 12-pounder. 'What think you now?' asked Hastings. 'I'll never use the other way as long as I live, sir', was the reply.

Excellent was also used as a boys' training ship, and a mizzen mast was kept rigged for sail training until as late as 1869. But the main practical training was firing the guns. A red flag was hoisted in *Excellent* to show that firing was about to start; none took place until the mudflats were uncovered by the ebbing tide. The shot was recovered by 'mudlarks', many of them from the same splendidly named local family of Grub, who sold the shot back to *Excellent*. The Grubs used to 'ski' about the mud with wooden boards strapped to their feet, had special implements for extracting shot from the mud, and on a good day could earn as much as eleven shillings.

Excellent might preach the word of naval gunnery, but for years it fell on stony ground in some quarters. There was much criticism of the need to have a gunnery school, by politicians who complained of the cost (about £35,000 a year in 1835) and by conservative officers who abhorred all new-fangled technical nonsense and loudly proclaimed that any time not spent in seamanship (which meant handling ships under sail) was sadly wasted.

In 1834, when Admiral Sir Charles Rowley read an examination paper for gunnery lieutenant, he found he could not understand the words 'impact' and 'initial velocity'. He asked another member of the Admiralty Board, Sir John Beresford, who said 'I'll be hanged if I know, but I suppose it is some of Tom Hastings's scientific bosh; but I'll tell you what I think we had better do—we'll just go at once to Lord de Grey and get the *Excellent* paid off.' However, Lord de Grey (the First Lord at the time) told him he could not sanction it, 'for you have no idea how damned scientific that House of Commons has become.'

There was also prejudice against *Excellent* at sea. When Cooper Key, a future Captain of *Excellent*, was gunnery lieutenant of the frigate *Curaçoa* (24) in Rio in 1843, he asked the captain to let him exercise the men firing at a target. The ship had been in commission eight months but this was the first gunnery practice on board.

A target was anchored 500 yards away and, though the ship was rolling heavily, Cooper Key was pleased with the results. 'The firing was excellent,' he wrote. 'We were obliged to send out four new targets during the forenoon. But what pleased me most was that the *Excellent* men we have on board were the four best shots in the ship. I may say, I had a good laugh at the fellows who run them down.'

Gun Drill at HMS *Excellent*. In the 1880s, *Excellent* began to move ashore to new buildings and facilities on Whale Island, made by convict labour from mud and spoil from new docks and extensions to the dockyard. The old ship *Excellent* was paid off and all hands came ashore in 1891.

The mention of a library in the letter from *Excellent* showed that some reforms were under way. In July 1838 the Admiralty approved the supply of libraries to seagoing ships; large ships were issued with 276 books, small with 156, the books being mostly of a religious or 'improving' nature.

The Admiralty also addressed the question of drink, which throughout the navy's history has resulted in more punishments, more accidents, more opportunities lost, and more careers spoiled, than any other cause. In 1825 the rum ration was halved and the custom of having 'banyan' or meatless days was abolished. The beer ration was abolished in 1831.

An attempt was made to standardize uniforms. *Instructions to Pursers* in 1824 listed the clothing to be carried in every ship, including blue cloth jackets and trousers, knitted worsted waistcoats, white duck trousers and frocks, shirts, stockings, hats, mitts, blankets, and black silk handkerchiefs.

Each reform moved the navy a little further towards a permanent state of manning. In June 1827 an Admiralty circular announced that henceforth petty officers of first or second class could not be punished by flogging, except by sentence of a court martial. Captains could, however, still disrate petty officers by summary punishment on board, as in the past. When a ship paid off, the petty officers were to be discharged to the flagship as supernumeraries and, after reasonable leave, they were to be given another seagoing appointment, at the same pay and rate as before. Also, to improve the status of the petty officers, they were permitted to wear the badge of their rate, the second-class petty officer a white cloth anchor on his sleeve, the first-class the anchor surmounted by a crown.

In the 1850s the government at last made an attempt to solve the long-standing problems of how to man the navy properly in peacetime, and how to increase the navy's strength rapidly in wartime without, if possible, resorting to the press gang.

It was high time something was done. The Syrian War of 1840 had demonstrated the navy's shortage of manpower, and for most of the 1840s the navy's manning was in a chronic state of imbalance between the number of men available and the number the navy required at any given time.

In 1835 the then First Lord Sir James Graham introduced a Register of Seamen which was, in effect, a gigantic roster of *all* mariners for service in the navy. Those who reached the top of the list were required to serve for five years, after which they were released and replaced by others on the register.

This was essentially a short-term arrangement. There were still no long-term engagements, no barracks to house men ashore, indeed no means of holding men permanently. Men were lodged in hulks until their ships were ready. The period of five years' service was rarely observed; normally it was three or four years, the length of a ship's commission, just as it had always been.

By 1839 the register had 175,000 names on it, with about 22,000 apprentices, but men left the navy at the rate of over 1,000 a year throughout the 1840s, and in 1852 the register had dropped to 150,000. Over 50,000 of those were exempt, and only about one man in 25 on any list had ever served in the navy. *Excellent* had been in existence over 20 years but fewer than 3,000 men had passed through it.

In 1852, Lord Derby's government appointed a committee of naval officers, with Admiral Sir William Parker as chairman, to examine the question of recruitment and to report.

When the Committee reported, Lord Derby's government had fallen and Lord Aberdeen was in power. It was his First Lord, Sir James Graham, and the Second Sea Lord, Rear-Admiral Maurice Berkeley, one of the great reforming figures of the Victorian navy, who put the recommendations of the report into effect.

An Order in Council of 1 April 1853 and the Continuous Service Act of the same year embodied the Committee's main recommendations. From 1 July 1853, all new boy entrants to the navy engaged for ten years' continuous service, their 'time' to start from the age of 18. Boys and seamen already serving were encouraged to transfer to the new continuous service. Standards of entry were tightened up. The chances of promotion from boy to ordinary seaman were improved. The pension was now payable after 20 years from the age of 18, instead of the old 1831 period of 21 years from the age of 20.

Pay in general was increased, for ordinary seamen from £1 6s. 0d. to £1 13s. 7d. a month and for able seamen from £1 14s. 0d. to £2 1s. 4d. A new rate of leading seaman carried an extra 2d. a day, and senior ratings could go on to become chief petty officers, at an extra 3d. a day. Men who had passed the *Excellent* could qualify for warrant gunner, or for 'captains of guns'.

There was now to be paid leave between commissions, sick pay conditions were broadened and improved and, most important for the older sailors, their right to choose their ship was still preserved. As for discipline, the act stipulated a more uniform and temperate scale of punishments, urging captains and officers once again to moderate their language and their behaviour, and to employ corporal punishment with more restraint. The 1853 act laid down, in fact, the broad foundations of the system which is in use in the navy of today.

TRAINING IN THE MID-NINETEENTH CENTURY

The secret was to 'catch them young'. Training brigs had been in commission for many years, going on training cruises in home waters and doing some discreet but useful recruiting. The Committee extended the principle of training afloat. The old two-decker *Illustrious* was established as a training ship for boys at Portsmouth in 1854, followed by the *Implacable* at Devonport a year later. The *Illustrious* was commanded by a very able officer, Captain Robert Harris, who instituted a year's course of seamanship instruction, with some general education.

The new boys, known as 'Jemmy Graham's Novices', grew up (despite some ridicule from older hands) to prove themselves in the Crimea, China, and the Indian Mutiny by the end of the decade. One of them wrote an account of his experience in the 'old Guardho' *Illustrious* in *A Sailor Boy's Logbook, from Portsmouth to the Peiho*, published in 1862.

He does not give his full name but refers to himself throughout simply as 'John'. He joined in 1855, much against the will of his family. When he arrived on board *Illustrious*, a lieutenant asked him if he could read. 'Most certainly I could. "Well, then look at that," and he handed me a bill pasted on a board, stating the advantages of serving one's country in the Royal Navy, scale of wages, provisions, and a list of necessary clothing. I read the bill, and gave it back, telling him I fully understood it: "Then you'll enter for ten year's continuous service, will you?" "Yes sir," I unhesitatingly answered.'

'John' of HMS *Highflyer*, as he was photographed for the frontispiece of his memoirs *From Portsmouth to the Peiho*, showing the sailor's uniform of the 1860s.

He was medically examined. The dispensary was a small cabin on the lower deck where the 'man of medicine looked first at my mouth and teeth, then felt the muscles of my arms and legs, telling me to take one step backwards or forwards; then, "Cough, will you? but don't do it in my face. Cough again—again. Were you ever ill? Did you ever break an arm or a leg?" and "Have you been vaccinated?".'

He was issued with his clothing: one pair of blue cloth trousers, two blue-serge frocks, two pair of white-duck trousers, two pair of white jumpers, two pair of stockings, two white frocks, three flannels, two caps, one knife, and a marking type, all of which cost £3 10s., including bed and blanket. His 'shore-going togs' were made up into a neat bundle. He was offered sixpence for them by a shark from shore, so he gave them away.

John's account of his training shows what a seaman of the 1850s was taught, and the training remained broadly the same for the rest of the century. His 'First Instruction' was three days of elementary knots and hitches, and the proper way to lash up a hammock. 'Second Instruction' was boat-pulling, how to dip, toss, and feather oars. After ten days he went to cutlass drill, in a class of 20. After three days they went on to rifle drill, how to march and present arms, and finally how to aim and fire. They fired at some stakes in the mud when the tide was out, firing out over the ship's quarter in the direction of Portchester Castle, at ranges of about 200 to 250 yards.

'Fourth Instruction' was exercise on the big guns: the different parts of the gun, its weight, the different charges, the uses of the train and side-tackle, and how to dismount a gun. The drill was 'four rounds quick firing, second and third round sponge, load and shift breechings'. John thoroughly enjoyed this drill, which was much better than handling a musket all day.

'Fifth' was more rope work under 'Old Pipes' the bosun: all sorts of splices, and complicated knots such as 'Matthew Walker' and 'Turk's Head'. After three weeks they went on to 'Sixth', under a very good-tempered middle-aged seaman who gave them instruction with a model-rigged ship in the schoolroom, to learn the names and uses of all the ropes, and 'such things as putting an eye into a hawser, making sword, thrum and paunch mats, and turning in a dead eye'.

The seventh and last instruction was in handling the lead-line and learning the points of the compass, after which they were passed as Boys. They had spent six months under instruction and now went to the training brig *Sealark* which made short cruises in the Channel. After three months in the brig, John went back to *Illustrious* where he and his mates were now of some real use, given some responsibility, were exempt from all drills and 'were allowed to improve ourselves in our profession in any way we pleased'. After a year they were drafted to the Fleet as Boys First Class. John himself joined the steam corvette *Highflyer*, commissioning for the China Station (one of her officers was Jackie Fisher).

MANNING PROBLEMS IN THE RUSSIAN WAR, 1854–1856

The 1852 Committee's reforms went some way to solving the navy's manning problems in peacetime, but did not address the problems of war. All their solutions were overtaken by events when the war against Russia broke out in 1854 and the navy had somehow to man two fleets, one for the Baltic and the other for the Black Sea.

The ships only got to sea through the efforts of a small nucleus of trained men on board, many of them coastguards, elderly by lower-deck standards. At the first Sunday church service in the *Nile*, for instance, 'off caps' revealed many balding heads, and a good few pairs of spectacles were adjusted. They resented being sent back to sea and objected to wearing 'pusser's slops'—service clothing—instead of their own. However, without them the fleet could not have sailed.

As it was, men had to be recruited from every available source. Even Norwegian and Swedish volunteers were signed on in Stockholm. The *Cumberland*, back from three years on the North American station, was sent direct to the Baltic (needless to say, without consulting or informing her officers or crew) where her experienced men, many qualified for higher ratings, were distributed as petty officers amongst other ships.

In April 1856, after the Russian War, there was a fleet at Spithead of 25 sail of the line and some 200 other classes of ships, with combined ship's companies of more than 26,000 men. Within two months more than half of the ships had been paid off and 15,000 men discharged to shore.

The Admiralty took advantage of this situation in May 1857 by inviting all petty officers, seamen, and boys in ships in home waters or arriving home from foreign service to take their discharge, without payment, from their continuous service engagements. Men of poor character or physique were compulsorily discharged. This was a return to the bad old days of casual recruiting and was bitterly resented, especially as many sailors took their tickets and then regretted it. After all the fanfares for a new deal for seamen, all the promises of a permanent career, it seemed the Admiralty's thinking had progressed little beyond the press gang stage.

In June 1858 the First Lord, Sir John Pakington, set up a Royal Commission on Manning which reported in January 1859. It recommended more training ships and brigs for boy seamen; the old *Illustrious* was paid off in 1859, but five more large ships were eventually employed: the *St Vincent* at Portsmouth and the *Impregnable* at Devonport in 1862, the *Boscawen* in Southampton Water until about 1867, when it moved to Portland, the *Ganges* at Falmouth in 1866, and the *Wellesley* at Chatham. Other recommendations were: free bedding and mess utensils for new recruits, free uniform for continuous service entrants; an improved scale of victualling; improved pay for seamen gunners; and, most important, the formation of a Royal Naval Reserve.

A NAVAL RESERVE

The Statute authorizing the Reserve was passed in August 1859. Its intention was to make use of the thousands of merchant marine sailors who made short voyages, returning to home ports three or four times a year. The reservists had to be of British descent; physically fit; under 35 years of age (although they were allowed to join up to 40 until 1860); with five years' sea service in the previous ten, at least one year in the rate of AB; of fixed abode, if possible; personally known and selected by their shipping master; and approved by their local coastguard officer.

They enlisted for five years at first, trained for 28 days a year, were paid naval rates, and could qualify for a pension from the age of 60. They could only be called out by a Royal Proclamation and in times of 'sudden emergency' for an initial period of three years.

About 20,000 men were expected to join. In fact, the response initially was almost nil. The merchant seaman still distrusted the Admiralty, with thoughts of the press, doubts about what would qualify as a 'sudden emergency', fears of being literally shanghaied to the China Station, and recent memories of the Admiralty's reactionary behaviour in discharging continuous service men in 1857.

The Royal Naval Artillery Volunteers at Exercise. The Royal Naval Reserve was founded by Statute in 1859, but made a very slow recruiting start because of the merchant seaman's suspicion of the Admiralty. The Naval Artillery Volunteers, one of the forerunners of the RNVR, and consisting for the most part of yacht-owners and professional men, were set up by Act of Parliament in 1873 to provide gunners for service in home waters.

However, numbers did eventually build up to some 17,000 in the Royal Naval Reserve by 1865.

In the meantime, the Admiralty's plans were once again overtaken by events. In 1859, there was an anti-French 'scare' and a bounty had to be offered to men who enlisted: £10 for ABs, £5 for ordinary seamen, £2 for landsmen. Some good men did join but the majority of the 'bounty men' were criminals, ne'er-do-wells, mental deficients, unemployables, debtors, and family black sheep, reminiscent of the old days of the press.

Thus, in many ships there were virtually two ship's companies, the 'bounty men' and the regulars, who understandably objected to having to do their own jobs and also the jobs the bounty men were supposed to do, all for the same pay.

But all discontents were aggravated by the main grievance of leave. Until the navy gave leave on a more regular and more humane basis, it could not even begin to approach a solution to a crisis of discipline in the fleet at the end of the 1850s. Discontent, exacerbated by harsh, old-fashioned, and sometimes downright stupid handling by the officers, led to serious unrest in some ships.

New regulations in April 1860 authorized captains to grant regular leave, of 48 hours or four days, depending upon the recipient and the circumstances. The first experiments were watched 'in an agony of suspense' by senior officers, and at first there was mass leave-breaking, with desertions reaching almost epidemic proportions.

Although a deserter forfeited everything—all pay, prize money, bounty, annuities, pensions, medals, decorations, clothes, and any effects on board—between 2,000 and 2,500 men were formally noted as deserters every year. The Naval Discipline Act of 1860 defined desertion for the first time: a man was marked 'Run' on his papers, with an 'R' on the ship's books, after 21 days' unauthorized absence from the ship.

In 1866 leave was granted in three categories: a 'special' class for men of good character; 'privileged' for less reliable men who were given leave when convenient; and 'regular' for anyone else, including men in the Second Class for Conduct. Not until 1890 did leave become a right and not a privilege.

The First Class for Conduct was introduced in 1859. No sailor or marine in this class could be flogged except by sentence of court martial. Parliament kept a close watch on the numbers of men flogged and the Admiralty strenuously issued strict injunctions to captains to refrain from excessive punishment. Publicity, constant pressure and surveillance in parliament, urgings by the Admiralty, and changing social attitudes to flogging at last had their effect. The navy suspended flogging in peacetime in 1871 and altogether in 1879. But the punishment is still suspended to this day, and theoretically the cat, like the press, could be reintroduced in wartime. Seamen boys and cadets were still punished with the cane until after the Second World War.

The success of the *Illustrious* 'Novices' encouraged Captain Harris to suggest a similar training ship for officers, but he was opposed by conservative officers who believed that a lad should be sent to sea straight from school. But in January 1856, Harris entered his own son, Robert Hastings Harris (another future admiral), on *Illustrious*'s books, to learn seamanship alongside the novices, and navigation and nautical astronomy on board *Victory*, under the Revd Robert Inskip.

· This was effectively the start of the system of training young naval officers which was to lead to the *Britannia*, Osborne, and Dartmouth. It was also the end of a period of some 20 years when there had been no formal training for officers.

One of the results of the opening of HMS *Excellent* had been the closure of the Royal Naval College in 1837. With new and better qualified instructors and with facilities for the further education of gunnery specialists, the college was thought no longer necessary as a boys' school; another powerful factor against it was the continuing hostility of officers who had themselves never passed through the college. It reopened in 1839 as an institution of higher education, as 'an additional means of scientific education to the young gentlemen and officers of the fleet'. For this purpose it was amalgamated with *Excellent*, but it quickly deteriorated into a 'crammer's' for mates (later sub-lieutenants) studying for their lieutenant's examination.

Entries for officers remained haphazardly defined for years. In 1838, an Admiralty circular laid down that a Volunteer First Class 'must not be under twelve years of age. He must be in good health, fit for service, and able to write English correctly from dictation, and be acquainted with the first four rules of arithmetic, reduction and rule of three.' However, though the term 'College Volunteer' was officially discontinued in 1838, the circular remained unaltered until the term 'Naval Cadet' appeared in 1843. Not until 1849 was the maximum age for entry fixed at 14. In 1851, a circular required all entrants to undergo an examination (as Fisher did) at the college within two months of the date of their nomination.

Since the closing of the college in 1837, cadets' education had been carried out by the Schoolmaster, whose title was changed in 1840 to 'Naval Instructor and Schoolmaster', 'Schoolmaster' being dropped in 1842. The Naval Instructor received the light blue distinctive colour on his sleeve in 1879, and was officially designated 'Instructor Officer' in 1919.

But the Harrises, *père et fils*, must have impressed the Admiralty, for in February 1857, only weeks after young Robert had finished his year in *Illustrious*, a circular was issued announcing an entirely new regime for the education and examination of young naval officers, which included a period of compulsory training in a stationary ship before being sent to sea.

The first batch of 23 cadets joined *Illustrious*, moored off Haslar Creek in Portsmouth Harbour, in August 1857. By the end of that year, the ship was devoted exclusively to cadets' training, the novices being abolished for the time being, because it was decided that officers' and seamen's training should not be carried out in the same ship. By the end of 1858 it was also clear that a bigger ship was needed. Captain Harris shifted his pennant to the three-decker *Britannia*, the fourth of that name, in Portsmouth on 1 January 1859.

In February 1862 the *Britannia* moved to Portland and then, in September 1863, to Dartmouth, a place forever associated with her name, where she was moored in the river half a mile above the town. The number of cadets had increased in a few terms from 230 to over 300. Extra accommodation was provided in 1864 by the old two-decker *Hindostan* which was brought round from Devonport and moored upstream of *Britannia*, with a walkway connecting her stern with *Britannia*'s bow. In July 1869 *Britannia* was replaced by another, the fifth of the name, a much larger ship launched in 1860 as *Prince of Wales*, a screw-ship of 131 guns.

From time to time the Admiralty issued circulars defining the age-limits for entry, changing the syllabus, and making the entrance examination competitive, but the general pattern of life on board *Britannia* was established for the rest of the century. All *Britannia*'s masts and rigging had been removed except the foremast, which was still rigged for the cadets to practise on, with a safety net stretched beneath it. The space of the upper deck was almost completely built over with class-rooms, model rooms, and special accommodation for the ship's officers. The guns had all been removed and the gun-decks were used as dormitories, mess-rooms, bathrooms, and promenades. The long triple tier of empty gun-ports had glass panes framed with wooden sashes.

Two terms of cadets joined every year, and each term spent two years in *Britannia*. Every cadet kept all his belongings in a sea-chest and slept in a hammock, the two junior terms in *Hindostan*, the two senior in *Britannia*. All four terms ate on the four long tables in the large cadets' messroom which was also used as a general assembly space for morning prayers, inspections, and evening preparation, and as a mess where the cadets spent their spare time reading, writing letters, and ragging about.

Theirs was a spartan regime of hard exercise and fierce discipline. The day began with a swim in a cold salt-water bath and proceeded on its laid-down course, the passing hours marked by the ship's bell: prayers, inspection, classes, meals, boat-work, exercise. Food was plain but plentiful: slices of stale bread with a scraping of butter and weak tea for breakfast, beef and beer for dinner, cake and milk in the afternoon, cold meat for supper, roast chicken on Sundays once a month.

Work fell into three broad parts. There was 'study', consisting of mathematics and navigation; 'out of study', which was French and drawing (so arranged by Their Lordships, it was said, that naval officers could accurately sketch headlands and other conspicuous navigational marks as they sailed past them); and seamanship, which comprised sailing lore and signals.

After school the cadets changed into sports gear or flannels for football or cricket, walks, and cross-country runs ashore. 'Out of bounds' orchards and game coverts were always strictly defined and there was generally a Royal Marine sergeant waiting to pop up unexpectedly from behind a hedge. A flotilla of four-oared gigs and pair-oared skiffs (known then, and still to this day at Dartmouth, as 'blueboats') was available, and launches and cutters for sailing. Occasionally there were longer sailing trips in the *Ariadne*, although these were discontinued in 1873, or in the *Dapper*, a 300-ton barque-rigged gunboat dating from the Crimean War.

Discipline and the general ruling of the ship, apart from academic studies, were in the hands of a captain RN, assisted by a commander, three lieutenants, a master at arms, four ship's corporals, and half a dozen cadet captains and chief cadet captains chosen from the senior term. The academic staff had no disciplinary powers but a report of slack or inattentive work by a cadet was punished, with no appeal.

Punishments were severe. Cadets could be caned with their trousers on or birched with their trousers off. They could also be confined to cells, on a reduced diet of bread and water. Offenders in the third (or worst) class for conduct wore a white stripe on the arm and ate in the cockpit or in a special messroom. They also had extra drill, with bar-bells or heavy Brown Bess muskets. Other punishment regimes involved offenders being shaken earlier in the morning and being kept on their feet after the rest had turned in, often being made to stand to attention facing the main deck bulkhead for an hour at a time (a punishment for the lower deck which was not abolished from the fleet until 1912).

Britannia was an almost completely enclosed world of its own. The cadets led an introspective, insular existence, which the navy actually maintained at Dartmouth with very little change up to the outbreak of the Second World War. There was nothing in Dartmouth harbour to excite the cadets' interest in naval affairs. The only seagoing men-of-war they ever saw were the small sailing-brigs attached to the seamen's training ships at Plymouth or Portsmouth which occasionally anchored for a day or so near the harbour entrance.

After passing their examination, the cadets went to sea as midshipmen, where, in the gunrooms of the fleet, they lived a life even harder and more energetic and in conditions even more spartan than in *Britannia*. They kept watches at sea and in harbour, learning how to take sights, how to handle parties of men, how to deal with various events and emergencies of shipboard life. They also ran the ship's boats. Every forenoon and on occasional afternoons they received instruction in navigation from the naval instructor on board. But watches and boat-running took priority over school, which the midshipmen themselves disliked and found every pretext to avoid. With their other duties, the loss of sleep that watchkeeping entailed, and the various commitments and distractions of life in a man-of-war, it was very difficult for any midshipman to concentrate on studies.

The sail training brig HMS *Martin*, a tender to the boys' training ship HMS *St Vincent*, entering Portsmouth harbour. Such brigs remained in service until 1903.

After five years at sea, a midshipman was examined in seamanship by a board of captains. This was the first major step towards lieutenant's rank. The standard demanded by the board naturally depended upon the captains who sat on it. John Jellicoe, who would have got a first-class pass anyway, said that 'The mail came in during the exam and the Captains were more interested in their letters.' A first-class pass brought advanced seniority.

Aged about 20, the young gentlemen went as sub-lieutenants to the Royal Naval College, Greenwich, which opened in 1873 as the 'university of the navy'. The Royal Naval College, Portsmouth, and the Royal School of Naval Architecture and Marine Engineering in South Kensington were both transferred to the new college, which aimed to cultivate 'the general intelligence of officers, to improve their aptitude for the various duties which a naval officer is called upon to perform'.

It was some years before this aim was achieved. Many of the sub-lieutenants had forgotten nearly everything they had learned at 15 and were not amenable to academic studies or to college discipline. In December 1878, a writer calling himself 'A Naval Nobody' published an article in *MacMillan's Magazine* scathingly critical of naval education. Of his time as a midshipman, he said, 'I can think of no single advantage that I have therein, no advantage whatever which I could not equally have gained by serving that time (or a great part of it) on shore at a college, going to sea occasionally for a sailing cruise in some small craft in the Channel.'

Greenwich he called simply 'a farce. For that that college is a farce no one who has studied there will deny.' He complained of the 'discipline fit only for boys and ship-board life. Your "harassing legislation" worries and sickens us.'

He also complained that naval officers were not taught foreign languages: 'How often have I seen two naval officers of different nationalities bowing and grinning to each other idiotically, comprehending each other less than two monkeys would . . .'. And naval officers showed no interest in geology or natural history: 'A man of war visits an unknown country, say New Guinea. And what information do we bring back? Can we describe what the special characteristics of the country are, what the botany, what the geology, what the fauna? Scarcely one scientifically intelligible word: a tree is a tree, a palm a palm, a bird a bird, an insect an insect!'

There was some substance in those complaints. There certainly was a Service suspicion of fields of knowledge outside the navy, a distrust of officers with 'outside interests', and, above all, a mild contempt, not always concealed, for the 'clever' officer. The native British distrust of the man who was 'too clever by half' was translated in the navy into the expressed belief that the intellectually bright officer, the man 'who chased x, y and z', the 'three-oner' who got first-class passes in his seamanship board, his Greenwich time, and his *Excellent* course, must also of necessity be a duffer at seamanship. The 'x-chaser' might know all there was to know about algebra, it was believed, but would be utterly useless on deck in a gale at sea. There was even a song about it. 'A three-oner' was one who 'shouted when the ship was flat aback, "Let go the starboard alpha cosine theta stunsail tack."'

A Naval Nobody had some strong views on the navy's branch specializations. 'It sounds impossible, inconceivable,' he wrote, 'that it is only a privileged few who are allowed to make a *study* of gunnery, practically and theoretically; only a privileged few who are initiated into the mysteries of torpedoes; only a privileged few who are taught thoroughly the all-important knowledge to a sailor of surveying and navigation; not even a privileged few who are taught—with any practical result—that science which has displaced the science of utilising the winds—the science of steam; and yet all this is so!'

Scientific development in weapons did, however, bring forth a response in the way of training arrangements. In the late 1860s, Mr Robert Whitehead, manager of an engineering factory in Fiume, developed a self-propelling, depth-keeping torpedo, which the Admiralty bought a licence to manufacture after successful trials at Sheerness in 1870. In 1872, the frigate *Vernon*, which was then a hulk serving as a coaling jetty at Portland, was brought round to Portsmouth and fitted out as a torpedo instructional ship. She was still a tender to *Excellent* and thus a branch of the gunnery school. Commander J. A. Fisher, an important figure in the formation of the new Branch, was appointed to *Excellent* in September 1872, for 'torpedo instruction'. *Vernon* broke away from *Excellent* and was commissioned as an independent command in 1876.

Through the years, *Vernon* accumulated various other hulks around her for accommodation and instructional purposes which all moved to Portchester Creek in 1895. Finally, in 1923, *Vernon* moved ashore to the Gunwharf in Portsmouth Harbour, where the Mining School was already established. The Torpedo Branch became responsible for all aspects of torpedo and mine warfare, and also for ships' electrical supply and installation. Training and instruction, for both officers and ratings, including ships' divers, was carried out at *Vernon*.

Seamen under instruction at HMS *Vernon*, the torpedo school at Portsmouth, which was first commissioned as a tender to *Excellent* in 1872 and became an independent establishment in 1876.

Sail training continued until the end of the nineteenth century, and had it not been for the outbreak of the Boer War in 1899, when the training ships were paid off and their ship's companies of some 1,400 officers and men transferred to four modern cruisers, sail training would very probably have continued until well into the twentieth century. The sailing brigs attached as tenders to the ships served until 1903.

The training squadron of the 1880s and 1890s had four and sometimes six full-rigged second-class cruisers or corvettes which made long cruises to Norway, Madeira, and the West Indies, always under sail except in narrow waters. The sailing-sloop HMS *Cruiser* was used for training ordinary seamen in the Mediterranean for many years.

On 24 March 1878 the sail training frigate *Eurydice* was returning from a cruise to the West Indies and was close inshore off the Isle of Wight, carrying all plain sail with studding sails, when she was caught by a sudden squall. But though her crew, including over 200 seamen under training, were superbly fit and well trained and could be relied on to have all sail off her in double quick time, the ship capsized, and of the 300 people on board only two survived.

The loss of *Eurydice* sent a thrill of horror round the nation. It was one of those tragedies which made people remember where they were and what they were doing when they heard the news. A young Jesuit priest, Gerard Manley Hopkins, took the story of *Eurydice* and wrote one of the great religious poems in the language, a poetic parable linking this naval disaster, so many men drowned with no chance of absolution, to a greater disaster to the nation, the abandonment of the Roman Catholic religion. Hopkins's poem was indicative of a marked change in the public perception of sailors since the early nineteenth century.

The old insanitary hulks were banished at last and barracks built ashore. Agnes Weston and her devoted helpers in the Royal Naval Temperance Society and in her Sailors' Rests, the first opened in Devonport in 1876, did much to raise the sailor's esteem in his own and the public's eyes. The bluejacket ashore was no longer 'poor Jack', the drunken 'common sailor' of 1815, but 'The Handyman', hero of a score of Naval Brigade campaigns, celebrated in poems, songs, and on the stage. Senior officers such as Lord Charles Beresford realized the influence parliament and press could have on the navy's reputation, and funding. A great naval exhibition on the Thames embankment in 1891 was a huge success.

Lionel Yexley, who had himself joined the *Impregnable* as a boy seaman in 1879, and underwent (and thoroughly enjoyed) a gunnery course at *Excellent* when Jackie Fisher was captain, became editor of a paper called *Hope, The Bluejacket and Coastguard Gazette* when he left the navy, and set himself to improve the bluejacket's lot, ashore and afloat, inside the navy and out. In 1905 he started the *Fleet*, the first true bluejacket's newspaper. Whenever the sailor was abused afloat or slighted ashore, the *Fleet* retaliated with a full account of the offence.

In the 1890s, after many years of internal administrative slumber and outward public indifference, the navy was suddenly fashionable. The sailor became a potent selling symbol on advertisements, especially for tobacco and soap. The

'Too proud, too proud, what a press she bore!' Gerard Manley Hopkins's poetic description of the boys' training-ship *Eurydice* capsizing in a sudden squall off the Isle of Wight when returning from a cruise in the West Indies on 24 March 1878.

Navy Records Society was founded in 1893, for the printing of papers and documents of naval biography, history, and archaeology, and a year later the Navy League was founded to secure 'as the primary national policy, the command of the sea'. Naval history was studied at Oxford and Cambridge (though not in the *Britannia*). Board games of naval tactics were on sale. Rudyard Kipling went to sea with the Channel squadron and published *A Fleet in Being* in 1898.

One board game of the 1880s was called 'From Sailor Boy to Admiral', a variation of 'Snakes and Ladders', in which sailor pieces progressed along numbered squares, being advanced for good work, put back for bad, until the first player to reach admiral's rank won the game.

But such progress could only be accomplished in a game; it was not possible in the navy. Promotion from the lower deck to the quarterdeck was the last great reform needed in the Victorian navy, and it was the most difficult to achieve.

The last half of the nineteenth century was a time of enormous technological change in the navy. New ships were built and commissioned, new machinery fitted, and new guns designed and fired. But ironically, and almost paradoxically, despite the tremendous progress in propulsion machinery, guns, mountings, and armour plate, the navy of the late nineteenth century was in some ways at its most reactionary. Its officers were recruited almost entirely from a self-defining, self-perpetuating section of society, as though they were all members of some semi-aristocratic yacht club, electing and re-electing each other over and over again.

The 1852 Manning Committee had recommended that warrant officers be promoted to the rank of lieutenant for gallantry and daring in action, but no such promotions were made until Queen Victoria's Golden Jubilee honours list in 1887, when two warrant officers were promoted, inevitably becoming known as 'Jubilee Memorials'.

But there were no more until the end of the century. Henry Capper joined the *St Vincent* as a boy seaman in 1869 and qualified as a gunner in 1879. When he attended his first dinner in *Excellent* as a young and ambitious warrant gunner in 1881, the president announced him as a new member and the 60 officers present clapped. Capper rose to thank them and said how much he appreciated their welcome 'but had done my best to reach this mess only as a halfway house to a commission', whereupon there was a great roar of derisive laughter. The president explained that there was no further promotion open to him until in some 30 years' time he might get a chief gunner's stripe a few months before retirement. 'No naval warrant officer has had a commission these last 70 years: if you want one you must get the system changed first.

That was precisely what Capper determined to do. For some years he contributed anonymously (although the Admiralty knew of it) to the press, and in 1888 founded the monthly *Naval Warrant Officers' Journal* to publicize their case. In 1890, the First Lord, Lord George Hamilton, announced that the navy urgently needed 100 more lieutenants. The warrant officers issued 'An Earnest Appeal for Promotion from the Ranks, Royal Navy' showing that they were already doing the duties of lieutenants and sub-lieutenants and that some of them were prepared to take the examination for lieutenant.

Cutlass Drill on the quarter-deck of HMS *Resolution*. The navy had not fought a fleet action since Trafalgar, nor a ship-to-ship engagement since Navarino in 1827, but naval brigades took part in many campaigns ashore all over the world throughout Victoria's reign.

'The Black Brigade' showing various punishments: for spitting on deck, carrying a spittoon around; for dirty clothing, carrying the clean article lashed to an oar; holystoning or scrubbing the deck; for slack hammock, carrying the properly lashed hammock over the shoulder.

They had strong support in the press and parliament and a motion was put down in the House. But the government opposed it. So too did some serving officers, whose families regarded the navy as a private preserve. As one otherwise kindly and sensible lady told Henry Capper, 'I have the greatest sympathy with you personally in your desire to rise, but you have chosen the wrong service. The Navy belongs to us, and if you were to win the commissions you ask for it would be at the expense of our sons and nephews whose birthright it is.' The extra officers, 'The Hungry Hundred', were therefore all RNR officers, who transferred to the navy from the Merchant Marine.

During the Boer War, Gunner Thomas Lyne was promoted for gallantry in the field and was granted a lieutenancy in King Edward VII's Coronation honours list. He eventually became a rear admiral, the first man from the lower deck to reach flag rank since John Kingcome, Admiral of the Red, in 1818. General lieutenancies for warrant officers were introduced in 1903.

Finally, in 1912, the First Lord, Winston Churchill, announced the appointment of a committee on discipline and the institution of the 'Mate' Scheme, whereby young petty officers of any branch could reach the wardroom, thus reopening a path to promotion which had been blocked for 93 years. The scheme was extended to the Royal Marines in 1913 and to Mates (E) in 1914.

THE ENGINEERS

That Naval Nobody of the 1870s was perhaps being unrealistic in complaining that all officers were not specialists in gunnery and torpedoes, but he did have a good point about knowledge of navigation, and an even better one about steam. The navy had had steam vessels for years, but engineering had no place in the *Britan-*

nia syllabus, although the Royal Naval College at Portsmouth did give instruction in 'steam'.

The earliest steamships in the navy, *Monkey* of 1821, *Comet* of 1822, *Lightning* of 1823, and *Diana* of 1824, all had engineers who, so to speak, came with the ship, like a walking book of operating instructions. Known simply as 'engineers', they were literally engine-drivers. They had no ranks, no uniform, no status, no terms of service, and could come and go much as they pleased.

Training for 'engineer boys', a five-year stint, began as early as 1828, but generally engineers had no formal training or terms of apprenticeship. It seems that engineering in the Royal Navy never properly recovered from this informal start.

Steamships saw very early action. *Lightning* went with the bombships to attack Algiers in 1824, and *Diana* served on the Irrawaddy in the Burma War of the same year. There were engineering hazards even in those days. One of *Diana*'s midshipmen went on deck one morning and saw a human leg in a bucket. It belonged to the engineer, who had gone to stop the engines during a sudden squall in the night, slipped, and crushed his leg in the works so badly it had to be amputated.

Many seamen officers were baffled as to how to treat engineers, who had vastly different upbringings and outlooks from themselves. Engineers were not officers. Yet they were not ratings. Nor were they warrant officers, indeed the boatswains and the carpenters very much looked down on the engineers. There was to be a long and heated debate down the years over whether engineers were 'military branch', and if they were officers, whether they were engineer officers or officer engineers.

Steamers were first commissioned as warships in 1833. A few naval officers realized that steam was here to stay and took steps voluntarily to improve their knowledge. But in 1844, when Lieutenant Cooper Key volunteered to change with a lieutenant in the paddle steam frigate *Gorgon*, it was regarded as a sensational and foolish action, likely to jeopardize his career. 'Steamer lieutenant' was a term of contempt.

An Order in Council of 1837 gave engineers warrant rank (below the carpenter) but by the 1840s the navy was plunged into the first of several manning crises over engineers. There was a shortage of them, and very few applicants. So, fatally, the entry standards were lowered. Even so, the poor pay and poor status did not attract the right sort. Those who did enter did not deserve better pay and status. It was a vicious circle.

The navy's need for engineers always exceeded its regard for them. Improvements were introduced, not out of any concern for engineers, but always as measures of desperation—because, unless things improved, there simply would be no engineers.

Thus, in the 1840s, when commissioned rank for engineers was being discussed, many seamen officers could not bring themselves to look upon engineers as fellow officers (some could hardly look upon them as fellow human beings). A letter to *The Times* put the widely held view that engineers 'were a most useful class of men but they are not "gentlemen"'. 'Gentlemen' or not, another Order in

Council in 1847 gave engineers commissioned rank, equal to the old navigating masters, and their names first appeared in the Navy List. But they still had to mess by themselves. Nobody wanted oily overalls in the wardroom.

The victory of the screw propeller ship *Rattler* over the paddle-ship *Alecto* in 1844 consolidated the engineer's position in the navy, and the Russian War of 1854–6 confirmed the steamer's versatility and usefulness. It also confirmed that engineers were combatants. Stokers had already played honourable parts in naval brigades earlier in the century. The earliest-dated Victoria Cross awarded to the lower deck was won by Stoker William Johnstone in the Baltic in August 1854. In an exploit straight out of *The Boy's Own Paper*, he and Lieutenant Bythesea of HMS *Arrogant* went ashore and captured the Tsar's mails. Johnstone had been chosen 'because he spoke the language' and was almost certainly not a British citizen, but a German or a Swede.

Engineers were given uniforms in 1837 and an 'engineer's button' with a 'side-lever' engine motif in 1841. The 'curl' for executive officers was introduced in 1860 and the purple stripe for engineers in 1863. Chief engineers with 15 years' seniority were equal to a commander in rank. But until 1883 only a chief engineer messed in the wardroom.

In 1868 the Admiralty introduced 'engine-room artificers' in an attempt to establish a supervisory role for engineer officers, thus elevating them above manual work. But junior engineers took this as a threat to their livelihood. Feelings on this and other matters ran so high that in the 1870s the Admiralty faced another crisis over engineers' recruiting.

A committee was set up in 1875 under Cooper Key to examine the position. Their recommendations about pay, rank, promotion, and 'military status' were all turned down. However, in 1877 an old 121-gun screw line-of-battleship, *Marlborough*, was allocated as an accommodation ship for engineer students in Portsmouth.

Students from the RNEC Keyham, under instruction in Devonport Dockyard Smithy. The Royal Naval Engineering College, Keyham, opened in July 1880. Students underwent a six-year course, a 'wedding of theory and practice', with lectures on mathematics and mechanics in the college and practical work in the dockyard workshops, foundry, pattern-makers and boiler-makers shops.

There had been an earlier engineers' training ship, the brig *Sulphur*, for a very short time in the 1840s. The dockyard schools trained the engineer boys and the main training of engineers was still carried out in the dockyards. A training school for engineer students, which became known later as the Royal Naval Engineering College, was built at Keyham, Devonport. The first students transferred to it from the old *Marlborough* in July 1880.

The most junior students (in later years the midshipmen) lived in six large dormitories. The more senior (sub-lieutenants) had cabins, which were really small cubicles partitioned off, open at the top. Students entered between the ages of 14 and 17, and entry was by competition, each new intake joining in July. They were divided into six sub-divisions, three to each corridor, the students in a division being taken from all ages, with a house captain in charge.

It was a somewhat spartan life of hard work and hard play. Keyham and *Marlborough* existed together until 1888, when classrooms and studies were built at Keyham, and *Marlborough* closed down. Initially, the standards of training were not good, but they improved when Professor W. M. Worthington arrived from *Marlborough*.

Under Worthington the students undertook a six-year course, a 'wedding of theory and practice', with lectures on mathematics and mechanics in the College and practical work in the dockyard workshops, foundry, pattern-makers and boiler-makers shops. The students built auxiliary machinery and carried out steam engine trials. From time to time they went to sea in HMS *Sharpshooter* for practical steaming experience, cruising up and down the Devon and Cornish coasts.

Nevertheless, by the end of the nineteenth century there was yet another crisis. At the time of the Cooper Key Committee the view was that 'too many engineer students are being taken from the sons of dockyard artificers, of seamen and marines and of others belonging to the same class of society'. This had become less and less true as the years passed but there was still an attitude amongst seamen officers that an engineer was really no better than a 'lascar with a bottle of oil'. The engineers themselves complained of their lack of status and the denial of the right to handle disciplinary matters within their branch.

Recruitment fell, and the Admiralty was forced to desperate measures again, taking in 'emergency engineers' to fill the gaps. These men had even less in common with the wardroom than the officers who had passed through Keyham and recruitment slumped again. It was the same old vicious circle. Nearly 80 years after the first steamships in the navy, the seamen officers who still made up the whole Board of Admiralty had learned nothing from the past.

The solution was the Fisher–Selborne Scheme of 1903, jointly devised by Fisher, who was then Second Sea Lord, and Lord Selborne, First Lord of the Admiralty. Its essence was that all officers, seamen, engineers, and Royal Marines, should be of one company, and would join, be educated, and train together. Engineering would be another specialization, like gunnery, torpedoes, or navigation.

THE FISHER–SELBORNE SCHEME

277

Initially, the scheme was not intended to give cadets anything like a public school education, although, chiefly for reasons of sanitation and discipline, the old *Britannia* would be replaced by a new college ashore, plans for which were announced in 1896. Fisher believed that officers should go to sea at 12 or 13, just as he had done. 'Nothing is worse,' he said, 'nothing is more retrograde than trying to ape the Public School system for the Navy.'

However, the great naval historian Julian Corbett suggested that the chief obstacle to effective naval training was that 'You cannot train except at sea and at sea you cannot teach.' The solution was the basis of the Dartmouth scheme, in which cadets entered at 13 years of age and were given a general education, equivalent to a public school, of four years.

The foundation stone of Aston Webb's new building at Dartmouth, to be called the Britannia Royal Naval College, was laid by Edward VII in March 1902. The new College would not be ready until September 1905, so a 'junior' College was opened at Osborne, on the Isle of Wight, where the first cadets arrived in September 1903.

Entry was by nomination after an interview, with a qualifying examination and a medical examination. Cadets spent two years at Osborne followed by two years at Dartmouth, and then went to sea for six months in a training cruiser before joining the fleet as midshipmen.

The syllabus was a full one of mathematics, mechanics, heat, and electricity; the science and practice of engineering; French, English composition, and literature; general and naval history; geography, navigation, seamanship, and religious knowledge. There were also boat sailing and rowing on the river, trips to sea in the sloop *Racer*, and the usual sports.

Under the scheme, the RNEC Keyham would no longer train boys from 14 to 19 but take lieutenants aged 22 instead, for specialist training, who would be known as lieutenants (E) instead of the old engineer lieutenant.

As Captain Rosslyn Wemyss, the first captain of Osborne, wrote, 'There is no reason why the cadets entered under the New System should not (indeed there is every reason why they should) *all* be eligible to attain to the very highest ranks in His Majesty's Navy, those who may become Lieutenants (E) or Marine officers thus finding themselves exactly on a par with Lieutenants (G), (T) and (N).'

It was a noble intention, but it was not to be. The Fisher–Selborne Scheme did not last. Engineering still bore some of the stigma of lower social status. Only about one midshipman in 20 actually volunteered for engineering on leaving Dartmouth. Parents, especially naval parents, were not generally best pleased if their offspring announced that he wanted to be an engineer. Wemyss suspected some parents of discouraging their sons in their engineering studies.

The First World War demonstrated to the Royal Navy (though not to the United States Navy) that engineering was not just another specialization, and it was impracticable (again, except in the US Navy) to amalgamate deck and engineer officers' duties.

After the war the system was changed. Midshipmen were selected for engin-

Sir Aston Webb's magnificent Britannia Royal Naval College, Dartmouth. The foundation stone was laid by King Edward VII on 7 March 1902 and the College opened in 1905. The old ship *Britannia*, which the college replaced, was broken up in 1916 for the copper sheathing on her hull.

eering at the age of about 19 and sent to Keyham for a four-year course—the so-called 'Long E' course. But they were still equal in status, as Fisher had stipulated. They were still 'military officers'.

But in 1925 it was all changed again. An Order in Council, using the phrase 'executive officer' for the first time, defined the branches, such as engineering, and laid down that engineers were to ship the purple stripe again. This 'great betrayal', as it was called, was as bitterly resented as anything in the nineteenth century. Furthermore, it came as a very severe jolt to many officers who had chosen engineering in good faith, reassured that it would not jeopardize their long-term career prospects. Now, at a stroke, their promotion ceiling had been reduced by two or three ranks. They could never now be full admirals. There was only one vice-admiral (E), the engineer-in-chief.

There were indignant meetings. But there was no service machinery for putting forward such grievances. The Captain of the College agreed to put forward a selection of letters of complaint to the Commander-in-Chief. But nothing came of them.

There would inevitably have been yet another shortage of engineer officers had it not been for Winston Churchill who foresaw the coming of war and the navy's need for more junior officers. In Churchill's 'special entry' scheme, introduced as early as 1913, cadets entered at the age of 18, having completed their education at public school. They had two cruises in HMS *Highflyer*, to learn some seamanship and navigation, before joining gunrooms in the fleet. Although a few Dartmouth officers continued to volunteer for engineering, eventually the great majority of engineer officers were 'special entries'.

Nevertheless, the 'great betrayal' suggested that in some ways the Board of Admiralty, like the courtiers who surrounded the Bourbon Louis XVIII, had forgotten nothing and learnt nothing. Although there were enormous changes in the Royal Navy during the nineteenth century, the feeling remains that the Admiralty normally responded to pressure rather than initiated reforms and in general followed closely Alexander Pope's philosophy to

> Be not the first by whom the new are tried
> Nor yet the last to lay the old aside.

10

THE BATTLESHIP FLEET: THE TEST OF WAR 1895–1919

JAMES GOLDRICK

*I*f the 1897 review had been a celebration of British maritime power, it also marked the sunset of the Pax Britannica. The strategic conditions which underpinned Britain's dominion were changing rapidly. Britain was no longer the greatest industrial power. Germany was overhauling her lead and the United States was following suit. The same march of industrialization was taking place elsewhere in Europe and in Japan.

The development of new weapon systems was having its effects, equally profound even if less immediately apparent. The classical British strategy of the steam age had been investment of an enemy's ports by light units under the protection of the battle fleet with a judicious distribution of cruisers along the trade routes to ensure the passage of British trade and the strangling of that of the opposition. The strategy would work because the Royal Navy was confident that no strong enemy forces could pass the blockade and interfere with the relatively weak trade protection units. The advent of the locomotive torpedo in the 1870s presented a challenge at first, but the short range of such weapons and the development soon afterwards of the quick-firing gun restored the advantage to the heavy ships.

General Payment on the quarterdeck of a pre-'dreadnought' battleship. Witnessed by the Officer of the Watch (in sword belt and frock coat), the Paymaster dealt out each rating's pay into his cap.

The introduction of the first gyroscope-equipped torpedoes in 1896 tilted the balance away from the battleship and cruiser by doubling the effective ranges of the underwater weapons. Despite contemporary efforts to improve gunnery and increase the fighting range, there could be no surety of safety for heavy units, particularly at night or in restricted visibility. The implications were obvious for any attempts to maintain a blockade against an enemy who could deploy large numbers of torpedo-armed vessels, enjoying the advantages of proximity and surprise. The introduction of the first relatively effective submersibles in the last years of the old century emphasized the difficulty.

Britain's political situation was increasingly awkward. Russia and France had been in alliance since 1894. Germany in 1898 enacted the first of a series of Navy Laws which presented a direct challenge by proposing construction of a fleet far greater in size than the Royal Navy had been accustomed to maintaining in operational condition in its home waters. With few friends elsewhere in Europe, Britain had also to deal with a United States of America intent upon strategic and commercial domination of the western hemisphere and robust in defence of its perceived interests. Only in Asia could Britain see the prospect of friendship, and this with the opportunist Japanese.

NAVAL COMPETITION AND POLITICAL ISOLATION, 1895–1905

The United Kingdom had faced isolation before, but without such implications for its maritime survival. A combination of the European powers now had the potential to match the British battle fleets in their own waters and to mount a world-wide commerce war which the Royal Navy would find very difficult to counter.

Germany extended her Navy Law in 1900 and France and Russia produced large armoured cruisers which forced Britain to respond in kind. Following the outbreak of the South African War in 1899 and accompanying increases in the army vote, the Exchequer began to warn of financial ruin if expenditure was not reined in. Nevertheless, as long as the Boer War continued and Britain remained isolated within Europe, the production of sufficient battleships and armoured cruisers to match any two European powers 'plus a margin' had to continue and with it expansion of naval manpower and the organization to handle the new ships. To pay for modernized dockyards to support the expanding fleet, the government used the device of loans which had to be repaid from future naval estimates. Although this measure was a sensible short-term palliative, the corollary was that the discretionary budget in future years would be reduced. This would have serious implications when progressive increases in ship size required further expansion of facilities and when the axis of the primary threat shifted from the south—France—to Germany in the east.

Even diplomatic action could produce little short-term comfort. Conciliation of the United States was an obvious and early step, but the United States had never been the real problem. The Anglo-Japanese Treaty of 1902 provided an ally in the Far East but its provisions required that the British keep a fleet in the region at least the equal of that of any other local European force.

The relief of the Boer War's end was short lived because of the collapse of the economic boom which had been in progress since the late 1890s. This rapid and unexpected reduction in receipts made the need for economies in military spending all the more urgent. The direction such economies were to take was largely determined by one man, Admiral Sir John Fisher, who was the personal selection of the First Lord, the earl of Selborne for the newly retitled office of First Sea Lord in October 1904. Perhaps alone amongst the senior flag officers, Fisher did not need Lord Selborne's reminder that the Royal Navy now had to force the greatest possible fighting value out of every sovereign it spent. With experience as Controller, Commander-in-Chief, Mediterranean and Second Sea Lord, Fisher had long nurtured revolutionary ideas for the reorganization of the navy.

ENTENTE AND NAVAL
REFORM, 1905–1910

It was also true, even as an alarmed government sought to make massive reductions in the 1905 estimates, that the strategic situation was beginning to improve, or at least to simplify itself. With the Russians in deep trouble in the Far East from the outbreak of war with Japan in February 1904, systematic British diplomatic effort brought about the signing of an *entente cordiale* with France, thus reducing markedly the threat in the Mediterranean. This left Germany as Britain's major naval rival.

Fisher's view of the need for flexibility in response to strategic demands and his assessment of the implications of technological change combined to give him a clear view of the future shape of the navy, even if it sometimes exceeded the capacity of contemporary technology to meet his demands. Fisher's command in the Mediterranean had coincided with the radical improvements in gunnery and the consequent increases in fighting ranges set in train by Captain Percy Scott in the cruisers *Scylla* and *Terrible* between 1898 and 1900. Fisher was quick to enforce Scott's standards throughout the Mediterranean and also insisted upon much improved machinery performance, converting the fleet from one, as even his arch enemy Lord Charles Beresford was later to admit, of 12 knots with numerous breakdowns to one of 15 knots without them. He had also watched the French development of submarines with acute, perhaps illicit interest.

Fisher's temperament and the complexity and uneven pace of technological development combined to confuse some of his own ideas. He was quick to believe that the dominion of the battleship was over, that flotilla craft in the form of torpedo boats and submarines would dominate coastal waters and the narrow seas and that the trade war would be fought by fast armoured cruisers hunting down commerce raiders around the world. This was essentially a vision developed in the context of the French and Russian threat and it was one complicated by tactical considerations such as the new fighting range, which favoured big guns such as the 12-inch over quick-firers like the 6-inch, but which was not yet long enough to give a battleship the full advantage of its heavier armour in resisting the effect of shot.

Before Scott's revolution in gunnery, Fisher had been convinced that fast cruisers armed with medium-calibre quick-firing guns had the potential to overwhelm battleships from ranges at which neither their slow firing and inaccurate big guns nor torpedoes would be effective. With the new potential of the heavier weapons, Fisher took the fast armoured cruiser a step further when he conceived the idea of marrying to it an all-big-gun armament. Such a ship would dictate the fighting range and overwhelm with its weight of fire.

Lord Fisher (right) shown with Winston Churchill (left) in 1913. The relationship between the older and the younger man is nicely shown in this photograph. Churchill was to derive much of his inspiration for naval development from Fisher and they achieved a great deal during Churchill's term as First Lord. Churchill's enthusiasm for the Dardanelles expedition in 1915 eventually proved too much for Fisher, bringing about the First Sea Lord's resignation and Churchill's dismissal from office.

Fisher's Committee on Designs, formed as one of his first acts in office, took a more cautious, albeit still revolutionary approach. Its work produced the first all-big-gun battleship, the *Dreadnought*, whose design was approved in May 1905. She was laid down the following October and, by the diversion of material from other ships, completed in a year and a day. The new vessel combined ten 12-inch guns and turbine propulsion, which gave her an operational speed at least two knots faster than any other capital ship in the world. She was an immediate sensation. Although other navies were working towards the concept, none had succeeded in producing such a combination of fighting qualities, nor in so little time. *Dreadnought*'s commissioning caused an effective holiday in battleship construction which, if short lived, gave the British government welcome breathing space.

Fisher's true children, the three 'Invincible' class which were soon to be known as battle cruisers, did not emerge until 1908 and met with a mixed reception.

Admiral Lord Charles Beresford as Commander-in-Chief, Channel Fleet. Mercurial, politically active, and deeply opposed to Fisher and many of the reforms made in the Royal Navy between 1904 and 1910, Beresford became Fisher's principal critic within the service, enlisting both politicians and press in an attempt to bring down the First Sea Lord.

Britain's strategic situation was by this time very different from that of 1903. Russia had been decisively defeated by Japan. The annihilation of her Far East fleet allowed the Royal Navy to withdraw its battleships to home waters. The mutual interest Britain and France now saw in the maintenance of the *entente* against Germany made war between the old rivals unthinkable and British *rapprochement* with Russia practically inevitable. Neither Italy nor Austria–Hungary, even in combination, could yet pose a serious threat to Anglo-French security in the Mediterranean. The greatest single problem was Germany and Germany was building battleships.

If the German empire meant to wrest dominion of the sea from Britain it could not do otherwise, since the British fleet lay across the German access to deep water. Despite Fisher's prescient assessment of the power of flotilla craft, securing the sea required battleships. In this context, the commerce protection role which originally spawned the *Invincible* concept was less relevant and the prospect of such lightly armoured ships fighting in the line or acting as a fast wing of the battle fleet worried many. The Admiralty, however, was now examining the possibility that experiments in predictive fire control, conducted in great secrecy from 1905 onwards, held the potential of marrying the speed and heavy armament of the battle cruiser with the ability to hit the enemy at such ranges that the latter had no chance of replying effectively. That potential was sufficient to sustain Fisher's faith in the battle cruiser concept, but poor comprehension of the issues, jealousy of the civilian Arthur Pollen, whose brainchild the system was, and inadequacies in the physical capacity of the undermanned Admiralty to deal with technical change brought most hopes to nothing. The Royal Navy was to enter the First World War with an inferior, partly plagiarized system that severely restricted the ability of ships to fire accurately while manoeuvring. This proved a critical limit on the capability of the Grand Fleet at Jutland in 1916.

The undermanning in the administration of the Admiralty was a hidden weakness of the Royal Navy, not remedied until well into the First World War. While the size of the active fleet increased steadily and the complexity of the units within it by orders of magnitude, the administrative infrastructure did not. The Board of Admiralty was poorly supported, not only through the absence of a proper naval staff, but in the limited numbers of clerical, financial, and technical personnel. Senior officers of the era were forced to become chronic over-centralizers, as much as a result of circumstances within the Admiralty as the culture of the fleet at sea.

Fisher's ideas coincided with the British government's need for retrenchment. To provide manpower for the new battleships and big armoured cruisers without increasing the ceilings on manpower, he struck from the effective list more than 150 ships, which he believed were too weak to fight and too slow to run away. Fisher substituted for the old and inefficient reserve units a force of fully manned units as a nucleus for additional three-fifths manned ships which he designated the Home Fleet. Significantly, the torpedo craft in British waters were allocated to the commander of this new force. Fisher's rationalization extended to an effective ban on cruiser construction in favour of destroyers. This derived from his assessment of the tactical weaknesses of the former and his conviction that large destroyers could do the work of scouting and countering an enemy's torpedo craft just as adequately, without being so vulnerable to the enemy's heavy units.

The nascent submarine arm received the new First Sea Lord's enthusiastic support. Britain had been cautious in adopting the concept of the submarine and other navies were permitted to experiment with them undisturbed. The near-simultaneous introduction of several efficient designs changed this situation and the 1900–1 estimates included a secret order for five of the American 'Holland' type. Development from then on was rapid and influential figures in the government began to see that submarines might possess the potential to provide all the coast defence and anti-invasion forces that Britain required. This would remove

The battleship *Dreadnought*, shown in 1914 as flagship of the Fourth Battle Squadron. Combining an all-big-gun armament of ten 12-inch guns and turbine propulsion, she revolutionized capital ship design, yet, by the outbreak of the First World War and only eight years after commissioning, she was outclassed by much new construction. She went into reserve in July 1918, her greatest war service being the ramming and sinking of U-29, whose captain had sunk the *Aboukir*, *Hogue*, and *Cressy* in the U-9.

the army's expensive commitment to coastal artillery and garrison troops. Although the pre-1904 Board of Admiralty had been dubious about the diversion of resources from the battle fleet this would imply, Fisher saw an opportunity for the priority for funds in the straitened budget to remain with the navy. The development of diesel engines and the increasing seagoing capabilities of the boats also opened up possibilities for their offensive use, which fitted closely with Fisher's evolving concepts of sea warfare. By 1908 a true ocean going submarine in the form of the 'D' class was in service, giving Britain a lead over the rest of the world.

The First Sea Lord's other reforms extended to wholesale reduction of the forces allocated to overseas stations and the accompanying closure of dockyards both abroad and in the United Kingdom which he did not see as necessary to face a threat from Europe. He cast a jaundiced eye over the stores organization and succeeded in making considerable reductions in holdings.

These economies were a godsend to the embattled government, but they left the Admiralty open to bitter and increasing criticism, which Fisher made few attempts to mollify. Some of that criticism was well judged; some was not. The danger lay in the way in which developing controversy was returning the Royal Navy to internecine political disputes which it had not seen since the eighteenth century. Fisher was soon more or less opposed by many flag officers on the retired list and by a good number who were still serving. There was much support for what Fisher came to label the 'Syndicate of Discontent' in Conservative political circles and this was even more apparent after the Liberals won the General Election in 1906. Although Fisher had little sympathy with Liberal attempts at naval arms control and was unsurprised when the Hague Peace Conference of 1907 made no progress on the subject, he could work within their policies of continuing restraint on military expenditure.

Despite Fisher's personal view that war should be fought on a total basis, the Liberal government forced the Admiralty to acquiesce in Britain's acceptance of the Declaration of London in 1909. This circumscribed the power of a belligerent to blockade to the point that it struck at the core of Britain's historical strategy of mercantile war. The Royal Navy's difficulties in producing a war fighting strategy which worked within these limitations were to result in much confusion in the years ahead, even though parliament failed to ratify the treaty in 1912.

Discontent came to focus around the ample form of Admiral Lord Charles Beresford, commanding first in the Mediterranean from 1905 and then the Channel from 1907. The feud between Fisher and Beresford had its origins in personality and it was fuelled by personality, but there were seeds of truth in Beresford's complaints. In particular, the creation of the Home Fleet as an entity independent in peacetime from the Channel Fleet went against the principle of concentration. Beresford argued that it was too weak to represent a reliable defence against a German surprise attack in the absence of the Channel and Atlantic Fleets from home waters. The Admiralty's reply was that the Home Fleet had to be built up slowly in order to prevent the Germans using it as a pretext for further expansion of the Imperial Navy. Beresford regarded the redistribution as a personal slight but the Admiralty was to wait until 1909 and Beresford's departure, by which time all the newly completed dreadnoughts and battle cruisers had been assigned to the fully operational Nore Division of the Home Fleet, to absorb the Channel Fleet into the latter. Only the Atlantic Fleet, now based in Gibraltar, would operate independently of the Commander-in-Chief, Home Fleet.

The same measures had been applied to the Mediterranean. From 14 battleships in 1902, there was a progressive reduction to eight in 1904 and six in 1907. The Atlantic Fleet, the Admiralty intended, would be available as a mobile reserve in the event of a crisis in the Mediterranean. With only a handful of ships scattered around the other overseas stations, the Royal Navy was now focused almost wholly upon the threat in the North Sea.

The concentration had its side effects. The Foreign Office complained about the effects on British prestige of the absence of permanently stationed warships. The Dominions, for their part, were becoming increasingly thoughtful about the implications of the new policies for their naval defence. They were moving towards the idea of creating services for their protection of their own ports and trade. The Admiralty, suspicious of the diversion of resources implicit in such ideas, was initially slow to take advantage of the developing enthusiasm for local navies but even it came to realize that such development promised an easing of the fiscal burden.

Fisher's flaw lay in his methods. Although his espousal of favouritism as a primary method of promotion was by no means unique in the Royal Navy, his use of selected juniors to report on the activities of the navy at large was. While the First Sea Lord was a good, if inconsistent judge of men and, in the absence of a coherent staff in an undermanned Admiralty, a relatively effective policy-maker, he made enemies of the majority of the senior flag officers. His pragmatic courting of the press aroused considerable bitterness. Fisher consistently refused to espouse the creation of a naval general staff, despite increasing evidence of the inadequacies in Admiralty organization. His sop in the form of the Naval War Council did not prove satisfactory, because it possessed only advisory powers and was too small and undermanned to do the work required. The trick which he had adopted from the outset of ignoring administrative paperwork began to redound to his disadvantage as the nature of his work inevitably changed from the initiation of policy to its implementation.

Fisher's insistence on retaining his war plans in his breast was in accord with precedent but had the effect of stifling analysis and inviting unquestioning acceptance of his concepts. He had abandoned close blockade and intended that the North Sea be dominated by destroyers and the submarines to which he was devoting increasing resources, despite the constraints on the naval estimates. This 'flotilla defence' would incapacitate the German battle fleet without risking Britain's capital ships. Fisher was largely free of the heresy of the 'decisive battle' and logical in acknowledging that control of the sea allowed Britain its free use as a mechanism for asserting power against land enemies. This approach strengthened the Admiralty's hand against the army's expensive proposals to create a

British Expeditionary Force (BEF) in readiness for a continental war. Distrusting the commitment to France which the existence of a BEF would imply, Fisher was in whole-hearted agreement with Edward Grey that the British army was a projectile to be fired by the navy—at a time and place of British choosing, not that of a European ally.

The real difficulty was that neither the intellectual culture nor adequate planning machinery existed to allow systematic examination of the strategic and technological issues. Despite the undoubted range and vision of his own thinking, Fisher did little to assist in bringing either culture or machinery to maturity.

Faith in Fisher's administration was progressively undermined from several directions. The primary strategic issue on which the government and the Admiralty were assailed was the battleship building rate. The Liberals had not accepted the policy of the departing Conservatives in 1905 that up to four capital ships should be laid down each year. In 1906, to pay for social reform and the required repayments on old naval works loans, they reduced the number in 1906–7 to three. The 1907–8 programme was likewise reduced and the government even held over the third ship in the expectation that success at the Hague would allow the pace of construction to be slowed. That hope soon foundered and shortly afterwards it became public that Germany intended to increase her annual order to four capital ships for the next four years. Rumours of increasing ordnance and shipbuilding capacity arose, while Germany resolutely refused to allow inspection of her yards. There were soon fears that the British lead in dreadnought construction would be lost; the 1908–9 programme included only two ships.

As public indignation mounted, the Admiralty pressed for increases in the 1909–10 programme. By this point it wanted not four but eight units. The government was prepared only to insert the extra vessels as conditional ones, to be ordered if thought necessary for national security. The Liberal hand, despite the

A destroyer (HMS *Ardent*). From introduction of the first of the type in 1893, Britain led the way in destroyer design, producing a succession of increasingly capable and weatherly vessels that performed well in the test of war.

Above: Manning the navy. For much of the nineteenth century the Admiralty wrestled with the problems of manning the navy in peace and in war. This picture, by George Barnard O'Neill (1828–1917) shows a naval recruiting party in about 1859 on London Bridge after a regatta trying to persuade a young oarsman to accept the 'Bounty' for enlistment.

Left: A popular board game of the 1880s. In fact, to rise from sailor boy to admiral was impossible in the Victorian navy. Promotion from the lower deck to the quarterdeck was blocked for 93 years until Churchill's 'Mate' Scheme, whereby petty officers could gain commissions, was introduced in 1913.

Above: HMS *Queen Elizabeth* shelling Turkish fortifications in the Dardanelles, 18 March 1915.

Right: The light cruiser *Southampton* at the Battle of Jutland, 31 May 1916, a painting by Oscar Parkes. Flagship of the Second Light Cruiser Squadron, her record of scouting and reporting was unmatched by any other ship in the Grand Fleet and was in outstanding contrast to the performance of most British units during the battle.

protests of the radical wing, was forced by the news that Austria–Hungary—and, in reply, Italy—would be embarking upon the construction of dreadnoughts. In the spring of 1910, the four conditional vessels were ordered and to them were added offers already made by Australia and New Zealand to cover the cost of a capital ship each.

The Dominions' generosity resulted in an Imperial Conference on the defence of the empire in July 1909. At this meeting the Admiralty finally gave way to their aspirations to maintain local services and proposed the creation of a Pacific Fleet which would be built around fleet units led by battle cruisers. Canada, however, was to play little part in this scheme. Although a Royal Canadian Navy came into being in 1910, domestic concerns prevented it operating more than a pair of training cruisers before the outbreak of war. Australia, on the other hand, formed the Royal Australian Navy in 1911 and immediately set about creating a complete fleet unit. Its flagship, the battle cruiser *Australia*, commissioned in 1913. New Zealand was still too small to sustain a separate naval service but paid for a battle cruiser, the *New Zealand*, which was intended for the fleet unit in China. The agreements made at the 1909 Imperial Conference depended upon the local navies being integrated with the Royal Navy in organization, administration, and operation. The concept, despite Admiralty misgivings, was to stand the test of war and provide the basis for all future nurturing of the emergent navies of the empire.

The massive building programme calmed public sentiment but 1909 also saw the formation of a committee to deal with Lord Charles Beresford's volley of accusations as to the inadequacy of Admiralty policy. Although few proved to have substance, the ill-feeling apparent within the Service undermined the First Sea Lord to the point where he knew that early retirement was necessary if he was to ensure a succession in his image.

Fisher's departure with a peerage in January 1910 was no consolation to his enemies because he achieved the appointment of the only senior flag officer of sufficient eminence to be a credible successor who was not a member or sympathizer of the 'Syndicate of Discontent' and who was acceptable to both government and navy. Admiral of the Fleet Sir Arthur Wilson came unwillingly to the post after three years in semi-retirement. His strategic views, particularly his distrust of the flotilla concept as a substitute for the battle fleet and his espousal of the concept of close blockade, were not in accord with those of Fisher, but he could be relied upon to support most of the reforms of the previous five years. Most importantly, the two years available before his compulsory retirement would allow a suitable successor to emerge.

APPROACH TO THE FIRST WORLD WAR, 1910–1914

Wilson was not a success. Reticent in the extreme, he refused to consult with the other Sea Lords or senior officers afloat and he proved unable to delegate authority. Having seen no sea service since 1907, Wilson was out of date and this was reflected in his insistence that submarines be confined to harbour and coastal defence until persuaded otherwise by the Commander-in-Chief, Home Fleet, Sir Francis Bridgeman. He failed, too, to push ahead with the development of new systems for long-range gunnery because he could only conceive of naval actions taking place at medium range—around 5,000 yards—at which sophisticated fire control was not really required. Wilson rejected attempts to improve the quality of armour-piercing shell, which had been proved inadequate at long range, believing that high explosive shells would be more effective in the early stages of battle. That he could adapt to technological change was demonstrated by his interest in wireless, but his principal achievement in this area was the exercise from his office in Whitehall of tactical control of the Second Battle Squadron—off Portugal. This was perhaps the first instance of an Admiralty tendency to misuse the capabilities of improved communications by over-control of commanders on scene that was to manifest itself at intervals over the next forty years.

Admiral of the Fleet Sir Arthur Knyvet Wilson VC, Fisher's successor as First Sea Lord. Reticent in the extreme, Wilson was ill fitted to deal with either the political problems of his office or the technological and strategic changes which the Royal Navy was facing.

Wilson's regime did not inspire confidence and the Admiralty's political position was not strong at the time Germany's intervention in Morocco sparked off the Agadir Crisis in July 1911. Naval readiness at the height of the crisis was questionable, with the Home Fleet scattered around its base ports and the Atlantic Fleet in northern Scotland. But the key problem which the government focused upon was the divergence in policy between the War Office and the Admiralty. While the army was firmly committed to the dispatch of an expeditionary force to France, Wilson declared his intention to capture Heligoland as the key to a close blockade of German ports. This was serious enough. Even more alarming was the Admiralty's apparent inability to staff the army's sea transport requirements. Herbert Asquith, the prime minister, eventually decided that radical improvements were necessary and forced Reginald McKenna to exchange his office of First Lord with Winston Churchill, then Home Secretary.

Churchill's mission was to set up a naval general staff within the Admiralty. He arrived with a reputation as an opportunist radical devoted to social reform at the expense of military expenditure and ruthless in achieving his ends. He soon proved so by dismissing Wilson, with whom he found it impossible to work, and installing Sir Francis Bridgeman. Churchill also had the benefit of Fisher's constant advice and the old admiral's drive for technological innovation was reflected in many decisions made in Churchill's tenure in office. Unfortunately, the naval general staff did not enjoy a healthy birth. The old Naval Intelligence Division was given a few extra personnel, redesignated the War Staff and placed under the direction of a flag officer. The division of responsibility between the War Staff and the First Sea Lord which this implied was to prove disastrous in 1914 and inefficient at all times until reform in 1917–18. The War Staff, if it reported to anyone, worked for the First Lord. Although Bridgeman had refused to constitute himself as Chief of Staff to anyone else, the failure to combine his office with the new meant that he was not a full player in policy-making. The one clear gain was the formation of the Staff College in Portsmouth to train junior officers in staff duties. While this attracted some highly talented officers and did good work in its brief existence before suspension at the outbreak of war, its effects were necessarily incremental and the full results would not be obvious until 1939.

Bridgeman himself did not last at the Admiralty. He felt that Churchill interfered in the running of the navy at the expense of the corporate authority of the Board. After a series of clashes, Churchill seized on the admiral's admission that he had suffered a period of ill health to demand his resignation. Despite his belief that such treatment amounted to blackmail, Bridgeman acceded to the First Lord's order and retired quietly, relieved by Prince Louis of Battenberg. Churchill, however, could not avoid considerable controversy over Bridgeman's departure and was probably saved only by the admiral's refusal to press the issue in public.

The technical achievements in the last years before the war were more encouraging. The pace of innovation was resumed, helped as much by the improvement in government finances resulting from the succession of radical budgets brought down by Lloyd George, as Fisher and Churchill's influence. The calibre of battleship armament had already increased from 12-inch to 13.5-inch; it now went to 15-inch with the order in 1912 for the first 'Queen Elizabeth'-class battleships as a reply to Japanese and American adoption of the 14-inch gun. These ships were also fully oil fired, which allowed both increased speed and range while providing spare displacement for the improvement of other military qualities in what was to be an outstandingly successful class of ship. Supply of the necessary fuel would be guaranteed by the British government's £2 million purchase of a controlling share in the Anglo-Persian Oil Company. Combining such strength in both offence and defence, the 'Queen Elizabeths' were the first true fast battleships and as such marked the expensive consequence of the failure of the battle cruiser concept through the lack of fire control. It was only financial restrictions which forced a return to coal for the successor class, the 'Royal Sovereigns', a step to which Fisher bitterly objected.

Construction of large light cruisers had been resumed in 1908 and the designs were improved to produce all 6-inch gun versions of the 'Town' class which proved ideal for both fleet and trade protection work. Destroyers developed immediately before the First World War showed a similar balance of fighting qualities. Churchill's administration also made up some of the time lost since Fisher's departure in submarine development, but continued with muddled attempts to produce types capable of operations with the fleet.

Churchill galvanized the Admiralty's approach to aviation, which had not been helped by the abortive project of 1909 to build a rigid airship, the *Mayfly*. Although recruiting and experiments in flying aircraft from ships were already in train, the First Lord was enthusiastic in forwarding expansion of the promising aviation arm. Within a few months of his arrival plans were in hand for naval air stations around the British coast and the first flight took place from the battleship *Africa* in January 1912. The technical sub-committee of the CID reported favourably on the need for military and naval aviation the following month and, as a direct result, a Royal Flying Corps was established with both naval and military wings. In March 1913, the first parent ship for aircraft, the converted cruiser *Hermes* was commissioned for the annual manœuvres. In reality, the naval wing soon broke away from the RFC and enjoyed a *de facto* existence as the Royal Naval Air Service, a separation confirmed in July 1914. Churchill was, however, an enthusiast for lighter than air craft, which the majority of his senior advisers were not. Although non-rigid airships were to prove particularly useful against the U-boat, Britain did not achieve the progress made by the Germans with their Zeppelins and the airship programme, inevitably, consumed resources which would otherwise have gone to heavier than air craft.

The continuing increases in the estimates which resulted from all this activity were justified because the pressure from Germany continued. A reduction in the rate of construction had been planned from 1912 under the existing German Navy Law, but it became clear that the Germans intended to lay down at least three extra ships in the next six years and put a third squadron of battleships into full— and thus high readiness—commission. To this had to be added the Italian and Austro-Hungarian dreadnought programmes which would put up to ten ships in commission by the end of 1915. Significantly, 1912 marked the end of the two power standard and its replacement by one which required Britain to match the next most powerful European navy, plus a 60 per cent margin. Although there was little actual difference between the resultant figures, it marked a clear indication of the Royal Navy's focus on the German threat.

The Admiralty was forced to propose a further concentration in home waters, bringing the Atlantic Fleet north and moving the battle squadron in the Mediter-

The exhausting routine of coaling ship aboard the battle cruiser HMAS *Australia* in 1916. Coaling ship was an 'all hands' evolution conducted immediately on returning to harbour. Introduction of oil fuel only ships did away with the need for this manpower intensive and dirty work—it also did away with an evolution that did much to bring officers and men together through a shared physical task.

ranean from Malta to Gibraltar. The latter would be replaced at Malta by a squadron of battle cruisers. The Home and Atlantic Fleets were reorganized into three groupings, the First, Second, and Third (collectively described as the Home) Fleets. The First, the strongest, was fully operational, the Second nucleus manned, and the Third retained in care and maintenance.

The overseas aspects of the scheme were not well received by the Foreign Office or within the country at large. British prestige was held to be at stake; but underlying this lay an uneasy feeling, amply justified in the coming war, that the empire's communications in the great oceans were vulnerable and threats to them difficult to locate. The Admiralty's failure to realize this was one of the complex factors leading to the rejection of convoy as a protection measure.

The Admiralty's response to pressure to revise its plans was itself complex. In the Far East, it relied on the recently renewed Japanese Treaty, effective until 1921, to justify cancellation of the plan for the dispatch of two battle cruisers to the region. Churchill importuned New Zealand to permit its gift battle cruiser to remain with the Home Fleets. The Australians proved impervious to similar suggestions. In the event, the Admiralty's revision of the 1909 plans left barely adequate forces to face the highly efficient cruisers of the German East Asiatic Squadron.

For the Mediterranean, an offer from the Canadians to provide three gift battle-ships seemed to give the opportunity to deploy a dreadnought squadron there from 1915. This would provide a long-term capacity to match Italy or Austria–Hungary without the aid of France, or both if Britain was in alliance with the latter. The Canadian offer, however, foundered in internal controversy and the interim solution of a weak Mediterranean battle cruiser squadron became perforce a continuing policy.

This was yet another factor in the increasing pressure which the Admiralty was experiencing in matching naval requirements to finance. Despite the new taxation regime, by 1914 the government was unwilling to increase the naval vote any further. The 1913–14 naval estimates represented over a quarter of the total budget. The last likely new contribution from the empire had come in the form of a gift 'Queen Elizabeth' from the Federated Malay States. Britain could not expect any more such additions to the battle line.

The pressure of technology on tactics and strategy was manifesting itself in progressive revisions to the war plans and in ever more radical concepts. Yet there was little consensus at any level of the Royal Navy as to the way ahead. Abandoning close blockade, the Admiralty initially favoured an observational blockade in the North Sea, but the 1912 manœuvres, together with steady pressure from the Commander-in-Chief, Home Fleet, Sir George Callaghan, forced the War Staff to think again. Fixed observation lines were simply too vulnerable and a distant blockade the only practicable alternative. This espousal of a doctrine which admitted the new vulnerabilities of surface forces and heavy ships in particular, inevitably led to reflections within the Admiralty and by the First Lord on the means by which a financially restrained Great Britain could deny the seas to its opponents. If battleships were too expensive, then the Royal Navy would have to rely upon the cheaper and more numerous units of the flotilla to make ineffective an enemy's battle fleet. This thinking was only embryonic in 1914, but it fore-shadowed great changes in the future composition of the navy.

A 'Queen Elizabeth'-class battleship shown from the air, c.1918. This photograph shows the balanced design, the standard five turrets of previous classes having been reduced to four to provide capacity for more powerful machinery. Combining a heavy and accurate gun with good protection and relatively high speed, the 'Queen Elizabeths' were the most effective battleships of the war.

Ironically, 1913 and the early months of 1914 saw a gradual easing in tensions with Germany, despite the forthcoming completion of improvements to the Kiel Canal which would allow the rapid passage of the largest dreadnoughts between the Baltic and the North Sea. To prove the machinery of mobilization, the Admiralty planned a test activation of all reserve ships in July 1914, with the intention of combining a fleet review with short exercises as an alternative to the annual manœuvres. This was largely an economy measure and was seen as such by Germany, whose High Seas Fleet was scheduled to conduct its summer cruise to Norway. Before that event the reopening of the Kiel Canal took place in the presence of the German Kaiser and a British battle squadron. Anglo-German naval relations seemed on a more amiable footing than they had been for over a decade and it was only the assumption of court mourning for the murdered Archduke Franz-Ferdinand of Austria–Hungary which cut short festivities. The British fleet review went on as intended on 20 July, a much more sombre event than that at Kiel. Perhaps the clearest shadow of the future was that cast by the sixteen seaplanes which flew in formation over the Royal Yacht *Victoria and Albert*.

Neither navy was quick to see hostilities coming. Britain was preoccupied by the prospect of civil war in Ireland over home rule and the possibility that the armed forces might have to be used against the Ulster Unionists. The refusal of army officers at the Curragh to serve in such operations caused bitter political controversy and there were deep divisions within the Royal Navy which would probably have manifested themselves in similar action if put to the test.

While some active units deployed to Ireland, British reserve ships returned to their home ports to pay off on 23 July. The High Seas Fleet was still in Norwegian waters. Only when the implications of the brutal Austro-Hungarian ultimatum to Serbia began to sink in did the respective admiralties move. On 26 July, Prince Louis of Battenberg issued orders which prevented the operational ships of the First Fleet from dispersing. The following day reactivation of the reserve units began, together with patrols around the naval ports and the issue of a warning telegram to overseas stations.

On 29 July, the First Fleet, newly renamed the Grand Fleet under Sir George Callaghan, left Portland in secrecy for its war base at Scapa Flow in the Orkney Islands. The Germans, in the belief that Britain would not enter the developing conflict, were slower to mobilize and at first concentrated in the Baltic against Russia. Not until 31 July, by which time most British preparations were complete, did the High Seas Fleet transfer to the North Sea and only the next day was a total naval mobilization allowed. This prevented the dispatch of converted merchant ships as commerce raiders and meant that Germany would be forced to rely on the eight cruisers operating outside the North Sea and Mediterranean, supplemented by a handful of local conversions.

The efficient British mobilization and the smooth transfer of the British Expeditionary Force to France under heavy escort marked both the apogee of pre-war naval planning and its limit. The Admiralty's recognition that there were hard decisions to be made was indicated in the sudden supersession of Callaghan by Sir

John Jellicoe, lately Second Sea Lord and one of the men most closely associated with Fisher in the technical and operational reforms of the previous decade. Long considered the ideal Commander-in-Chief of the wartime fleet, Jellicoe had deep experience as an administrator and operational commander and shrewd ideas as to British weaknesses. He quickly determined that he would not risk fighting the High Seas Fleet on its own terms in the southern North Sea, where the Germans might succeed in leading him on to submarine and mine traps, but rely on patrols to wear down the German strength while hoping that the latter would venture so far from home that they could be cut off. The Grand Fleet would remain at Scapa Flow or in northern waters, covering the distant blockade of cruisers between the Orkneys, Shetlands, and Faeroes.

Jellicoe's estimate was a fair vision of the German policy. The latter would not risk their heavy ships against the Grand Fleet. Although the Germans soon realized the error of their expectation of a close blockade, they too pinned their hopes on flotilla craft. Only when the Grand Fleet was weakened to the point of inferiority would the High Seas Fleet venture forth.

Admiral Sir David Beatty (*right*) and King George V (*left*) on board HMS *Queen Elizabeth* at Scapa Flow in June 1917, twelve months after the Battle of Jutland and after Beatty's appointment as Commander-in-Chief, Grand Fleet.

The Grand Fleet's situation was not an easy one. The strain of war service showed up deficiencies in British machinery and many of Jellicoe's best ships were soon in need of repairs. The old cruisers of the Northern Patrol suffered heavily as winter drew on and they required replacement by converted merchant ships. In fact, the Grand Fleet's margin of superiority in 1914 was not great in any class of ship. Its problems were magnified by the lack of a dockyard on the east coast and the absence of anti-submarine defences at the anchorages with access to the

North Sea. The strain on the naval budget caused by the flow of new construction had not permitted sufficient emphasis on new naval works. Rosyth, the intended dockyard, would not be ready for years; Scapa Flow had no fixed facilities at all.

Jellicoe's fears were not shared by all his juniors; the limitations on surface ship operations in the North Sea had not yet sunk in. A false sense of optimism was created by the little victory in the Heligoland Bight on 28 August 1914. An ill-considered destroyer-sweep into the Bight by light forces from Harwich under Admiralty orders was saved from disaster by the intervention of battle cruisers under Vice-Admiral Sir David Beatty. These had been added to the operation on Jellicoe's initiative and it was well that he did so, because the local German light cruisers and torpedo boats had proved more than a match for the ships of the Harwich Force under Commodore Reginald Tyrwhitt. Unsupported by their own heavy ships, which remained at anchor, caught by the tide within the Jade and Weser estuaries, the Germans lost three light cruisers and a torpedo boat.

The action brought to the public eye the flamboyant Beatty, hero of the 1898 Nile campaign and the Boxer Rebellion and the youngest flag officer since Nelson. Married to a divorced American heiress and himself very much the socialite, Beatty was nevertheless a highly professional officer who was to become one of the dominant personalities of the naval war. Although he had his faults as a tactical commander, Beatty was a natural leader. He possessed shrewd judgement and an ability to grasp key issues which was not confused by the tendency to over-centralize so common in his contemporaries.

Beatty's grasp of fundamentals was not shared by many. Much of the wishful thinking of 1914 resulted from the general strategic and tactical immaturity enforced by over-centralization within the Admiralty. Matters were not improved by the obvious defects within the War Staff machinery or by the lack of hard knowledge about the operational implications of the new technology now at sea. Most seagoing officers had also to overcome the effects of a fixation upon the American historian Alfred Thayer Mahan's ideas of the decisive battle and the overweening need to destroy the enemy's fleet as the primary aim of naval power. More complex ideas of sea control and trade protection, espoused in part by Mahan himself and developed by other thinkers such as Sir Julian Corbett, were poorly understood both within the Royal Navy and in the country at large. The protracted realities of the blockade would take time to manifest.

The optimism did not last. Once the Germans had nerved themselves for U-boat operations away from local waters, U-21 sank the old cruiser *Hermes* off St Abb's Head on 5 September. The British submarine E-9, under Lieutenant-Commander Max Horton, torpedoed and sank the old German cruiser *Hela* one week later. These warnings failed to save three armoured cruisers, *Aboukir*, *Hogue*, and *Cressy*, which were sunk in quick succession by U-9 off the Netherlands coast on 22 September.

Jellicoe was forced to withdraw the Grand Fleet to Lough Swilly on the coast of Ireland while Scapa Flow was given proper defences. Shortly after this, the German auxiliary minelayer *Berlin* broke through the blockade and laid a

Sir Julian Corbett, naval historian and strategist, who evolved sophisticated ideas about the strategic roles of navies which were not properly understood within the United Kingdom. Corbett's views about the use of navies to influence the war on land focused on combined operations and the protection of trade at the expense of wishful thinking for a decisive action between opposing battle fleets.

minefield across the shipping routes in the Irish Sea. This caught the super dread-nought *Audacious*, one of the Grand Fleet's most valuable units, which sank within hours. Despite the spectacle being witnessed by many Americans aboard the liner *Olympic*, the Admiralty was so concerned by the implications of the sinking that the loss was not admitted. What was intended as a temporary measure of censorship remained in force for the duration. Neutral—and Allied—knowledge of the lie undermined Britain's credibility for propaganda claims in the maritime war.

The difficulties of the British position were underlined by a successful raid on Yarmouth on 3 November 1914 by the battle cruisers of the First Scouting Group under Rear-Admiral Franz Hipper. Although the operation had little military significance, it forced the British to redouble their efforts on the North Sea anchorages so that the Grand Fleet could be brought back. Meanwhile, the Admiralty began development of a signals intelligence network that would prove increasingly valuable as the war progressed. Based on Room 40 in the Admiralty Old Building (Room 40 OB), the organization under the former Director of Naval Education, Sir Alfred Ewing, combined with direction finding stations established by the Naval Intelligence Division under Captain Reginald Hall to develop a sophisticated interception and decryption service. Its beginning benefited immensely from the successive capture in 1914 of the three principal German ciphers—from a merchant ship in Australia, by the Russians from the wreck of the light cruiser *Magdeburg* in the Baltic, and through salvage from a torpedo boat sunk in the English Channel.

The first concrete product from Room 40 OB was warning of an intended raid on Scarborough and other east coast ports in December 1914. Only Beatty's battle cruisers and a battle squadron under Vice-Admiral Sir George Warrender were dispatched to deal with the German Scouting Groups because the Admiralty did not realize that the German Commander-in-Chief, von Ingenohl, had brought the High Seas Fleet out as far as the Dogger Bank. Poor visibility and excessive caution on von Ingenohl's part saved the British from an embarrassing retreat or the defeat in detail which would have achieved the Germans' immediate aim of reducing the opposition to parity. The British did little better and the poor co-ordination, defective communications, and failures of initiative demonstrated in the affair were indicative of the problems that lay ahead.

The Royal Navy replied on 25 December with a seven-strong seaplane raid on German hangars near Cuxhaven. Launched from three converted carriers escorted by the Harwich Force, this sortie was only a limited success but it marked the first ever offensive use of shipborne aircraft.

Hipper, worried by the accurate British intelligence, which he thought originated in reports from fishing vessels, next determined to attack trawlers on the Dogger Bank and sortied on 23 January. Out to meet him came Beatty with five battle cruisers. The interception at dawn the next day proved too exact and Hipper was able to turn for home with Beatty hard on his heels. In the chase which followed, the battle cruiser *Seydlitz* was damaged and the large armoured cruiser *Bluecher* crippled. At his apparent moment of triumph, Beatty's flagship, the *Lion*,

The battle cruiser *Lion*, first of the 13.5-inch gunned 'splendid cats' and Beatty's flagship at Heligoland Bight, Dogger Bank, and Jutland. Improved internal anti-flash precautions, instituted after the Dogger Bank action, saved her at Jutland but were not in force aboard the three battle cruisers, including her sister ship *Queen Mary*, which blew up and sank.

was forced out of line through accumulated damage to machinery. Confusion over signals and the Admiral's intentions meant that the remaining British units concentrated on sinking the *Bluecher*, while the First Scouting Group ran clear. *Lion* was brought in only with great difficulty. The battle was a useful propaganda victory, but all concerned were acutely aware of the opportunity missed. There were material lessons, too, many of which the British did not absorb. While the Germans took new anti-flash precautions after the near loss of the *Seydlitz* in a turret fire, little attention was paid to similar measures after damage to the *Lion*, whose inadequate protection should have been manifest to thoughtful observers.

Beatty's forces were reorganized into the Battle Cruiser Fleet, based on Rosyth as the advanced scouts for the Grand Fleet battle squadrons from Cromarty and Scapa Flow. The Germans, however, would not offer battle and there were no more capital ship encounters in 1915. Von Ingenohl was replaced by the even more cautious Hugo von Pohl and Kaiser Wilhelm II was loath to risk his ships, preferring to hold them as a bargaining counter for peace negotiations.

The Mediterranean proved a more active theatre for the first twelve months of the war. The confusion which surrounded the attempts to destroy the battle cruiser *Goeben* and her consort, the light cruiser *Breslau*, set the pattern for the coming months. The British Commander-in-Chief, Sir Berkeley Milne—no Nelson—was hamstrung by Admiralty orders which appeared to emphasize protection of French troop transports and an excessive respect for Italian neutrality. As a result, the Germans escaped from his battle cruisers. An armoured cruiser squadron, commanded by Rear-Admiral E. C. T. Troubridge, refused action because Troubridge believed that *Goeben* could outrange and outgun the British units and that the Admiralty had instructed him not to engage a superior force. Troubridge's judgement had some tactical merit, but it was a sorry start to the

DISTANT WATERS, 1914–1915

The cruiser HMAS *Sydney*, shown in the North Sea on the Dogger Bank in 1918. *Sydney*, which destroyed the *Emden* in 1914, is shown here with a shelter for a Sopwith Camel. She was the first light cruiser fitted with a rotating flying-off platform.

war for the Royal Navy. The extent of the opportunity missed was rammed home by the German squadron's passage of the Dardanelles. The escape of the *Goeben* and *Breslau* not only gave Germany considerable prestige in Turkey, but the Germans hit on the masterly stroke of presenting the ships as replacements for the two battleships which Britain had just confiscated as they completed in British yards. While the newly renamed *Agincourt* and *Erin* were useful reinforcements for the Grand Fleet, the bitterness with which Turkey took the news was a factor in bringing the Ottoman empire into the war against Britain a few months later.

Confusion in the Mediterranean was matched on other overseas stations, with Churchill and the War Staff too often attempting to override local commanders-in-chief. The German East Asiatic Squadron under Vice-Admiral Graf von Spee escaped from Chinese waters and, after detaching the *Emden*, made its way towards South America. Despite assessments in both Singapore and Australia that this was the direction von Spee had taken, no pursuit was allowed across the Pacific. The Admiralty was content to let a poorly equipped squadron under Rear-Admiral Sir Christopher Cradock meet von Spee off Coronel on the coast of Chile on 1 November 1914. Both Cradock's armoured cruisers were destroyed without loss to the Germans. Meanwhile the *Emden* was pursuing an active career as a commerce destroyer in South-east Asia and the Indian Ocean and neither she nor the *Karlsruhe* in the West Indies had been located by the time the news of Coronel came through to Whitehall.

Fortunately for the Admiralty, Lord Fisher had been reinstalled as First Sea Lord only two days before. Battenberg proved unable to bear the strain of war, compounded by ill health and public disaffection over his German birth. His resignation was sensibly engineered by the government and Fisher brought in at Churchill's request. The combination of the two personalities was to prove unstable but it produced immediate results in the form of a revived building programme that focused on large numbers of destroyers, escorts, and submarines which were to prove vital later in the war. Fisher was quick to realize Cradock's weakness and tried to reinforce his squadron with an additional, much more powerful armoured cruiser. The defeat off Coronel determined him to strip the Grand Fleet of battle cruisers and dispatch the *Princess Royal* to the West Indies and *Invincible* and *Inflexible* to the South Atlantic. In command of the latter the Admiralty placed Vice-Admiral Sir Doveton Sturdee, formerly Chief of the War Staff and, in the new First Sea Lord's opinion, responsible for much of the trouble. Jellicoe and Beatty protested bitterly about the weakening of the battle cruiser squadrons but Fisher paid little attention.

Sturdee's ships were spectacularly successful at the Battle of the Falkland Islands on 8 December 1914. Sturdee used the superior speed and armament of the battle cruisers to pick off von Spee's ships at long range. Only the light cruiser *Dresden* escaped the British and she was to be caught and sunk in March 1915. *Emden* overreached herself by attacking the cable station at Cocos Island in the Indian Ocean on 9 November at the same time as the first Australian–New Zealand troop convoy was passing on its way to Egypt. She was destroyed by HMAS *Sydney*. As *Karlsruhe* had suffered an internal explosion five days before (it was some months before the news reached the British) and the remaining German attempts at commerce raiding had been easily dealt with, the sole threat was the *Konigsberg*, now bottled up in the Rufiji River in East Africa. She would be destroyed in July 1915.

The cramped conditions in a First World War submarine. This photograph shows the coxswains of HM Submarine E2 at diving stations in the Mediterranean in 1918. The relatively pristine uniforms of most indicate that this is a posed photograph taken in harbour.

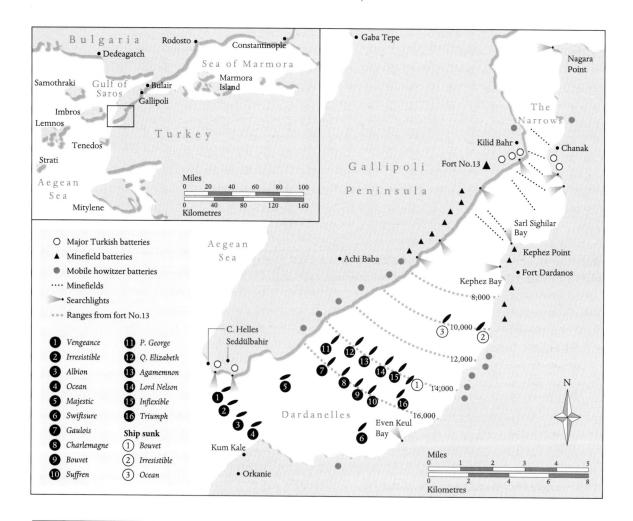

The map includes the following labels and legend:

Legend:
- ○ Major Turkish batteries
- ▲ Minefield batteries
- ● Mobile howitzer batteries
- ···· Minefields
- ▸ Searchlights
- ··· Ranges from fort No.13

#	Ship	#	Ship
1	Vengeance	11	P. George
2	Irresistible	12	Q. Elizabeth
3	Albion	13	Agamemnon
4	Ocean	14	Lord Nelson
5	Majestic	15	Inflexible
6	Swiftsure	16	Triumph
7	Gaulois		**Ship sunk**
8	Charlemagne	1	Bouvet
9	Bouvet	2	Irresistible
10	Suffren	3	Ocean

Black Sea and Dardanelles, February–March 1915.

German efforts at commerce destruction by surface units enjoyed some success later in the war, particularly through the efforts of the disguised raiders *Moewe* and *Wolf*, and they proved a drain on British patrol resources world-wide. Nevertheless, the Falkland Islands battle marked the end of any serious possibility that the British might lose control of the trade routes in distant waters. That strategic situation was only underlined by the entry of Japan into the war and the progressive reduction of Tsingtau, the main German base in China, and other possessions.

Perhaps the most useful move outside home waters in 1914 was the decision in October to dispatch submarines to operate from Russian bases in the Baltic. Remaining until after the 1917 revolutions, these craft were to prove a thorn in the Germans' side and went a little way towards redressing the toll taken by the U-boats.

Unfortunately, Churchill and Fisher soon began to work at cross purposes within the Admiralty. While Fisher devised emergency building programmes,

Churchill was inspired by Turkey's entry into the war to propose that the Dardanelles be attacked. Control of the passage to the Black Sea would, he reasoned, force Turkey to surrender while opening an all-year-round access to Russian ports. Because the War Office would not provide the troops, the First Lord turned to the idea of an attack by ships alone. This concept received sufficient support from the local Commander-in-Chief to convince the Cabinet. In the atmosphere of frustration over the stalemate on the Western Front, doubts as to the efficacy of ships against forts and the task of the fleet once it had forced the Dardanelles were suppressed.

The bombardment began on 19 February 1915 against well-prepared defences and went on for weeks. It proved an expensive failure, with the loss of the battleships *Irresistible* and *Ocean* and the French *Bouvet* to mines and damage to the battle cruiser *Inflexible*. By late March it was apparent that the Turkish defences were too much for the ships alone and that troops would be required. The navy played an integral part in the abortive land campaign which followed, developing considerable local expertise in amphibious operations after some expensive early lessons. Ironically, the most successful achievement of the Dardanelles campaign was the 1916 evacuation from the beaches of the entire Allied force without alerting the enemy. Submarines on both sides had the most active role. German U-boats dispatched to the theatre in 1915 disposed of the battleships *Triumph* and *Majestic*, while British submarines penetrated the Dardanelles and conducted attacks off Constantinople itself.

Fisher's doubts over the Dardanelles eventually overflowed in May 1915 in a dispute with Churchill over the need for naval reinforcements. The First Sea Lord's resignation and his refusal to reconsider his position brought little credit on himself, but also served to bring down Churchill. The combination had brought on the disaster which many had predicted but the Admiralty did not do well in the exchange. Arthur Balfour and Admiral Sir Henry Jackson's cautious and studied methods were the antithesis of their predecessors and their regime lacked the creative energy so necessary for success in war. Ironically, Fisher's light craft ordered in 1914–15 rather than efforts initiated in the following eighteen months were to prove critical in the U-boat crisis of 1917.

The submarine E11 returning to Mudros on 7 June 1915, having completed the second submarine patrol in the Sea of Marmora. Her captain, Lieutenant-Commander Martin Nasmith, was awarded the VC and promoted Commander.

CHANGES IN THE
PATTERN OF THE NAVAL
WAR, 1915–1916

It was inevitable that the Germans should turn to the submarine as the alternative weapon for the commerce war, if only in response to the increasingly efficient—and ruthless—British efforts to cut off supplies to Germany and extend the definition of contraband. The Germans were, however, slow to realize that the conflict was not only a matter of supply lines but one of achieving or losing international support. British insistence on the right of search in a designated port was the source of much bitterness, particularly in the United States. But the German declaration on 18 February 1915 that waters around the British Isles were a war zone, with a 'sink on sight' policy for merchant ships, returned the moral advantage to the British. Increasing American fury over losses of ships and lives mounted to the point where the Germans were forced to return to a policy of warning and search. A brief revival in February 1916 was abandoned only two months later after fresh protests from the United States.

These interruptions to the German campaign, although they did not mean an end to merchant ship losses, took the pressure off an Admiralty which had devised no means for protecting shipping. Experiments in locating and destroying submarines, some of them begun a decade or more before, had been abandoned with the outbreak of war. In the middle of 1915, work was resumed on the development of sensors and weapons which would bear fruit in the coming years, but the Royal Navy, preoccupied with the need for light craft to operate with the Grand Fleet in home waters and the lack of escorts for commerce protection, would not yet consider the strategy of convoy. The merchant ships seemed too many and the difficulties of organization too great.

1916 saw a flurry of activity in the North Sea. The new and energetic Commander-in-Chief of the High Seas Fleet, Vice-Admiral Reinhard Scheer, adopted an all arms approach in the belief that the integration of U-boats and Zeppelins would give the Germans sufficient tactical advantage over the Grand Fleet. Given Beatty's geographic location and his known intent to hold contact at all costs, the strategy had possibilities. The key would be entrapment of the battle cruisers before Jellicoe's battle fleet could come up in support. A bombardment of Lowestoft and Yarmouth in April gave a clear demonstration of the new approach. Thoroughly alerted, the British mastery of signals intelligence, still unrecognized by their opponents, virtually guaranteed an encounter if the Germans emerged again.

THE BATTLE OF
JUTLAND

This came on 31 May when units of the Battle Cruiser Fleet encountered the Scouting Groups out in the Skagerrak on a 'coat trailing' operation. Unknown to Scheer, his plan had already misfired to the extent that none of the submarine ambushes set outside the British bases had achieved success. He also had no idea that Jellicoe was already at sea. Confusion did not only exist on the German side. Misinterpretation within the Admiralty of call-sign identities resulted in the assessment that Scheer's flagship—and thus the High Seas Fleet—had not sailed. Both sides thought that they were dealing only with scouting forces or, at the most, detachments at squadron strength.

The first rounds went to the Germans. Hipper turned away to lead Beatty's six battle cruisers and four battleships towards the main body of the High Seas Fleet. In the ensuing gunnery duel, the British lost the *Indefatigable* and *Queen Mary* to catastrophic magazine explosions resulting from inadequate armour and poor anti-flash precautions. *Lion* was nearly lost from the same causes. Poor co-ordination between the battle cruisers and the attached fast 15-inch gun ships of the Fifth Battle Squadron also resulted in the heavy armament of the latter coming late into play. By the time that Hipper's ships began to suffer heavily, the roles were reversed. The appearance of the High Seas Fleet forced Beatty to retreat to the north towards Jellicoe.

Little over an hour afterwards, the first units of the Grand Fleet joined the action, forcing Jellicoe to deploy his battleships from cruising formation into line of battle. Poor reporting procedures and inevitable navigational inaccuracies combined with progressively worsening visibility to give him little clear idea of the location of the enemy. Yet Jellicoe's decision proved masterly. The deployment of his squadrons on the port wing crossed the T of the German advance and, by allowing him to turn to a south-easterly course, held the potential to cut the High Seas Fleet off from its bases.

Further British disasters in the loss of the armoured cruiser *Defence* and the battle cruiser *Invincible* could not disguise the increasingly tenuous situation of the High Seas Fleet. Scheer's heavy ships found themselves facing the entire British line and suffered heavily, while Hipper's battle cruiser force was progressively reduced to only one fully effective unit, the *Moltke*. Scheer was forced to turn the German battle squadrons together away to the south-west.

The British would not follow. Jellicoe had long expected that the geography of the North Sea would force any battle fleet engagement to take place in the late afternoon in poor visibility with the prospect of darkness. Even in the fine early summer conditions of the day, mist and the smoke of the two coal-burning fleets were reducing visibility to a few miles at best. Jellicoe feared that the German fleet would have left torpedo craft astern to deal with any British units in pursuit. He knew, too, that the inadequate fire control systems of his heavy ships would be ineffective while the ships were manoeuvring, heightening the risks which unexpected encounters held. He would hold his course until his ships were well clear of the German track.

Nevertheless, Scheer gave the Grand Fleet a second opportunity. Even as Jellicoe turned south with the intent of placing himself across the enemy line of retreat, the High Seas Fleet turned north-east. Scheer's belief was probably that this would bring him clear of the British but he was soon disabused of the idea when the leading ships of the column came under heavy fire. A rush at the Grand Fleet by the battle cruisers provided some respite while Scheer ordered a second turn away and this time unleashed his torpedo boats. Jellicoe's doctrine was absolute and the battle squadrons turned away to avoid the torpedoes. In this they were successful, but the manoeuvre resulted in the loss of contact between the battle fleets.

Battle of Jutland, 31 May 1916.

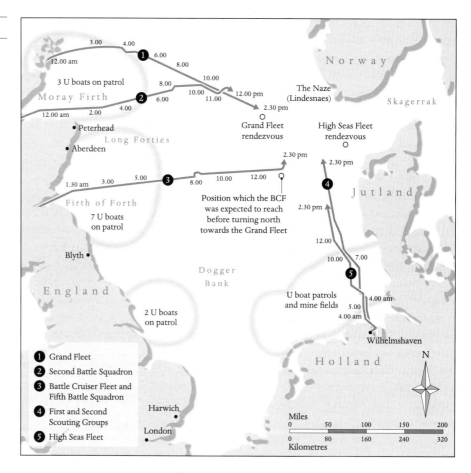

It was never fully regained. Jellicoe eventually turned west-south-west and then south-west to keep the High Seas Fleet from its bases. Despite some early evening indications of the proximity of the German fleet, not all of which were reported to Jellicoe, he put the battle squadrons into night formation, disposing the destroyer flotillas astern to block any attempt to pass behind the Grand Fleet. It was not his policy to bring on action at night. Jellicoe and most of his senior officers believed, with some justice, that the uncertainties of night fighting gave the advantage to the weaker side. Unfortunately, the doctrine of not provoking night action had resulted in there being little training in night fighting at all. Thus, when Scheer judged the moment right and turned across the British rear, the Grand Fleet light forces were ill prepared to deal with him. In particular, they lacked not only the skills of co-ordinated night fighting but systematic methods of enemy contact reporting. Over-emphasis on wireless discipline bred a reluctance to report which had already served the Grand Fleet ill that day. From the series of brief but bloody encounters which went on throughout the night, no information reached Jellicoe to alert him of the German retreat. The destroyers were not the only culprits. Several heavy ships sighted German units and either thought them friendly or did not wish to reveal their own positions by engaging or reporting the contacts. The one other chance which Jellicoe might have had to remedy the situation was lost because the Commander-in-Chief would no longer rely upon Room 40 decryptions. An accurate estimate of Scheer's intent to make for the Horns Reef passage, which would have given a clear indication of his planned track and timing, was ignored.

Between sunset and dawn the British lost a flotilla leader and four destroyers. The Germans lost the pre-dreadnought *Pommern* and three light cruisers, and thought the price cheap. Daylight found the sea around the Grand Fleet empty of Germans and the High Seas Fleet only hours from its anchorages. In sinking 14 ships, including three battle cruisers, for the loss of a battle cruiser, a pre-dreadnought, and nine other vessels, the Germans had some justification in claiming a tactical success, despite the heavy damage suffered by their battle cruisers and many of the battleships. But the strategic intent of Scheer's operations had not been achieved and the passing months would make such achievement steadily less possible. The weaknesses uncovered in British organization and material were immediately addressed. The inflexibility and lack of delegation in the Grand Fleet battle orders, the poor night-fighting techniques, and the defective communications organization were all dealt with. The design defects of the British heavy ships in their fire control, armouring, and subdivision, as well as in armour-piercing shells were not so easily resolved, but improved drills and internal safety procedures and damage control produced a sea change before the end of 1916.

Only once more before 1918 did the Germans emerge. Scheer attempted a sortie to bombard Sunderland on 19 August 1916, with U-boat traps set outside the British ports. Once again, forewarned by Room 40, the Grand Fleet was early to sea but a renewed main fleet encounter did not occur. Scheer took a Zeppelin's report which described light craft of the Harwich Force as heavy units at face value and beat a hasty retreat, while the loss of two light cruisers to U-boats reinforced British fears about operations in the southern North Sea.

THE AFTERMATH OF
JUTLAND IN THE NORTH
SEA, 1916–1918

Scheer's assessment was that the High Seas Fleet could not achieve a decisive victory against the British. The submarine arm would have to take the lead with limited action against merchant shipping. Although the U-boats were tactically constrained by this deference to concepts of prize warfare, the results were immediate. The destruction of merchant ships rose to over 120,000 tons a month and began to present real difficulties to the British. There was little consolation in a monthly sinking rate of U-boats of less than three units from all causes.

Some of those difficulties were self-imposed. The Admiralty effectively monopolized the British shipbuilding industry and the merchant building rate and, equally important, the repair effort for merchant ships had declined disastrously. Furthermore, the allocation of merchant tonnage to wartime priorities was not well co-ordinated. The requirements of the Mediterranean theatre and of the Grand Fleet were significant factors and the fleet itself also consumed a dispro-

HMS IRON DUKE (1916)

KEY

A Admiral's bridge.

B Admiral's chart house—Assistant Flag Lt maintaining flag plot.

C torpedo control—LCdrCT and assistants.

D conning tower—Helmsman and OOW at action stations. In practice ship was conned even in action from exposed bridges.

E boiler rooms—18 coal fired, all steaming. Over 100 Stokers plus supervisory staff.

F engine rooms—port and starboard. Engine room Officers of the Watch, Artificers, Stokers.

G torpedo rooms—two topedo tubes each side under the supervision of Gunner (T).

H transmitting station producing fire control for 13.5″ guns—manned by Royal Marine Band.

I 13.5″ director and spotting top—gunnery contol personnel under supervision of Gunnery Officer (9 men aloft in all).

J after director and emergency conning position—the Commander (Executive Officer) usually had his action station here but in reality had a roving commission.

K 3″ anti aircraft guns: *Iron Duke* originally carried 3 pounder AA weapons, the first ship to be so fitted.

L searchlights— yet to be under fully centralised control. Each searchlight was generally under the supervision of a midshipman.

M 'Q' turret—in most ships this or Y turret was manned by Royal Marines.

N 'A' turret—The four other turrets were manned by seamen (21 per turret) under the supervision of a Lieutenant (the Turret Officer) and a Chief Petty Officer.

O 13.5″ shell rooms—crew of 19 per shellroom.

P 13.5″ magazines (for propellant)—crew of 19 per magazine.

Q tiller flat—at action stations manned by personnel ready to take over local control.

R signal bridge: manned by seamen ready to execute orders from Yeoman of Signals, and reply to communications from other ships.

S navigating bridge, normal position from which the ship was conned in practice—C in C, Flag Captain, Navigator, CINC's staff and OOW were here at action stations with supporting personnel.

T 6″ gun batteries—the ship's defences against enemy torpedo boats. Each gun had a crew of 8 plus magazine and shellroom crews.

U sick Bay—3 medical Officers, 5 rating medical staff.

Admiral Sir John Jellicoe aboard his flagship, HMS *Iron Duke*. Photograph taken *c*.early 1916 as the Commander-in-Chief makes his way from the upper deck to the forward superstructure. Jellicoe exercised command of the Grand Fleet from the *Iron Duke* for over two years.

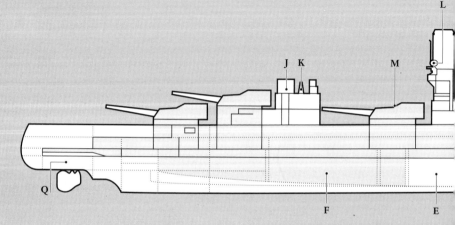

Note: In this diagram it is not possible to detail more than the salient features of manning in action. The *Iron Duke's* war complement by department in 1914 was approximately as follows, but it may have been considerably enhanced by the time of Jutland:

Seamen: 30 officers, 80 chief and petty officers, 310 junior rates
Communications: 2 officers, 22 ratings (the great majority flag signalmen)
Engineering: 7 officers, 20 artificers, 30 senior rates, 180 junior rates
Artisans: 36
Medical: 3 Officers, 5 ratings
Supply: 3 Officers, 40 ratings
Royal Marines: 3 Officers, 90 other ranks, 24 bandsmen

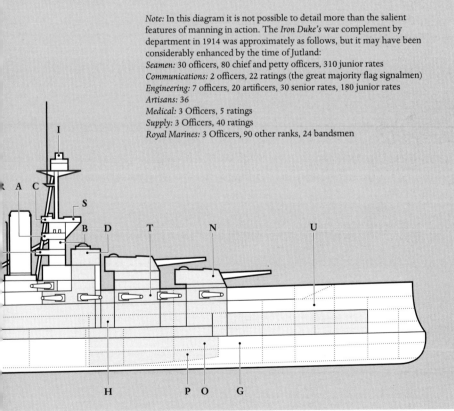

portionate component of the British domestic transport effort in sustaining the supply of coal and stores to the northern ports. Thus, even the restricted German submarine campaign could turn a critical situation into disaster.

The continuing inadequacies of the Admiralty's staff and administrative capacity had not been addressed by the Balfour–Jackson board, whose lacklustre performance resulted in a progressive erosion of confidence in the navy. In November 1916, Jackson went most willingly, with Jellicoe translated from the Grand Fleet as his successor. Beatty was advanced to the acting rank of Admiral and appointed Commander-in-Chief over the heads of most of the senior flag officers of the battle squadrons. The appointment was viewed with mixed feelings, but Jellicoe's favoured nomination, his Chief of Staff (and brother-in-law) Charles Madden, was actually junior to Beatty and possessed neither the command experience nor the public fame. In the event, Beatty proved highly effective. Notably, he brought the battle cruisers much more closely under the control of the Grand Fleet and did not allow his successor, William Pakenham, the autonomy that he himself had been permitted. Although Beatty breathed an air of initiative and decentralization into the Grand Fleet's tactics, he made few changes to Jellicoe's strategic doctrines. By far the most important was the relocation of the Grand Fleet from Scapa Flow in 1918 to join the battle cruisers at Rosyth, a move which placed the battleships nearer to the Germans and considerably eased the supply problem.

THE U-BOAT CAMPAIGN, 1917–1918

The fall of Asquith and the installation of David Lloyd George as prime minister provided the catalyst for a change of First Lords. Balfour was replaced by Sir Edward Carson, another Conservative politician, but a much warmer personality. Despite good professional relationships within the Admiralty, however, the administration's performance was little improved. Jellicoe attempted to co-ordinate measures with the establishment of an Anti-Submarine Division and he regarded as his own first priority the defeat of the U-boats. But the essential need for the direct escort of merchant ships went unrecognized as literally thousands of armed craft, large and small, were allocated to patrol duties. Jellicoe, too, was a tired man, depressed at leaving the Grand Fleet, still prone to over-centralize and increasingly dispirited and pessimistic as the result of years of over-strain.

Attempts to block the Straits of Dover to U-boats had yet to succeed. The Germans were making free use of the Belgian ports of Zeebrugge and Ostend to ease their access to the Atlantic. Offensive and defensive minefields were ineffective, partly through the continuing poor quality of British mines. Attempts to use disguised craft, known as 'Q-ships', to attack U-boats enjoyed temporary but diminishing success. The Germans soon became more cautious and would torpedo anything suspicious without warning.

The Germans were also under pressure. Increasing economic crisis and stalemate in the trenches brought the army High Command out in favour of Scheer's proposals to renew the unrestricted U-boat campaign, whatever the risks of American intervention. On 1 February 1917 the policy was renewed, and it brought immediate dividends in the doubling and redoubling of sinking rates to the point

where, in April, more than 500,000 tons were sunk. It also brought the United States into the war in the same month, but the new partner could offer little short-term assistance to the Allies. The German reply was to extend the campaign to the eastern seaboard of North America and the rate of sinkings again increased.

The Admiralty finally accepted convoy after considerable pressure from within the junior ranks of the naval staff, from the Grand Fleet itself and after experience with convoys to Scandinavia and other experimental ventures. Perhaps the key was not so much the increasing impatience of Lloyd George, who himself visited the Admiralty at the end of April to examine the prosecution of the anti-submarine war, but the discovery that the statistics for ocean-going merchant arrivals and departures in the United Kingdom had been grossly exaggerated. The capacity existed to give convoys adequate escorts, and escort numbers would soon be greatly supplemented by the US Navy.

The convoy strategy produced immediate and increasing dividends as it was extended throughout home waters and into the Mediterranean. Monthly shipping losses were practically halved in both theatres. Nevertheless, adoption of the measure did not disguise the basic inefficiencies of the Admiralty and the need for reform. With his own experience of the munitions problem, Lloyd George was

One of the host of small ships engaged in anti-submarine or minesweeping work during the First World War: the mine-sweeping trawler *John Davis*, completed just before the Armistice.

shrewd enough to perceive that the fundamental requirement was administrative reorganization. Sir Edward Geddes, a Scottish businessman and transport expert, was already installed in the Admiralty as Controller and making headway on ship-building and repair. In July 1917, the prime minister replaced Carson with Geddes.

Geddes's attempts at reform soon came up against the inability of Jellicoe to work within a decentralized organization. A new, much larger and better-organized staff machine was created, with the office of Chief of Naval Staff being combined with that of First Sea Lord. The subtle and likeable Vice-Admiral Sir Rosslyn Wemyss was brought in as Deputy First Sea Lord. His primary task was to relieve Jellicoe of some of his heavy burden but the latter would not delegate, not even to an appointed Deputy, and he bitterly opposed the division of Plans and Operations into separate entities.

Wemyss was on the point of resignation and Geddes, in equal frustration, had pleaded with the prime minister to be freed from the Admiralty when matters were brought to a head by events at Dover. Proposals from the naval staff to the local flag officer, Sir Reginald Bacon, to consolidate efforts to stop U-boats forcing the Straits had been subject to peremptory rejection. While Jellicoe retained faith in his old friend, the collective opinion of Wemyss and the Plans and Operations Divisions of the Naval Staff was that Bacon had to go if the situation were to improve. Jellicoe's support of Bacon proved the last straw for the First Lord. Geddes first sounded out Wemyss as to his readiness to serve as First Sea Lord and then dismissed Jellicoe on Christmas Eve 1917. The decision was inevitable, but its timing and manner caused deep offence within the Royal Navy and only consid-erable pressure prevented the other Sea Lords from resigning in protest. Geddes's apparent brutality was at least partly motivated by his unwillingness to do the deed, but it left a bitter taste and Wemyss was hard pressed to re-establish confidence in the Admiralty in the navy at large.

With no experience of the Grand Fleet, having spent the war in the Mediter-ranean and East Indies, Wemyss was much more a man of alliances and diplo-macy than Jellicoe and the appointment was a good one to meet the new exigencies of the naval war. While the holding strategy continued in the North Sea, made even more cautious by Beatty's acknowledgement that there could not be sufficient escorts for offensive fleet operations and for convoy work and that the latter had priority, it was the Mediterranean that now required considerable attention. Wemyss's experience of inter-service and international co-operation was unrivalled. After service in the Dardanelles campaign, he had commanded the East Indies Station and his ships in the Red Sea had played an important part in supporting the Arab Revolt against the Turks.

The Mediterranean problem was partly a matter of protecting shipping and partly of land strategy. Italy had entered the war on the Allied side as far back as 1915, but this had not resulted in any immediate lessening of British naval require-ments in theatre. On the contrary, while forces were maintained in the Aegean to prevent sorties by the *Goeben* (renamed *Yavuz Sultan Selim*) and *Breslau* (*Midilli*), the Italians demanded reinforcements to face the Austro-Hungarians. These went to Taranto in the form of a squadron of pre-dreadnoughts and accompanying light cruisers. The land campaigns in Italy, the Balkans, and the Middle East created further demands for naval support, particularly for the protection of shipping.

The central issue of the Mediterranean war became the failure to co-ordinate Allied action. French and Italian mutual suspicions repeatedly hamstrung efforts to create a supreme naval command. Italy's natural caution at risking its forces against Austria–Hungary at the expense of its future capabilities to match France further hampered efforts to achieve local superiority within the Adriatic. Although there were large numbers of escort craft, progressively reinforced from Britain, the lack of co-operation and a mistaken emphasis on setting up barrages across the access to the Mediterranean from Austro-Hungarian ports meant perennial shortages. These reached the point where the British were forced to call not only on the Australians for light craft but the Japanese, who by the end of the war were providing the lion's share of the destroyers based on Malta.

Wemyss's experience had made him the logical nominee for the revived post of Commander-in-Chief, Mediterranean in 1917, an appointment deemed necessary by the need to improve the dismal U-boat situation. His translation to Whitehall brought the selection of Vice-Admiral Somerset Gough-Calthorpe, although later efforts to impose a recalled Jellicoe as supreme Allied naval commander foundered on Italian intransigence. Gough-Calthorpe's command was nevertheless marked by much improved Allied co-operation within the theatre and increasingly effective measures, including convoy, against German and Austrian U-boats. Less satisfactory was a sortie by the *Yavuz Sultan Selim* and *Midilli* in January 1918 which caught British forces at Mudros without their heavy support. The two ships, still German manned, were able to make their escape without interference until entry to a minefield outside the Dardanelles resulted in the sinking of the *Midilli* and severe damage to the battle cruiser. The latter was aground for several days and subjected to a series of abortive attacks, including several from aircraft, but was eventually moved to the safety of the Bosphorous.

In home waters, a new energy entered the war in 1918. While the Admiralty sensibly accepted Beatty's decision to keep the Grand Fleet out of harm's way, the selection of Roger Keyes to command at Dover meant that plans already in train for attacks on the Belgian U-boat ports were pursued with zest. This was despite the fact that local forces were already heavily engaged in providing support to the troops who were protecting Allied held ports from the great German land offensive of March 1918. Blockships, light craft, and monitors were assembled and manned with specially selected volunteers, many from the Grand Fleet itself. Raids took place on Zeebrugge and Ostend on 23 April and 10 May 1918, but neither the blockships nor the accompanying efforts to disable the local port facilities resulted in more than temporary dislocation to the U-boats. Far more important was the fact that the sorties were a great propaganda success and a tremendous fillip for the Royal Navy itself, particularly the Grand Fleet. While the

CLOSING ACTIONS OF
THE WAR, 1918

withdrawal of key personnel from the heavy ships of the High Seas Fleet to man submarines was taking its toll on morale and efficiency, the use of Grand Fleet officers and ratings had quite a different effect on those who remained in Rosyth. The battle squadrons received a tonic that was to sustain them through the influenza epidemics and the ennui of the last six months of the war.

The High Seas Fleet emerged once more. In April 1918, after von Hipper had taken over the command from Scheer, who became the supreme naval commander ashore, the Germans sortied into the North Sea to intercept a Scandinavian-bound convoy. Their timing proved out by a day and catastrophic machinery defects in the battle cruiser *Moltke*, symptomatic of the poor condition of the fleet as a whole, forced Hipper to turn for home. Alerted too late because of much improved German communications security, the Grand Fleet had no chance of catching its opponent before the latter was inside the Bight.

In the meantime, the Americans injected their own energy with ambitious plans for a mine barrage to close the entrances to the Atlantic. Beatty was inclined to doubt both the technical feasibility and the tactical value of the scheme but the

Ratings at dinner in the mess deck of a light cruiser. The mess tables were secured to the deckhead after use to provide space to sling hammocks. Habitability was better than in some navies, but ventilation and recreation and washing facilities were generally poor. Reform was under way, led by the battle cruiser *Queen Mary*, but progress was slow in improving the lot of sailors.

The aircraft carrier HMS *Furious* shown in late 1918 in the Firth of Forth with a Submarine Scout airship on her after flight deck. A series of disastrous accidents caused by air turbulence around the superstructure had led to a ban on landings by fixed-wing aircraft. *Furious* later received a complete conversion to a carrier and served actively almost to the end of the Second World War.

Admiralty insisted on co-operation in the interests of solidarity. Like the raids, the barrage operation probably proved as valuable for the sense of purpose it instilled in the Allied naval forces involved as for any material results.

The air arm was at last coming into its own, although the creation of the Royal Air Force in April 1918 promised bitter disputes in the near future. Despite conversion of increasing numbers of merchant ships and warships to carry aircraft and the wholesale expansion of the Royal Naval Air Service, development had been fragmented. Not only had the Admiralty been assigned the air defence of the United Kingdom but there were ill-judged attempts to set up a strategic bombing element. Amalgamation of the air arms of the army and navy was made inevitable by poor co-ordination of procurement policies and lack of co-operation in many areas, much of which was due to Admiralty intransigence. The Board was slow to realize that closer direction was needed, and only in January 1917, after great delay, was the office of Fifth Sea Lord created and the responsibility for aircraft production passed to the Ministry of Munitions.

The value of naval aviation was appreciated too late to allow the Admiralty to mount an effective defence against the August 1917 decision to create the RAF. Sir Rosslyn Wemyss did his best to ensure that the Admiralty accepted the new arrangements and made them work, but there were second thoughts on the issue within a few months. Most critical for the long term, the vast majority of RNAS officers, particularly those who had achieved most in the new environment, transferred to the RAF and their expertise was lost to the navy.

As with many other areas, much of the initiative for the development of naval aviation came from the Grand Fleet, particularly after the establishment of the post of Rear-Admiral (Aircraft) at the end of 1917. The possibility of a massed attack by torpedo craft on the High Seas Fleet in its anchorages seemed to offer the first substantial hope for breaking the stalemate in the North Sea. This was made practicable by the commissioning of ships such as the carrier *Furious* and the prospective completion of the first clear-run flight-deck unit, the *Argus*, as well as improvements in the aircraft themselves. On 19 July 1918, seven were launched from the *Furious* to raid the Zeppelin base at Tondern, and they succeeded in setting fire to the airships L54 and L60. By late 1918, plans were well advanced for the torpedo bomber attack. Fighters carried on capital ships and cruisers had already proved an effective defence against Zeppelins, while seaplanes and airships were providing protection for coastal convoys which was so effective that no more than a handful of ships so escorted were ever sunk by U-boats.

Although the primary American contribution to the sea-war in the European theatre came in escort ships, a detachment of battleships joined the Grand Fleet in December 1917 and formed the Sixth Battle Squadron under Rear-Admiral Hugh Rodman. Both services made deliberate attempts to promote mutual harmony and both at Scapa Flow and at Queenstown, the principal base in Ireland for American destroyers on anti-submarine duty, relations were excellent. As well as senior officers on the spot, much of the credit for the success of the integration of US Navy forces into Royal Navy operations was due to Admiral William S. Sims, an Anglophile who quickly established close friendships with Jellicoe and with Wemyss.

Good relations at sea and in the United Kingdom did not prevent increasing concern over American intentions in the long term. Significant sections of the US Navy, including Admiral Benson, the Chief of Naval Operations, were deeply suspicious of the British Empire and a 1916 Act of Congress had declared the clear American intent to have a navy second to none. The possibilities for divergence in future strategic interest could not be ignored, particularly as the Americans were slow to accept that the priority should be for escort construction, not capital ships. The Royal Navy, which had by 1918 only one battle cruiser, the *Hood*, under construction, faced the prospect of being overtaken by the US Navy within a few years. The country could not afford a renewed naval arms race and the Admiralty had little stomach for it.

The differences in strategic aims became more apparent as the prospects for victory improved. The failure of the German offensive and the subsequent success of the Allied efforts on the Western Front in August 1918 were matched by the rapid collapse of the other Central powers. That collapse, together with the increasing anarchy in Russia, found Royal Navy units ranging as far inland as the upper reaches of the Danube. In Siberia and Murmansk, operations which began in order to protect huge amounts of British war material from German use after the treaty of Brest–Litovsk took on a more political aspect as the government determined to support anti-Communist forces. More encouragingly, in October 1918 British ships led Allied units through the Dardanelles to enforce an armistice with Turkey. Wemyss's quiet insistence on British pre-eminence forced the French

Naval Officers of the First World War in the old Admiralty Board Room. Among those in the painting are Reginald Tyrwhitt (11th from left), Roger Keyes (12th from left), David Beatty (9th from right), John Jellicoe (2nd from right), Charles Madden (3rd from right), and Rosslyn Wemyss (extreme right).

to give way to Gough-Calthorpe's flagship, the *Superb*, sent from the Grand Fleet only weeks before.

Wemyss had his hands full elsewhere when Germany, too, collapsed. The German High Command's admission of defeat was soon followed by rebellion in Germany. Leaders in the revolution were the capital ships of the High Seas Fleet. Their crews first refused to sail on a sortie which might have placed them in a hopeless confrontation with the overwhelming superiority of the Grand Fleet and then mutinied outright. Although the U-boats remained loyal, the army would fight no more. Wemyss's aim when the Germans asked for an armistice was simple: to reduce German naval power to the extent that hostilities could not possibly be renewed. In the hectic few days of negotiations in early November, he achieved much of what he sought, despite indifference from the triumphant army commanders and suspicion by the Americans that Britain wanted to supplement the Grand Fleet's strength with that of its adversary. The bulk of the High Seas Fleet and all the U-boats were to be interned. The Allied blockade would continue and the rump of the Imperial Navy was to be immobilized until peace was signed.

The emergence of the High Seas Fleet on 21 November for internment in Scapa Flow was a bitter disappointment to Beatty and to the Royal Navy as a whole. Despite the satisfaction at what was in every effective sense an abject surrender, there was an incompleteness about the victory in keeping with the navy's ambigu-

The light cruiser HMS *Cardiff* leading the battle cruisers of the High Seas Fleet into internment on 21 November 1918. The spectacle was described as 'looking for all the world like a school of leviathans led by a minnow'.

ous record during the war. The new Trafalgar expected in 1914 had never occurred; the Royal Navy had won the war against the U-boats but it had also nearly lost it. If the record on inter-service co-operation was one of great effort and steady improvement, it was nevertheless difficult to rate such operations as significant against the titanic scale of the Western Front. While the victory was at least partially due to Germany's industrial and economic collapse, it was arguable that the blockade had been no more critical in this process, particularly in reducing food supplies, than the loss of manpower to the German army from vital industries. If Germany had been destroyed as a naval rival, there were new and possibly more dangerous problems on the horizon, which Britain was now incapable of reinforcing the navy to meet. There was much comfort in the achievements of individual units and forces such as the submarine service, but the future of naval warfare and the direction of technological development remained uncertain. The navy had already lost control over one vital element of that future; it remained to be seen whether it would preserve the capability to keep up elsewhere.

RETRENCH-
MENT
RETHINKING
REVIVAL
1919–1939

GEOFFREY TILL

11

*T*he Royal Navy emerged from the First World War victorious and with what was by far the strongest fleet in the world. Its nearest rival, the German High Seas Fleet, was imprisoned and later scuttled at Scapa Flow. After that extraordinary event, the Royal Navy had nearly half, and sometimes more, of the effective warships of all classes to be found in the world's big five remaining navies. The Royal Navy also comprised over 37,000 officers and 400,000 men, experienced and hardened after four long years of war. Britannia did indeed appear to rule the waves (see Table 11.1).

Table 11.1 Strength of the fleets of the major naval powers, 1918

	Great Britain	United States	France	Germany	Japan	Italy
Battleships	61	39	20	40	13	14
Battlecruisers	9	—	—	5	7	—
Cruisers	30	16	21	3	10	7
Light cruisers	90	19	8	32	16	10
Flotilla leaders	23	—	—	—	—	8
Destroyers	443	131	91	200	67	44
Submarines	147	86	63	162	16	78
Aircraft carriers and seaplane tenders	4	—	—	—	—	—

Source: S. Roskill, *Naval Policy between the Wars*, 2 vols. (London, 1968 and 1976), i. 71.

ECONOMIC CONSTRAINT

But to a worrying extent such appearances were deceptive. The British had been exhausted by the war. Recovery, prosperity, even political stability depended on improving industrial efficiency and exports. This required major cuts in defence spending, the biggest item of government expenditure. The means chosen was the so-called 'Ten-Years Rule' approved by the Cabinet on 15 August 1919. This decreed that the services should base their preparations on the assumption that there would not be a major war in the next ten years. The rule became self-perpetuating in 1928, lapsed in 1932, and was finally abandoned only in November 1933, by which time, of course, serious war had effectively been ruled out until 1943.

This arrangement gave the Treasury the whip hand in negotiating the naval estimates every year and was instrumental in producing a progressive decline which
resulted in the 1932 estimates being over one-third lower than those of 1920, the
first normal post-war year.

Not surprisingly, naval officers of the time were inclined to attribute every
ailment in the navy to the parsimony of the Treasury, and there is some justice
in their complaint. But the other two services fared even worse, and until 1938,
much more was spent on the navy than on them. Moreover, until the fascist
dictatorships started their major rearmament programmes in the 1930s, large
decreases in naval and other defence spending were the international norm, so
in relation to other navies the position of the Royal Navy was not quite so bad.
Nevertheless, the fact remained that by 1932 the Americans were spending far
more on their navy than the British were on theirs. While it is undoubtedly true
that the Admiralty made mistakes in, say, its construction priorities, the constant
erosion of the naval estimates was a major problem.

There was a drastic slow-down in the quality and the quantity of naval
construction. In the 1920s the building rates in battleships tailed off completely
with the completion of only the *Nelson* and the *Rodney*. The refitting of older
battleships with new weaponry, especially Anti-aircraft (AA) gunnery, and
armour plating was very slow. It was the same story for the supporting fleet too,
with the scrapping of dozens of 'R' and 'S' class destroyers, slow progress in
cruisers, corvettes, submarines, and so on.

The effects on naval aviation were particularly significant. In 1918, British naval
aviation was far ahead of all its rivals, but by 1930, it had been overtaken by the
Americans and was being menaced by the Japanese. Older experimental carriers
were kept on far longer than originally anticipated, and the start dates of the four
new carriers the Admiralty wanted were repeatedly postponed. The number of
first-line aircraft fell from the 3,000 or so operating in 1918 to barely 100 in the early
1920s. For much of this decade, the Treasury was able to resist Admiralty requests
partly on the argument that the British had far more carriers than anyone else
(which was true so far as this went) and on the equally true argument that the Air
Ministry regarded this form of air power as much less important than their own.

More insidiously, financial constraint limited battle readiness. The value of
exercises at sea, including the large-scale Combined Fleet exercises of the
Atlantic–Home Fleet and the Mediterranean Fleet, was diminished by big reductions in the fuel allowance, which increased the time ships spent in harbour.
Controls on the expenditure of ammunition both for the main and secondary
armament, and for the all-important AA practices was severely curtailed. The
delay in providing a remotely controlled AA target, and exercise firing restrictions,
do much to explain the unreasonable optimism that many naval officers of the
period had about the prospects for warships defending themselves against air
attack.

The navy's people were affected by financial stringency too. Its personnel
strength slumped from 380,000 in 1919 to some 89,600 by 1932. Under the terms of

Above: Night action off Cape Matapan. The battleships *Warspite*, *Valiant*, and *Barham* in the act of surprising and destroying the Italian cruisers *Zara* and *Fiume* at a range of less than two miles on the night of 28 March 1941.

Right: 'The Greatest Raid of All'. The assault force sails into St Nazaire in the early hours of 28 March 1942. In the centre is the former American destroyer *Campbeltown* disguised as a German ship and crammed with explosives that will destroy the Normandie dock. The destroyer is escorted by motor launches (MLs) and the vessels are carrying between them 2,698 commandos.

the infamous 'Geddes axe' of 1921–2, some 2,000 officers, including about one-third of the captains on the active list, were removed. This savage pruning of the officer corps, especially in the lower ranks, was to create problems and consequent imbalances in the 1930s.

An ex-First Lord of the Admiralty, Sir Eric Geddes was Chairman of the Committee on National Economy, 1921–2, and responsible for swingeing cuts in naval personnel.

In some ways, the lower deck fared worse. The war had widened the gap in pay and conditions between workers ashore and their equivalents afloat in the immediate post-war period. This led to dissatisfaction, and even a degree of disaffection in the fleet. This in turn had led to the introduction of the reasonably generous 1919 pay rates. Together with the relative down-turn in the economy in the 1920s (which produced 2 million unemployed and a healthy flow of recruits into the navy), this led some to suppose that the 1919 rates had been too generous. As the Head of the Naval Personnel Committee remarked:

Undoubtedly the lower deck are very well off. Some of the higher ratings keep motor bicycles and can afford to take the more expensive seats at local entertainments and their meals at places which officers patronise. In some cases they are able to buy their houses...

Such views led to the introduction in 1925 of revised rates of pay which were, for new entrants, some 25 per cent lower than the 1919 rate. The idea of paying people doing the same job (and often in the same ship) at two quite different rates of pay might in itself seem hardly likely to encourage that sense of being 'all of one company' on which the navy had prided itself for the past several centuries. But worse was to come.

A particularly vicious economic crisis in 1931 led the government's May Committee to recommend swingeing pay cuts for all public sector workers, including those in the services. It was decided that the pay of the men on the 1919 pay rate should be reduced by a quarter to bring it into line with the 1925 rate. In fact, the financial consequences of this were somewhat mitigated by the 1919 men having protected allowances, but this was not fully appreciated at the time. The consequence was the so-called Invergordon Mutiny of September 1931. Often portrayed in the most lurid terms, not least by many senior officers of the time, this was in fact little more than a respectful, and respectable, 'down tools' of a few hundred sailors in a handful of ships of the Atlantic Fleet for two confused and worrying days. Inspired by the fear of bolshevik hordes pouring out of the naval bases, the Admiralty and the government backed down and the reductions were in fact limited to some 10 per cent across the board, more in line with equivalent cuts ashore.

In the long run, the Invergordon Mutiny was not important despite the shock it caused. It resulted in a run on the pound which forced Britain off the Gold Standard, but that in fact did much to facilitate Britain's steady economic recovery through the 1930s. As far as the men were concerned, their losses were largely recouped within two years anyway. The Mutiny's real significance lies in the light it sheds on the navy's personnel policy of the time. The fact that the men *seemed* to have been asked for greater sacrifices than the officers, the scant regard paid to the hire-purchase and other financial commitments of the 1919 men, the abysmal presentation of a cuts package which united the 1919 and the 1925 men in opposition, and the widespread and quite unfounded suspicion at the time that there were bolshevik activists beavering away in the fleet were all evidence of a widening gap between senior officers and their men.

Labour politicians were aware of this and sought to improve the situation by various schemes designed to widen the social background from which the officer corps was drawn. Although the Admiralty, chastened by the Invergordon experience, was prepared to improve conditions on the lower deck, it was extremely reluctant to tinker about with the officer corps until the exigencies of the Second World War forced it to. In other ways too, the navy's personnel policy showed itself to be less than perfect. In 1929, there was the farcical *Royal Oak* affair when with the greatest publicity imaginable, the careers of an admiral, a captain, and a commander were ruined amongst other things by an argument over the tunes to be played by a Royal Marine Bandmaster with the improbable name of Barnacle at a wardroom dance. All in all, the navy's policy towards its people is clearly one of those policy areas in which its deficiencies cannot simply be attributed to the parsimony of the Treasury. But, in this of course, the Admiralty's attitudes simply reflected the social mores of the time.

The effects of economic constraint become much clearer in other areas. During the 1920s, there were repeated Treasury assaults on the size and composition of the Naval Staff—the thinking and policy-making section of the navy. By 1927, the Naval Staff was down to a bedrock of some 60 officers from a total of around 330 in 1918. The need, on a day-to-day basis, to try to fight off programme cuts reduced the overall quality of planning and decision-making. The results of this are difficult to calculate, but it certainly reduced the efficiency of the fleet. To illustrate with just one example, the effects on Naval Intelligence were particularly severe, and this contributed significantly to Britain's misperceptions about developments in the German and Japanese navies.

Still worse in some ways was the effect that financial constraint had on relations between the three services. Competition for scarce resources became severe. Put most simply, the amount of money made available to the Admiralty was partly determined by the extent to which the Treasury found the countervailing arguments of the army and the Royal Air Force persuasive. For most of the inter-war period, the Admiralty tended to win this competition, but by 1938 the Air Staff's argument that the need to defend Britain against German air attack was paramount, and the very evident deficiencies of Britain's putative expeditionary force for operations in Continental Europe were so evident that the navy sank from top to bottom of the spending priorities in a space of twelve months.

At a lower level, this argument was played out in the form of a series of sometimes quite rancorous disputes about what was the best way of performing certain military tasks. How, for example, should important territory be defended, whether this was Singapore, Aden, or Britain itself? The navy pinned its faith on guns at sea, or failing that, guns ashore. The army supported the notion of guns ashore. The air force, on the other hand, had in mind a mobile air force shuttling cost-effectively along imperial air routes from one threatened site to another as the occasion warranted. Few advocates of their service went to the length of arguing that their competitors were unnecessary, but they did advance partisan views that made joint action in such matters as the defence of bases or coastal shipping, amphibious operations and so on more difficult. And much of this was due to the increased inter-service rivalry imposed by low ceilings on defence expenditure.

As 'Father of the Royal Air Force', Air Marshal Sir Hugh Trenchard was Chief of the Air Staff from 1919 to 1929 and involved in incessant controversy with the Admiralty over the relative priority of air power and sea power, and the ownership of the Fleet Air Arm.

As far as the navy was concerned, the worst example of this was probably in the long and ferocious dispute between the Air Ministry and the Admiralty over who should control the Fleet Air Arm. The establishment of the Royal Air Force in April 1918 deprived the navy of its 60,000 flyers and future naval air decision-makers, but a miniature Fleet Air Arm was recreated in 1921. For much of the inter-war period, responsibility for this was uneasily shared between the Admiralty and the Air Ministry on the so-called 'Dual Control System' eventually hammered out by the Trenchard–Keyes agreement of 1924. Personal relations between the sailors and the airmen of the Fleet Air Arm were very good but the fact that every advance in naval aviation had to be funded by two major departments of state which were in a state of acute and potentially mortal competition with one another clearly slowed progress.

The Admiralty found it difficult to attract naval recruits to the Fleet Air Arm in the number it wanted. The design of naval aircraft fell increasingly below American and Japanese standards and far below comparable carriers' standards where the Admiralty was master of its own house. Most worryingly, the effective departure of the first generation of naval flyers meant that naval aviation was inadequately represented in the navy's senior councils until the mid-1930s.

The Air Ministry had its problems with the system too, but felt it could not let go for fear of the effects this might have on its ability to conduct its main business, namely the air defence of Britain and the development of a strategic bombing capacity. The capacity of Britain's aeronautic industry was finite and the Air Staff thought that any expansion of an Admiralty controlled Fleet Air Arm could only be at the expense of the country's fighter and bomber forces, which the Air Ministry naturally thought more important.

The dispute dragged on until the government decided in 1937 to give the Fleet Air Arm back to the navy. Although from the Admiralty's point of view this was a significant improvement, it did not entirely solve the problem. This was partly because the hand-over itself was expensive and time-consuming (and was still far from complete when the Second World War began) but also because the government had fought shy of handing over shore-based maritime airpower in the shape of Coastal Command, which Admiral Chatfield, the First Sea Lord who led the final campaign, had clearly wanted.

THE BEGINNINGS OF
RECOVERY

The British economy moved into a period of steady recovery from about 1932. This, plus concern about the aggressive and expansionist military activities of the Japanese in China and Hitler's accession to power in Germany in 1933, led to the feeling that retrenchment had gone too far, given the unsettled and worrying state of the world. The Ten-Years Rule was formally abandoned on 15 November 1933 and in the following year the Defence Requirements Committee was set up to oversee Britain's programme of rearmament. It was generally agreed that the navy's task was to be able to field a force capable of affording cover against the Japanese navy, while at the same time keeping at home sufficient force to deter any conceivable European threat. For this the Admiralty concluded it needed a New Standard Fleet comprising:

20 capital ships
15 aircraft carriers
100 cruisers
200 destroyers

together with an appropriate flotilla of submarines and other support vessels. In theory, such a modernized fleet was to be ready for 1942. As the 1930s proceeded, however, senior naval officers began to conclude that this fleet was at the same time likely to be insufficient for the task, and was probably not achievable anyway.

There were two reasons for concern on the latter point. In the first place, the beginnings of rearmament in no way betokened a bonanza for the services. The Treasury was quite properly adamant that a sound economy was the fourth arm of Britain's defence and therefore insisted that the pace of rearmament should not be allowed to interfere with the recovery of Britain's normal trade and industry.

The fact that the American economy was now over four times as large as the British, and the Germans were out-producing Britain too in many key areas, emphasized the need for financial prudence.

After a brief but worrying mini-recession in 1937–8, the Treasury introduced a strict rationing system between the services, the proportions of which clearly disadvantaged the navy. So, despite the rearmament campaign and the fact that by the Spring of 1939 the British were spending more on defence than they had at the height of the First World War, things actually got worse for the Admiralty, because its future adversaries were rearming faster than it was.

Arguably the finest First Sea Lord of the inter-war period, Admiral Chatfield (*left*) presided over the naval recovery of the 1930s, and was succeeded in November 1938 by Admiral Backhouse (*right*). They are both seen on the deck of HMS *Nelson* during the Combined Fleet manœuvres of that year.

There was another, perhaps more serious, problem too. The long years of reduced defence orders had seriously diminished British industry's capacity to respond to increased defence orders when money became less of a problem. The number of yards with the necessary skilled workers and machine tools had atrophied. As a direct result of this the new 'King George V' class of battleship took some four to five years to complete where American equivalents took just over three, and the British contented themselves with 14-inch guns, where the Germans, French, and Italians had 15-inch, the Americans 16-inch, and the Japanese 18-inch. It was the same story in every other aspect of fleet development. The result of all this was, despite a huge construction programme in 1937, a much slower rate of expansion than the Admiralty had hoped for.

In the early 1920s, the urge to encourage at least partial naval disarmament was very strong. There were two reasons for this. In the first place, battleships were seen as the main weapon system of the period and were regarded as extravagantly wasteful of scarce resources. Secondly, there was a widespread view that arms races in general and naval arms races in particular had increased the probability of war in 1914, and could well do so again, unless checked. An immediate start should be made since in the immediate post-war period both Japan and the United States had embarked on huge programmes of naval construction that seemed likely to plunge the world into a violent naval arms race as ruinous as it was dangerous.

The result was the Treaty of Washington of 1921/2. This brought in a system of fixed ratios of naval strength for the major powers. The United States and Britain were both on 5; the Japanese on 3; and the Italians and French on 1.75. The treaty introduced a number of complex equations about how many battleships and aircraft carriers the powers could retain, build, or modernize. It was revolutionary in that it scythed through the expansion plans of all the navies concerned and set up a naval arms control system that was intended to absorb other countries in due course, and extend to other naval vessels, like cruisers, destroyers, and submarines. For the British, it was even more radical in that it signified the formal loss of the naval supremacy Britain had enjoyed for the past century, and the necessary abrogation of Britain's treaty relationship with the Japanese. Quite clearly, it was a historic turning-point for the Royal Navy.

In fact, though, this was naval disarmament's finest hour. Thereafter problems began to appear and the whole system petered out in disillusion in 1936, after the Second Treaty of London, which was beset with so many qualifications in its implementation and with so few signatories as to be more of a stimulus to naval construction than a preventive.

These essays in naval disarmament had, however, some beneficial consequences for the Royal Navy. They seemed likely to limit the increase in the future size and gunpower of the battleship, which the Admiralty wholly welcomed. Naval disarmament could also be a means by which the number of Britain's possible enemies and/or their naval strength could be reduced. Thus the Admiralty was enthusiastic for the Anglo-German Naval Treaty of 1935, which limited the German fleet to 35 per cent of the strength of the British, with the exception of submarines where parity was allowed in certain circumstances. As far as the British were concerned, this was a good deal better than what might have happened had German naval construction been unrestrained by treaty.

It was the same with regard to the Japanese and the Americans in the 1920s. In the United States, Congressional opinion was so enthused by the Washington treaty that the US Navy was never able to build up to the limits allowed. In consequence, the Royal Navy stayed far ahead in numerical terms at least, and was certainly in a better relative position than would have been the case if the Americans' original building plans had gone ahead.

None the less, naval disarmament had its detrimental consequences too. Most obviously, it strengthened the Treasury's hand in arguing against any accretion in naval strength. Any disarmament regime is only as effective as the extent to which its implementation is verified, and the Axis powers certainly bent the rules quite significantly, to the disadvantage of Britain and the United States.

Naval disarmament also had quite important effects on the shape and composition of the fleet, which sometimes posed particular problems for the British. The Washington Treaty, for example, limited the tonnage that could be devoted to aircraft carriers. In practice, this disadvantaged the British, who had a lot of carriers already, because the Treasury fastened on this as a reason for severe restraint in carrier building. In effect, the Treasury concentrated on the number of Britain's carriers and were unimpressed by the Admiralty's argument that they were old, experimental, and often very inefficient in operating aircraft. The Americans and the Japanese, on the other hand, had their work cut out to get up to treaty limits, but were able to do so with much newer and better vessels.

In a more general way, the treaties limited the power of capital ships more than they limited new forms of warfare that were developing to challenge them, particularly aircraft and submarines. In fact, the British continued to hope that their reward for participating in the naval disarmament process would be the banning of submarines altogether, or at least severe limits on their use. The first turned out to be completely impossible; while operational limits on submarines were agreed, these hardly survived the declaration of war in 1939.

One of the reasons why the Admiralty was sometimes less hostile to naval disarmament than might be supposed was, as we have seen, that it could reduce the number and power of Britain's possible adversaries. At the grand strategic level, the Admiralty like the other two service ministries became increasingly perturbed at what increasingly seemed to them to be the prospect of over-commitment. The popular image of the military sometimes seems to be of a force thirsting for war and glory. This could hardly be less true of the inter-war period. Britain's service chiefs were all worried at the open-ended commitment to collective security operations that Britain's leading role in the League of Nations seemed to make likely. And they were quite appalled at the prospect of war with Germany, Italy, and Japan simultaneously, with only France as an ally.

The point was made with the greatest clarity by the Admiralty under the stewardship of Chatfield as First Sea Lord. Chatfield argued strenuously that Britain had to reduce the number of its possible enemies, and at times advocated coming to at least temporary accommodations with each of the Axis powers. At the same time, the Admiralty had serious reservations about the reliability of both France and the United States as possible allies.

Strategically, the situation was really very difficult and the three services tended to have different security agendas, all of which were justified by subsequent events. The army focused on the immediate need to police the empire, and the more long-term requirement to prepare another expeditionary force to fight alongside the French on the continent of Europe. The RAF, naturally, was principally concerned about the prospect of an air assault on the heart of the empire,

STRATEGIC CONSIDERATIONS

In their bid to reduce the number of Britain's enemies, British statesmen sought constantly to peel Mussolini away from his understanding with Hitler. Chatfield, desperately concerned about the naval balance, enthusiastically supported this policy.

London, and Britain's other great cities. In the 1920s, the likeliest adversary seemed to be France, but from the early 1930s, the expanding German air force was seen as the main threat. This aerial menace demanded the development of fighter defence for Britain itself and of a strategic bombing capacity as a deterrent. The Treasury was, as we have seen, determined to keep Britain's economy strong, not least so that it could prevail in a long war, should one come.

The Admiralty's own preoccupations were equally pressing. In political terms, the US Navy was the main adversary through the 1920s. Britain did not take the prospect of war with the United States seriously, although such a prospect did provide the backdrop for some aspects of British planning, such as, for example, the Bonar Law Capital Ship Enquiry of 1921. The US Navy in effect simply provided a convenient measure against which the Admiralty could gauge its needs. In addition, both government and Admiralty wished to maintain at least parity against the American fleet, as a necessary condition of Britain's continued international power.

Until about 1935, naval planners were not particularly concerned about the German navy, severely constrained as it was by the Treaty of Versailles. It was thought that such naval aspirations as it had were largely confined to the Baltic, where it confronted Soviet Russia. The Anglo-German Naval Treaty of 1935, moreover, should, it seemed, limit the rivalry. Only towards the very end of the period did the Admiralty really begin to worry about the growth of the German navy.

Japan on the other hand was a major preoccupation from the beginning. Despite the Washington Treaty but perhaps partly because of the connected abrogation of the Anglo-Japanese treaty of 1902, Japan was plainly the expansionist up-and-coming power at sea. The Japanese navy might well menace Britain's imperial trade routes and territory in the Far East. A possible war against the Japanese navy though, would be a difficult one, fought far from home.

The Admiralty therefore concluded that it was essential to have a modern first-class fleet base in the vicinity. Singapore was chosen as the site and work on the base began in 1921. Since it was clearly impossible to keep the fleet out there permanently, the idea was that the Mediterranean Fleet would continue to be the navy's strategic centre of gravity, able to dispatch forces to support operations in the Far East or Europe, or both, as occasion warranted. The strategy therefore developed of building up the defences of Singapore so that it could look after itself until the Main Fleet arrived from European waters. British naval forces on the China and East Indies stations would engage in holding and raiding operations until they were relieved from the West. What would happen then, was a little on the vague side (not least because it partly depended on what the Americans were doing) but the Admiralty planned to impose an economic blockade on Japan, considered engaging in raiding operations, possibly involving carriers, and hoped eventually to defeat the Japanese battlefleet in a decisive engagement. It was all very reminiscent of the last war against Germany.

Geography and distance alone would have made this a difficult strategy even if all the expected forces were available. But, in fact, they were not. The construction of the naval base at Singapore and of its defences was repeatedly slowed down and reduced partly through shortage of funds and partly through virulent controversy between the services as to the form those defences should take. Moreover, as the European theatre grew more threatening, planners began to extend the period before relief during which Singapore would have to look after itself, from 42 days to 90, and eventually to 180. The size of the putative fleet went down too; in the end it was no more than a battleship, an old battle cruiser, and a handful of destroyer escorts.

One of the reasons for the decline in Britain's Far Eastern capacity was the view widely held in Britain in the mid- to late 1930s, especially in the Treasury and the Foreign Office, but to an extent in the Admiralty too, that longer-term European threats were more dangerous to the existence of the British empire. It was felt the Japanese were unlikely to move against Britain unless events in Europe provided them with an opportunity, and so a defence policy that focused on the need to counter, for example, German air attack, was in fact an effective indirect defence of Singapore. And, as we have seen, the essential primacy of European concerns was in fact accepted by the likes of Sir Oswyn Murray, the Admiralty's prestigious Permanent Secretary, and Admiral Chatfield when they urged a policy of accommodation with the Japanese. The problem was that the Japanese seemed to be unappeasable and the Americans were hostile to the whole idea. For these reasons, Japan remained an enemy, but the necessary forces to contain it were not provided.

PRESSING DISTRACTIONS

Before the start of the negotiations that led to the Washington Treaty, the First Lord, Walter Long, when dealing with the problem of identifying possible enemies, remarked that 'owing to the peculiar geographic constitution of the British empire, we run constant risks of unexpected and unforeseen troubles'. The accuracy of this remark was demonstrated almost immediately, upon the outbreak of the first of five major civil/international wars in which Britain became involved during the so-called inter-war period. Having to meet these urgent and necessary requirements naturally distracted the Admiralty from its long-term strategic planning.

None of these interventions was successful in the end, and British governmental policy towards these issues was often characterized by a level of indecision and muddle which considerably reduced the prospects of success, and which frequently left a good deal of initiative to local military and naval commanders. Usually, British interests and citizens were at risk, and there were often persuasive humanitarian reasons for a degree of intervention. But in truth it seemed impossible for Britain to stand aside from such major events if other countries did not.

This was particularly true of Britain's role in the War of Intervention in Soviet Russia. British motives were as complex as the situation, a fact which communicated itself to the men. Their exposure to a confused and undeclared war when their colleagues back home were being demobilized led to some disaffection among the Royal Marines ashore and the crews of the seaplane carrier *Vindictive*, and the cruiser *Delhi*, amongst others.

None the less, in this unwise affair, the Royal Navy was remarkably busy and chalked up some extraordinary successes. In the Baltic, the problem was how to prevent the Bolshevik fleet at Kronstadt from interfering with British sea-based support for the Baltic republics. An extensive and reasonably successful Bolshevik mining campaign had also to be countered. On the night of 16–17 June 1919, Lieutenant Augustus Agar, in a bold freelance operation, led a force of coastal motor boats from a small base in Finland into the vicinity of Kronstadt where they sank the 7,000-ton Bolshevik cruiser *Oleg* as it was bombarding a White fortress at Krasnaya Gorka. On the night of 18 August, the British launched an even more extraordinary combined operation against the Bolshevik fleet base itself. While aircraft from the *Vindictive* launched diversionary attacks, a number of coastal motor boats slipped into Kronstadt harbour and sank two battleships and a submarine depot ship with torpedoes.

In this theatre and elsewhere, British naval forces defended positions ashore,

The attack on the Russian cruiser *Oleg* at Kronstadt, June 1919.

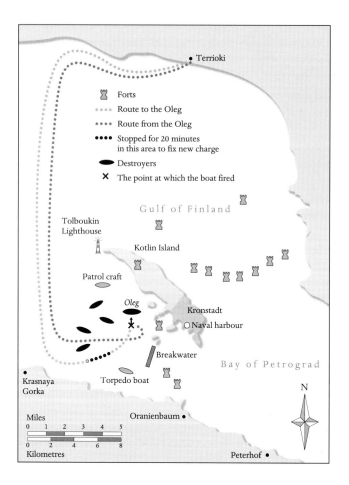

bombarded German and/or Bolshevik forces, and transported troops and refugees from a number of ports all round Russia's coastline. The Royal Navy discovered in itself a remarkable facility for taking to small gunboats and beating the Bolsheviks in vicious little encounters on the River Dvina, the Caspian Sea, and lakes such as Onega and Baikal in Siberia. Perhaps most amazingly of all, the crew of the cruiser *Suffolk* equipped and manned an armoured train which shuttled 3,000 miles inland along the Trans-Siberian railway, fighting and outshooting their Bolshevik adversaries of the Fifth Red Army on the Omsk front. Rarely indeed can naval gunners have fought so well, so far from the sea. But this was only one example of an extraordinary, and for the Royal Navy very unfamiliar, style of war.

Events in Turkey from 1919 to 1923 were very nearly as challenging and certainly as confusing. British policy was ill-advised and eventually failed. The local military commanders of the army and navy had very political roles to play and often exhibited more real understanding of the situation than did their masters in London.

Royal Navy battleships and aircraft carriers protected British troops ashore on both sides of the Bosphorus, and supported them when they seized key parts of Constantinople to prevent a nationalist take-over. British ships joined in humanitarian action evacuating thousands of terrified refugees (mainly Greeks) in the frightful aftermath of the burning of Smyrna in September 1922. When nationalist forces moved into the neutral zone at Chanak, menacing the British forces there, the navy prepared for war, which was expected daily in October 1922. It was avoided only by a last-minute peace conference and the fall of the Sultan. Symbolically, the navy spirited the deposed leader away to Malta in the battleship *Malaya* and the crisis ended a few months later. By the end of 1923, all British forces had left.

The navy also became embroiled in the long civil/international war in China which went on through most of the inter-war period, where British interests and citizens were threatened by local warlords and pirates, the fervour of the nationalists, the Communists, and finally the Japanese.

China was another area in which much of the naval effort had to be conducted on rivers by small gunboats. But when there were ferocious anti-foreign outbursts such as that at Wanhsien on the Yangtse in 1925 and at Canton, Hankow, Nanking, and Shanghai in the mid- and late 1920s, the British had to use larger warships, in the company of their allies, to help restore order.

But worse was to come with the expansion of Japanese activity in Manchuria and then China, especially after a clash between Japanese and nationalist troops on the Marco Polo bridge near Peking in July 1937. Incident piled on incident, and the Royal Navy became engaged in large-scale evacuation of British citizens from Shanghai in 1937. Shortly afterwards, British gunboats were bombed and bombarded by the Japanese. The situation deteriorated still further and by 1939 the British concession at Tientsin was being menaced. While the British sought to maintain reasonable forces on the China Station, which nearly always contained a carrier, these were badly outnumbered by the Japanese and overshadowed by the geographic size of the theatre. It is easy to see why the appeasement of Japan should have seemed so sensible an option to Chatfield and others.

The most serious danger of war, however, was in 1935–6, when Britain found itself leading the League of Nations effort to deter Mussolini's aggression against Abyssinia. The Mediterranean Fleet was extensively reinforced and went to full war readiness in its new temporary base at Alexandria. In point of fact though, nothing short of a wholesale sea-based sanctions campaign against Italy with its high attendant risk of war seemed likely to have the desired effect on the Italians. Neither the British nor the French were in fact prepared to take that risk, the sanctions were half-hearted, and Mussolini entered Addis Ababa in triumph on 5 May 1936, in effect ending the crisis.

Although it had not fired its guns in anger, this was an exhausting business for the Royal Navy and the whole episode had many implications for the future. Although the mobilization itself was well handled, the crisis revealed many deficiencies in the British fleet. The air defences at Malta were shown to be far too weak, making the base seem very vulnerable to Italian air attack. In consequence, the British were forced to retreat to the eastern end of the Mediterranean. The situation was not much better at sea, however. The Mediterranean Fleet's main armament stocks were thought good only for some 15 minutes of war operations, and the AA ammunition situation was little better.

All these lessons were learned, however, under the energetic and confident Admiral W. W. Fisher, Commander-in-Chief of the Mediterranean Fleet. By the end of the crisis, that fleet had been tremendously improved and was confident of success against the Italians in a set of planned operations which included a Fleet Air Arm attack on the Italian naval base at Taranto and aggressive sweeps and bombardments all along the Italian coastline.

Japanese troops entering Peking. Japanese policy in China put severe pressure on the Royal Navy and made a political *rapprochement* with Britain impossible.

Capital ships of the Mediterranean Fleet pictured in Malta. *Queen Elizabeth* is in the foreground, with the *Revenge* directly astern. Moored to their right is the battleship *Royal Oak*. To the rear may be seen the cruisers *London* and *Suffolk*. The vulnerability of Malta harbour to air attack, unless properly defended, is clear from this photo.

The harbour of Alexandria, filled with the ships of the Mediterranean Fleet. Its capital ships may be seen at the top of the photo. During the Abyssinian crisis, Alexandria was held to be much safer from air attack than Malta. During the crisis, the reinforced Mediterranean Fleet grew so large that the British had to use Haifa as an overflow.

The Admiralty back home was less sanguine, however. It had to tell Fisher that his plans to upgrade Malta's defences would have to be reduced by the need to improve Singapore's even greater deficiencies. The Abyssinian crisis demonstrated how tightly stretched the Royal Navy was, how much the good offices of the French were needed, and how important it was to improve relations with Italy. As Chatfield remarked, 'It is a disaster that our statesmen have got us into this quarrel with Italy, who ought to be our best friend because her position in the Mediterranean is a dominant one.' Unfortunately, any such attempts to improve relations with Italy were bedevilled by the last great crisis of the inter-war period in which the Royal Navy was involved, the Spanish civil war. Humanitarian considerations and the activities of other countries made it difficult for the British to stay out, whatever their preference. And, of course, once the British did become involved, considerations of prestige, and of deterrence, required that they must not seem to be defeated.

For all such reasons, the Royal Navy did indeed become involved in the Spanish civil war, paradoxically in the cause of non-intervention. Operations involved the use of several capital ships, including the battleships *Resolution* and *Royal Oak* and the battle cruiser *Hood*, and many cruisers and destroyers. These ships protected the passage of British merchant ships entering and leaving Spanish ports; saw off challenges by Republican warships, most notably the cruiser *Almirante Cervera* that sought to enforce the siege of Bilbao; rescued the crews of sinking ships; and, after the Treaty of Nyon in September 1937, frightened away Italian submarines harassing Republican shipping in the Mediterranean. The Spanish imbroglio only finally ended in 1939, by which time even more serious events were threatening Europe's peace.

These interim commitments, though, had been important for naval policy and plans. They had all shown that the threat of war could come with dramatic swiftness, much faster than the stately procedures of the Treasury might seem to suggest. They were all politically complex, with the admirals helping to make policy as much as they implemented it. All these tragic episodes were, in the last analysis, political failures.

At the level of grand strategy, the Admiralty's main concern was that each crisis showed Britain to be over-committed, with too many enemies and not enough reliable friends. In the 1930s in particular Admiral Chatfield was quite clear that Britain needed to reduce the number of its enemies, by appeasement if necessary. He was reinforced in this by the fact that the exigencies of the Abyssinian crisis, and the Spanish Civil War in particular, had thrown the Admiralty's plans for construction, refit, and careful preparation into confusion. The Royal Navy needed an unimpeded period in which to proceed with rearmament, but events conspired against it.

At the operational level, there were of course some off-setting compensations. These episodes provided a degree of practice in the business of naval warfare. Both the last two crises of the period exercised the navy in blockade, convoy operations, the naval control of shipping, minesweeping, and anti-submarine

The crew of HMS *Hood* peer down at the people of Barcelona begging for help during the Spanish Civil War.

A detached flight of Fairey IIID seaplanes operating at Wei-Hai-Wei in 1927. Reconnaissance aircraft like this helped the British keep an eye on the confusing events in the China of the time. A special force of them, operating from HMS *Argus*, went to the aid of the Shanghai Defence Force protecting British interests against Chinese insurgents.

operations. The Spanish Civil War also led to the creation of the Operational Intelligence Centre which was to have so important a role in the Second World War.

But above all, perhaps, these operations demonstrated the value, and to some extent the threat, of air power to naval operations. Aircraft had a vital role to play in all of these operations, especially in providing reconnaissance and in the attack of enemy shipping and military forces ashore: in China, Malta, and Spain, aircraft showed themselves to be a threat to warships. Close co-operation with forces ashore was a characteristic of the Russian and Turkish conflicts. In Russia and China, riverine and small coastal operations were the order of the day. In sum, while these operations may certainly have disrupted strategic preparations for war, they did at least provide a surprising variety of operational practice for the Royal Navy.

The battleship was still regarded as the navy's capital ship. However, the 16-inch gun *Nelson* and *Rodney* of 1927 were the only battleships completed during the inter-war period, although some older ones were quite extensively modernized. In the late 1930s, the Admiralty initiated the 'King George V' class, eventually completed between 1940 and 1942. These ships had only 14-inch guns, but the never-to-be-completed 'Lion' class were much nearer the Admiralty's real aspirations in that they would have had 16-inch guns. Britain's last battleship was the *Vanguard*. This was a singular design of 1939, intended to make use of some 15-inch guns left over from the First World War; the ship was only completed in 1946.

This rather modest programme shows that battleship modernization and construction was much affected by shortage of resources and the disarmament regime described above. It also reflected the fact that there was much technological, and sometimes near theological, dispute about the extent to which new technology in the shape of aircraft and submarines would threaten the battleship's traditional primacy. There were really two issues.

CONSTRUCTION PROGRAMMES

335

First, how vulnerable was the battleship to submarine and air attack? Innumerable experiments and exercises were conducted but largely thanks to exaggerated anticipations about the effectiveness of AA gunnery, British naval officers on the whole tended to underestimate the extent to which future battleship operations would be constrained by the threat of aircraft and submarines. The same was true in the world's other major navies, although at the very end of the inter-war period there were signs that the Americans and especially the Japanese were beginning to pull ahead in this regard.

Second, there was almost as much controversy about how big future battleships should be. The Admiralty was concerned about the battleship's apparently inexorable tendency to get bigger, and to have larger guns. This would make them expensive, reduce their number, make them too valuable to risk, and, since there were limited large dock facilities around the world, reduce their strategic mobility. But the Admiralty's hopes of using the naval disarmament process to delay this expansion came to nothing, leaving the British with a class of battleship more lightly armed than most of their contemporaries.

The other great controversy of the period was the extent to which airpower, perhaps even of the land-based variety, could take on roles traditionally carried out by the navy as a whole. As we have seen, at one level, this could lead to inter-service dissension over such issues as the defence of Singapore. But there was also the question of where the balance should be struck between the navy's own sea-based air power and other more traditional naval weapons systems. Would naval aircraft perform no more than ancillary duties (for example, spotting the fall of shot, reconnaissance, anti-submarine patrols, and so on) or might it take on the primary task of attacking the enemy's battle line?

Opinion on this matter was as divided in the Royal Navy as it was in most other navies. The poor quality of naval aircraft, due in significant measure to the Dual Control system described earlier, limited the potential performance of the Fleet Air Arm and therefore the expectations that the navy had of it. The anticipated place of air power in battle will be described later, but it is certainly true that the Admiralty thought naval aviation sufficiently important to fight a ferocious campaign against the Air Ministry throughout the inter-war period in order to regain control over it. The perceived importance of naval aviation was also demonstrated by as vigorous a carrier development programme as economic and disarmament conditions would allow.

In 1919, the Admiralty envisaged the maintenance of perhaps a dozen specialized carriers of various sorts, but this quickly proved to be over-ambitious, and after the Washington Treaty the plan settled down to a 20-year programme designed to produce a squadron of seven first-class carriers by 1939, with a number of others for the escort–trade protection role. But this programme was slowed down at every turn.

Immediately after the war, the Admiralty scrapped many of the first generation of experimental carriers. Its first effective carrier, *Argus*, was completed in 1918, followed by *Eagle* and the purpose-designed *Hermes* in 1924. Within a few years,

The carrier HMS *Glorious*, one of the Royal Navy's three modern carriers, entering the harbour at Valetta, Malta in 1932. Originally, *Glorious* was designed as a large cruiser.

none of these carriers could be regarded as first-line, but the completed conversions of *Furious*, *Glorious*, and *Courageous* provided vessels capable of performing that function. The Admiralty's intention to build four brand-new carriers from the mid-1920s was successively watered down, until in the end only the very useful *Ark Royal* of 1937 ever materialized. In 1939, the first of Britain's revolutionary armoured carriers, HMS *Illustrious*, appeared. Because they were heavily protected, these ships had limited aircraft complements, but they performed well in the Second World War and were in due course succeeded by a second class of armoured carrier, designed to the same principles, but with bigger aircraft complements.

The Royal Navy's carrier programme, and especially the gap between what it wanted and what it got, clearly refutes the notion that the Admiralty of the inter-war period was insufficiently air-minded. After all, the battleship programme suffered if anything proportionately more but this is not usually taken as evidence that the Admiralty was insufficiently 'battleship-minded'!

The Admiralty was in fact quite dismayed to have to go into the Second World War with only four first-line carriers and three obsolescent ones. The constraints of the inter-war period meant that the Royal Navy was not normally able to operate its carriers as a squadron, could operate far fewer aircraft at sea than it wanted, had no effective carriers for trade protection, and had a Fleet Air Arm with no more than modest capability when compared to its American and Japanese equivalents. The Admiralty was keenly aware of all this and at the very end of the inter-war period had the biggest current carrier construction programme in the world as it sought to catch up with the Americans and the Japanese.

The story was even less happy when it came to naval aircraft. The design and provision of these suffered from the Dual Control system described above. At the beginning of the inter-war period, British naval aircraft were better than most, but by the late 1930s they were significantly behind their Japanese and American equivalents both in quality and quantity. The development of Britain's naval aircraft in no way matched the rate of aeronautical advance, as measured by such things as the winning speed for the Schneider Cup (107 mph in 1920, 340 in 1931). British aircraft tended to be robust and, because so few could be embarked, multi-purpose. The Fleet Air Arm had no modern single-seater fleet fighter worthy of the name by the late 1930s. All of this tended to confirm the sceptics in their view that the Fleet Air Arm could be expected to perform no more than a supportive, ancillary role.

As far as carriers and their aircraft were concerned, trade protection was of secondary concern, and was a task largely relegated to obsolete vessels too old to serve with the battlefleet. This was not the case with cruisers. Here the protection of merchant shipping against surface raiders was an important priority. The length of the empire's shipping routes seemed to the Naval Staff to require a force of 70 cruisers, but they need not all be very large. The US Navy on the other hand concentrated on cruisers operating with the battlefleet and so wanted far fewer but much heavier vessels. This led to a great deal of Anglo-American discord during the 1920s and soured the disarmament process. Disarmament agreements were, none the less, one of the main influences on the Royal Navy's cruiser construction programme in the inter-war period.

In the event, the British built 13 large 10,000 ton 'County' class cruisers with 8-inch guns. These were very large comfortable ships, lightly armoured, splendid for a world at peace; the war was to show their deficiencies in armour and gun-power. The Admiralty preferred smaller cruisers and the last two 'Counties', *York* and *Exeter*, were 1,200 tons lighter, but kept the 8-inch gun. With the smaller

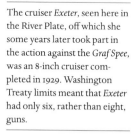

The cruiser *Exeter*, seen here in the River Plate, off which she some years later took part in the action against the *Graf Spee*, was an 8-inch cruiser completed in 1929. Washington Treaty limits meant that *Exeter* had only six, rather than eight, guns.

'Arethusa' and 'Leander' classes, however, the Admiralty reverted to its preferred 6-inch gun. Nevertheless, and just as was the case with capital ships, American and Japanese pressure forced the Admiralty into building bigger and more powerful vessels than they really wanted. The large 'Southampton' class of 1933–6 was the result; these were very fine ships, but kept the 6-inch guns. They were followed by the 'Fiji' and 'Dido' classes of 1939, the latter having only 5.25-inch guns. During the Abyssinian crisis, the Admiralty became more aware of the danger of air attack, and converted several old 'C'-class cruisers into AA cruisers. In the end, the Royal Navy had nowhere near the number of cruisers of all types that it needed to fight the forthcoming two-ocean war.

The need to participate in battlefleet operations, to protect merchant ships, and to respond to foreign building were also the determinants of British destroyer policy. The Admiralty scrapped many of the 'V' and 'W' classes of the First World War and then built a series of destroyer classes going from the 'A's of the mid-1920s to the 'I's of the mid-1930s. British destroyers tended to be of rather short endurance and were certainly insufficiently protected against air attack. They were, however, tough and reliable. Concern about the large Japanese types then building led to the much bigger 'Tribal' class of 1935 which proved itself in surface action, but it is probably true to say that destroyer designers paid too little attention to their anti-aircraft and anti-submarine roles, and rather too much to their torpedo attack activities in support of the battlefleet.

This leads naturally on to discussion about whether the Admiralty unduly neglected the defence of trade and concentrated too much on the needs of the battlefleet. As we shall see, there was something of a tendency to think that the submarine menace to merchant shipping had been satisfactorily contained, but nevertheless the Admiralty had a substantial programme of convoy escorts, sloops, and corvettes in place by the time war began, was busy fitting ASDIC (submarine sound detection equipment; the acronym was derived from the Anti-Submarine Detection Investigation Committee responsible for its invention) to its destroyers, and was converting trawlers to an anti-submarine role. Rightly or wrongly, there was a tendency to think that smaller classes of warship for such purposes could be improvised much more readily as war approached, or even after it had begun, than larger vessels. This being so it was sensible in times of peace to concentrate on the construction of larger vessels.

Connectedly, the British preference throughout the inter-war period was to ban submarines rather than build them, but they had few realistic hopes that this would be possible. During the period, therefore, the Admiralty kept up a steady programme of submarine construction. In the main, they were general purpose, and the last, failed attempt to build a submarine intended for close operations with the battlefleet was the three 'Thames' class submarines of 1929–30. None the less, in exercises and expectations, the main target of the British submarine was expected to be enemy warships. As strategic attention focused on Japan in the 1920s, submarine constructors began to emphasize the endurance so essential to Pacific operations—and this resulted in the 'O', 'P', and 'R' classes. Finally, the

Table 11.2 The Royal Navy, September 1939

	With the fleet	In reserve	In refit	Building
Capital ships	12	—	3	5
Carriers[a]	7	2	—	6
Fleet cruisers	33	2	3	⎤
Convoy and anti-aircraft cruisers	23	—	3	⎦ 19
Fleet destroyers	95	5	⎤	—
Escort destroyers, sloops, corvettes	101	—	⎦ 15	24
Submarines	38	—	—	11
Mine warfare vessels	52	11	—	44

[a] Includes two Seaplane Carriers, *Albatross* in South Atlantic Command and *Pegasus* in reserve.
Source: Roskill, *Naval Policy between the Wars*, i. appendix E.

Table 11.3 Strength of the fleets of the major naval powers, January 1939

	British Common-wealth	USA	Japan	France	Italy	Germany
Battleships	12	15	9	5	4	2
Battlecruisers	3	—	—	1	—	2
Pocket battleships	—	—	—	—	—	3
Cruisers	62	32	39	18	21	6
Aircraft carriers	7	5	5	1	—	—
Seaplane carriers	2	—	3	1[a]	—	—
Destroyers	159	209	84	58	48	17
Torpedo boats [small destroyers]	11	—	38	13	69	16
Submarines	54	87	58	76	104	57
Monitors, coast defence, and old armoured ships	3	—	1	—	1	—
Minelayers	1	8	10	1	—	—
Sloops and escort vessels	38	—	—	25	32	8
Gunboats, river gunboats, and patrol vessels	27	20	10	10	2	—

[a] aviation transport

Source: Roskill, *Naval Policy between the Wars*, i. 577

British built a number of submarines specifically designed for coastal operations, notably the successful 'Swordfish' and 'Shark' classes.

The result of this varied if constrained programme of construction was a balanced fleet which comprised on the eve of war the forces listed in Table 11.2. How the Royal Navy of 1939 compared with other navies of the time may be seen in Table 11.3.

Of course, the Admiralty's construction programme and general preparations for war had to be matched by parallel developments in the field of naval personnel. From the low of 1933–4, numbers went up by nearly one-third to 119,000 in 1938–9. Difficulties in filling various specializations led to a gradual liberalization in recruitment along the lines that Labour politicians had wanted. For example, the Admiralty's previous insistence that all pilots be officers was quietly dropped when it became clear that the existing pool was unlikely to satisfy the requirement. In the training world too, the demands of a bigger fleet hugely expanded the load. Technological developments were making naval warfare increasingly complex and this demanded higher-calibre personnel and enhanced training. In all these ways, the navy was gradually changing, a process that accelerated when the war began.

DOCTRINES OF WAR

The Admiralty of the inter-war period is often said to have concentrated too much on battlefleet operations, and to have neglected the sometimes competing and somewhat specialized requirements of trade defence and amphibious operations. Certainly, the doctrinal orthodoxy inherited from naval theorists such as Alfred Thayer Mahan and, to a lesser extent, Sir Julian Corbett, was that one had to win command of the sea before one could exercise it by the conduct of operations against the shore or the defence and attack of trade. The point seemed to have been demonstrated even in the minor military operations of the period. When the battle cruiser HMS *Hood* was present in all its naval majesty, Nationalist cruisers left British destroyers and merchantmen alone off Bilbao. But when British gunboats on the Yangtse were unsupported, they were bombed and shot at.

It followed, therefore, that preparing for battle as a means of securing command of the sea was by definition the Royal Navy's first duty. Only those countries which could not realistically aspire to command of the sea, or which did not need it, could contemplate any alternative. Certainly, the British knew that the Japanese were as orthodox about such matters as they were. When German naval expansion really began in the late 1930s, it seemed that their Admirals also thought the same. Only when the war started years before the German navy was ready for it, did the Germans, perforce, develop other ideas.

It is, though, true that the Admiralty persistently regarded surface raiders as the main threat to trade, rather than submarines. This was partly because convoy-and-escort had defeated the U-boat last time. Moreover, the development of ASDIC was held to have reduced the U-boat threat still further. In the idealistic 1920s and early 1930s there were some hopes that the barbaric and strategically counter-productive business of unrestricted submarine warfare would not be tried again. As late as 1935, Germany was, moreover, not supposed to have any submarines at all. Finally, there was no reason to suppose that a future Germany would be able to overrun France and Norway, thereby enormously increasing the potential reach of its U-boat arm. For all these reasons, the submarine threat to British shipping was underestimated, but it was understandable that it should have been.

In point of fact, as we have seen, the Admiralty did in the last years of peace embark on a very considerable programme of building convoy anti-submarine escorts of one sort or another. The formation of convoys, if mainly against surface raiders and aircraft, was assiduously prepared for. Such was the rationale for Britain's 70 cruisers, as we have seen. It also explains why the navy devoted so much effort to rehearsing the mechanics of the naval control of shipping, regulating the shipping industry, and so forth.

In some ways, the same could be said of preparation for amphibious operations. The Dardanelles experience had tragically demonstrated that amphibious operations were perhaps the most difficult form of war, especially if the landings were to be contested by first-class opposition. Nor was it very clear, unless planners were to envisage the loss of much of Europe and Britain's holdings in Southeast Asia, where such difficult operations would need to be conducted anyway. By definition, moreover, amphibious operations required the closest co-operation between the services, and all three of them had other tasks to perform which they thought had higher priority. Accordingly, and despite the early interest in the matter demonstrated by the combined staff courses at Camberley every year, little

Combined operations were neglected for much of the inter-war period. Here a landing craft is unloading its light tank during combined exercises conducted in South Devon in 1938.

progress was made in the 1920s at developing either a concept for amphibious operations, or the necessary equipment. None the less, the first Motor Landing Craft were slowly developed from the mid-1920s. The Japanese operation against Shanghai in 1933 stimulated interest. A large combined exercise was staged against the Yorkshire coast in the autumn of 1934 and a joint service team began to prepare a 'Manual of Combined Operations' in 1936. In 1938, an Inter-Service Trials and Development Centre was set up at Hayling Island, and progress in creating amphibious warfare vessels began to accelerate as the war approached. None the less, the disastrous conduct of the Norway campaign of 1940 showed how much there still was to learn.

The Admiralty is also often accused not just of devoting too much attention to the necessities of battle, but to those of the wrong kind of battle. In particular, the charge goes, the Royal Navy, unlike its main competitors, was wholly concerned with re-fighting the Battle of Jutland. In fact, such a charge almost entirely misses the point, since all of the Royal Navy's preparations were designed to ensure that Jutland would not happen again. Thus, all processes of battlefleet operations were meticulously exercised, divisional tactics were explored, night fighting skills developed, communications improved, and so on. Such activities could be justified, moreover, by the fact that they were also an excellent preparation for many of the other duties the navy might be called upon to perform. Finally, they were what Britain's most likely enemies were preparing for too.

THE APPROACH TO BATTLE

The Royal Navy's approach to battle can best be described by following through, staff college fashion, the various stages of a notional battle of the time. The navy's instrument would be a balanced fleet comprising a judicious mix of all types of war vessel co-operating in mutual support. The mix might vary from time to time, from assumed enemy to assumed enemy, but Figure 1 illustrates the assumptions of 1924. After that time, the role of the battlecruisers tended to lose its distinctiveness, the fleet submarine operating in quite such close support was found to be impractical, and the more common form was for divisions of the battlefleet to be the unit of concern rather than the battlefleet as a whole. Nevertheless, exploring this ideal battle of 1924 will establish most of the principles which held true of the inter-war period.

It was firstly generally assumed that the enemy might very well be running away in a limitless ocean and so the first task was to find him. There were various ways of doing this. Submarine picket lines might be established off his harbours, although these had not proved very effective in the First World War. Reconnaissance aircraft operating from Coastal Command land-bases might help, though the Admiralty tended to place more reliance on its own reconnaissance aircraft operating from fleet carriers. These were expected to range out up to 200 miles from the fleet, and were flown by FAA personnel who set world standards in such tasks. If the weather was bad, or further refinement was needed, the advanced cruiser screen on what was known as the A-K line (positions identified by letters of the alphabet) would keep the battlefleet in touch with its intended victim.

343

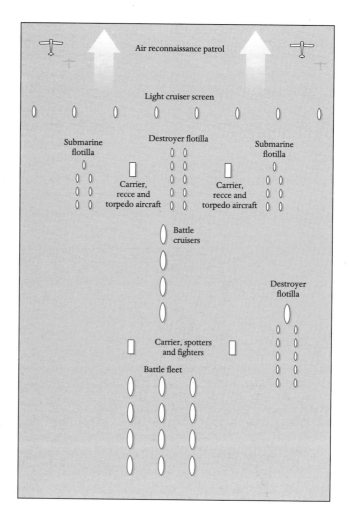

The fleet's next task was to 'fix' the enemy, or stop him running away. Submarine attack on some of his major units or aircraft operating from the fleet might be the way of doing this. Inter-war exercises and experiments tended to confirm the view that capital ships were in fact very difficult to sink. Accordingly, the expectation was that submarines and aircraft would probably only be able to damage and slow down the enemy's major units. The Royal Navy was slow to develop the dive-bomber (partly through difficulty in persuading the Air Ministry to develop the necessary dive-bombing sight) and was properly dismissive of the level bomber. Instead, the navy pinned its hopes on the torpedo-bomber, but these slow, lumbering aircraft were vulnerable to AA fire and their torpedoes inaccurate and lightly armed. The real problem though was the low number of any such attacking aircraft formation. But, none the less, with luck, the enemy would be fixed by such means. In the early part of the inter-war period, there was some expectation that the battlecruisers, operating as a squadron, might also have an important role to play in blocking the enemy's retreat.

Most of the world's naval officers believed that a major battle would culminate in a savage decisive artillery duel between battleships of the sort seen here. These are of the 'Royal Sovereign' class, were launched in 1915 with 15-inch guns, and were approaching obsolescence by the time war began.

As the fleets converged, both would move from a cruising disposition into one of a number of battle formations. The heavy guns of the battleships operating in a synchronized and co-ordinated fashion were expected to decide the battle, pulverizing the enemy's battle line at a range of perhaps up to 25 miles. But all other units of the balanced fleet were expected to have their vital supportive roles to play.

It was, for example, extremely important for both sides to get their battleships into a properly organized battle line, in which all the guns of fleet could be wielded as a co-ordinated whole. Given the importance of the first salvo, anything which disrupted the enemy's manœuvres might decide the outcome, even before the firing began. For this reason, the attacks of fleet submarines, or of destroyer flotillas, torpedo-bombers, or even of fighter aircraft spraying the bridges of enemy flagships with machine-gun fire might be all-important.

The enemy, desperate to escape his impending doom, could be expected to try to do the same. The British battlefleet, therefore, would need to be protected against enemy air attack by a fighter screen (though greater reliance tended mistakenly to be placed on shipboard AA fire) and against enemy submarines and destroyer attack as well.

Finally, the battlefleet would open fire with its 14-inch or 16-inch guns. A naval broadside was still very understandably regarded as one of the most awesome displays of military power possible in the inter-war period, and by dint of a good deal of practice, experimentation, and technological development the British hoped that their performance would be a good deal better than it had been in 1916. One of the most important developments of the 1920s was that of naval aircraft that could spot the fall of shot. These too would fly from the fleet carriers, and since they were operating over the enemy fleet, would need to be protected by fighters. Although the battlefleet of 1918 (with the battleship at its centre) is sometimes considered as a developed weapon system, in truth it changed through the inter-war period almost as much as any of its constituents, like the aircraft carrier, and certainly as much as did equivalent formations in the other services.

In any event, a long-range duel between opposing battle lines was expected to decide the outcome. Afterwards, further aircraft attacks, torpedo attacks by submarines, destroyers, and cruisers would finish the enemy fleet off.

This, of course, was a stylized battle which no one expected to unfold in quite this way (although both the battles of Calabria and Matapan against the Italians in 1940–1 were like it in many ways). In 1939, the German navy was in no position to offer this kind of opposition; later, the Japanese very well might, but here the sheer size of the Far East theatre was likely to be a complicating factor. None the less, this in broad outline was what the Royal Navy of 1939 thought, above all else, it should be prepared for.

After such an engagement the British would then be much more easily able to assert their command of the sea by defeating the enemy's attempts to interfere with British shipping, would be able to impose an economic blockade, deter or defeat enemy expeditions against British territory, and mount such amphibious operations as were necessary. These, then, were the expectations about the nature of battle and of naval warfare as a whole with which the Royal Navy went to war in 1939.

12

A SERVICE VINDICATED 1939–1946

ERIC J. GROVE

*T*f the war of 1914–18 had been a disappointment to the Royal Navy, that of
1939–45 was a vindication. The First World War had shown that there was room
for improvement in the Royal Navy's skills. As described in the last chapter, much
work had been done to improve equipment and operational and tactical doctrine,
especially in the areas of surface action and night fighting; the Royal Navy had few
equals in both by the late 1930s. Most importantly of all, the 1939 navy was imbued
with a new offensive spirit that the service of 1914 had conspicuously lacked.

Shortage of resources in the 1920s and 1930s had meant that naval planners had
been forced to make choices, investing in some areas and neglecting others. There
were two major weaknesses that would be vital in the forthcoming war. Naval
aviation was very weak, with small numbers of low-performance aircraft at sea
and land-based maritime aircraft still completely divorced from the naval service.
There was also insufficient capability to defend merchant shipping from
submarines. Doctrine in this latter area was flawed, with ocean convoy only to be
restored if unrestricted submarine warfare was reintroduced. The escorts under
construction were therefore short-range escort destroyers and corvettes intended
for coastal work. Neither proved good ocean escorts, although the latter were
often pressed into service as such. There had been little or no training in the
defence of mercantile convoys from submarine attack. Nevertheless, there was a
core of anti-submarine (A/S) expertise and equipment in the navy's A/S branch
that provided a foundation for the slow and painful construction of a successful
trade defence system. Moreover, the Royal Navy was able to cope with the threats
to the empire's shipping until the strategic situation changed in ways that pre-war
planners could not have predicted.

The strategic situation at the outbreak of the war was as favourable as could
have been expected for the Royal Navy. Neither Italy nor Japan joined the war and
the fleet could be concentrated against Germany, whose navy was far from strong.
There were 51 U-boats available in September 1939, half of which were small
coastal boats. Twenty-one of the larger submarines were at sea in the Atlantic and
the Channel. Their orders were to operate in a restricted manner and not sink
unarmed merchantmen without warning. One boat, however, U-30, mistook the
liner *Athenia* for an armed merchant cruiser and sank her without warning. The
112 passengers and crew lost did not die in vain for they convinced the Admiralty
that the Germans had indeed begun unrestricted submarine warfare. It was there-

Facing: The First Sea Lord,
1939–1943, Admiral Sir Dudley
Pound. His leadership and
relationship with Churchill
have been the subject of
considerable historical
controversy.

fore decided to introduce ocean convoys in the western approaches which did much to neutralize the U-boats. The ill-considered policy of hunting them with fleet carriers was abandoned after *Ark Royal* had a lucky escape and *Courageous* was sunk. Although outward-bound convoys were only escorted to 15° west, the convoy system was able to reduce losses from 41 ships in the first month of war to 21 in November. By the end of 1939, nine U-boats had been sunk, three by escorting destroyers, two by patrolling destroyers, one by a British submarine, and three by the Royal Navy's mine barrage in the Straits of Dover—which stopped German use of this path to the Atlantic.

In November and December 1939 it was the German mine that posed the main threat to British shipping. Fields of magnetic mines were laid off the east coast and the British had no answer; the port of London was almost closed. In December, the mining threat rose to a peak of 33 ships sunk. The threat then receded, first because the Germans ran temporarily out of mines, and then because of British countermeasures. In late November, Lieutenant-Commanders Ouvry and Lewis had dismantled a mine on the Shoeburyness mudflats. After examination at HMS *Vernon*, the Admiralty's torpedo and mine school in Portsmouth, its secrets were laid bare, and at the end of March 1940 minesweeping trawlers with special 'LL' magnetic sweeps began work. Ships were also fitted as rapidly as possible with 'degaussing' coils that neutralized the magnetic polarity of ships thus rendering the mines' firing mechanisms ineffective. The anti-mine campaign continued throughout the war with each German measure, such as the acoustic mine introduced in 1940, being met with an adequate countermeasure. Royal Navy personnel also distinguished themselves disposing of unexploded mines used as bombs in the attacks on Britain's cities.

The final prong of the German trident aimed at Britain's maritime trade was made up of the three 'Panzerschiffe', the 11-inch gun armed 'pocket battleships' specially built for long-range raiding operations. Two were at sea at the outbreak of war: *Deutschland* in the North Atlantic and *Admiral Graf Spee* in the South Atlantic and Indian Ocean. Both had strict orders to avoid British warships, which limited their impact as they could not therefore attack convoys. *Graf Spee*, however, decided to disobey instructions and win a significant naval action by sinking an escorted convoy expected off the River Plate. This sealed her fate. *Graf Spee* found not a convoy but Commodore Henry Harwood's raider-hunting group 'G', the heavy cruiser *Exeter* and two light cruisers, *Ajax* and *Achilles*. Harwood had guessed the mystery South Atlantic raider might be drawn to the Plate. He handled his ships brilliantly in action splitting his force in two, in the fashion of the divisional tactics developed pre-war. This divided *Graf Spee*'s fire and the German ship was so badly damaged in the ensuing action that she entered Montevideo for repairs. Harwood was reinforced by his fourth cruiser *Cumberland* and heavier ships still were on their way. Even if *Graf Spee* broke out she was low on ammunition and her chances of escaping heavier Allied forces were slim; she had little choice but to scuttle herself. Harwood was knighted and promoted Rear-Admiral, and full propaganda value was garnered from a truly famous victory. Allied command of the sea was unimpaired and surface raiders sank only 15 ships of 61,337 tons in the first four months of war.

On 3 September 1939 the Admiralty had signalled to the fleet, 'Winston is back', as First Lord of the Admiralty. This news buoyed up the morale of the junior ranks, as Churchill's arrival as political head of the navy seemed to confirm the prospect of decisive action. Churchill's restless energy always yearned for the offensive and he was far from satisfied with a policy of defending shipping and the distant blockade that was restored on the war's outbreak, with armed merchant-men backed up by the Home Fleet at Scapa Flow interdicting German traffic into and out of the North Sea. This was especially so as the Germans took the opportunity to make raids on the blockaders both at sea and in harbour. In the early hours of 14 October, the battleship *Royal Oak* was sunk by Gunther Prien's U-47 in Scapa Flow. Then the following month, the armed merchant cruiser *Rawalpindi* was sunk on the Northern Patrol by the battleships *Scharnhorst* and *Gneisenau*.

The Naval Staff, headed by Admiral Sir Dudley Pound, the First Sea Lord, strug- THE NORWEGIAN gled to mitigate the First Lord's wilder ideas. Much work went into dissuading CAMPAIGN Churchill from Operation Catherine, forcing the Baltic and cutting Germany off from her vital Scandinavian iron ore supplies. This plan, that would have risked serious losses to the fleet to air and underwater attack, was replaced by thinking about a Scandinavian offensive to capture the Swedish iron ore mines, or at least cut the supply line down the Norwegian coast. In effect, a race for Norway began between the Allies and the Germans which the Germans narrowly won. On 9 April 1940 a series of German landings took place at every major Norwegian port from Oslo to Narvik.

On 8 April the Royal Navy had begun minelaying in Norwegian waters. In the Clyde cruisers were ready with troops in case this provoked German reaction, but these forces were disembarked by Admiralty order when the German movements were misconstrued as an attempted break out by German heavy ships into the Atlantic. The forces covering the minelayers clashed with the enemy off the Norwegian coast. Lieutenant-Commander Roope's destroyer *Glowworm* came across the heavy cruiser *Hipper* of the Trondheim force, which Roope decided to ram, causing her considerable damage—though not enough to prevent the land-ing. Roope was awarded a posthumous Victoria Cross. Admiral Whitworth in the battlecruiser *Renown* fought a running battle with the two 11-inch armed German fast battleships *Scharnhorst* and *Gneisenau*, in which the modernized 15-inch gun British capital ship with its improved fire control equipment came off best. The Germans were forced to use their superior speed to get away in the heavy seas and bad weather. The Germans succeeded, however, in getting their troops ashore, even at Narvik, where eight destroyers slipped undetected into Vestfiord and destroyed the Norwegian coast defence vessels protecting the harbour. The Admi-ralty had uncovered the entrance by its orders to concentrate forces to deal with the German 'break-out'.

Nevertheless, this whole adventure should still have been strategic suicide for the Germans with their inferior fleet, but for one factor: Fliegerkorps X of the Luftwaffe. This was specially trained in maritime operations and was rapidly deployed to Norwegian airfields. The Admiralty prevented Admiral Sir Charles Forbes, the Commander-in-Chief, Home Fleet, mounting an immediate attack on Bergen before Fliegerkorps X arrived, and the events of the afternoon of 9 April showed that it was already too late. In an all too accurate bombing attack on the Home Fleet the large destroyer *Gurkha* was sunk. Other ships were hit but not seriously damaged. The attack persuaded Forbes that waters covered by bombers were too dangerous for his fleet. Even if the capital ships were safe, his cruisers and destroyers were not.

The viability of the surface ship against the bomber had been a matter of great debate in the pre-war period. The Royal Navy put much faith in medium anti-aircraft guns directed by a 'High Angle Control System' (HACS). This was a relatively crude device that only estimated the enemy aircraft's course and speed rather than measured it. The Naval Staff seems to have thought that the stabilized platform required by a full tachymetric system was beyond its suppliers' capabilities, but the performance of HACS against real attacks proved disappointing. The danger to his ships from the air was such that Forbes decided to withdraw northwards, leaving the main German maritime supply lines to submarines.

The Royal Navy also used its own aircraft to strike back, notably when Skua dive-bombers from Orkney sank the cruiser *Konigsberg* in Bergen on 10 April—the first large warship ever to be sunk from the air. The main British successes, however, were in the north where Fliegerkorps X was still unknown. On the morning of 10 April, the five 'H'-class destroyers of the Second Flotilla, led by Captain Warburton-Lee, an archetype of the aggressively minded destroyer commanders bred by inter-war exercises, made an attack on the German destroyers at Narvik. Two German ships were sunk, including the flagship, and three damaged. Six merchant ships were also sunk, including the supply ship for the troops ashore. Sadly, however, the attack had gone in unsupported because of more Admiralty-originated confusion and Warburton-Lee was caught by the five remaining German ships on his way out. Warburton-Lee's *Hardy* was hit and the Captain (D) killed. She had to be run aground and the destroyer *Hunter* was also lost, colliding, before she sank, with her sister *Hotspur*. The latter was able to get away with the other two ships, but the award of a posthumous Victoria Cross to the heroic Warburton-Lee could not wipe out the slightly sour taste of an operation that could have gone better. Whitworth blamed himself for not having used more initiative to go against the orders from London and led an overwhelming force up the fiord on 13 April to finish the job. This consisted of the battleship *Warspite* and nine destroyers, which annihilated the eight remaining German destroyers at the cost of serious damage inflicted upon the destroyers HMS *Cossack* and *Eskimo*.

The two battles of Narvik were a disaster for the German navy, involving the loss of the backbone of its destroyer force. They were, however, the only bright spots in an otherwise dismal record for the Royal Navy. In mid-April troops were landed under naval escort at Harstad near Narvik, but neither here, nor at the other areas in central Norway chosen to mount counter-offensives, was the navy able to make much progress against the twin enemies of German air power and back-seat driving from a disastrous combination of over-enthusiastic First Lord and over-centralizing First Sea Lord. British aircraft carriers without, as yet, the means of fighter control, were incapable of using their own aircraft to interfere with German air attacks. From early May, these were being carried out by the formidable Stuka dive-bomber which proved its anti-ship potential by sinking both a French destroyer and HMS *Afridi* during the evacuation of Namsos at the beginning of May. In the end Narvik was briefly taken only to be evacuated again as a result of the opening of a life and death struggle in France and Flanders. It was the final overflow of misfortune that yet another valuable fleet carrier, HMS *Glorious*, was thrown away during the evacuation when it was separated from the main fleet and caught by *Scharnhorst* and *Gneisenau*.

The German conquest of France in May–June 1940 had a fundamental impact on the Royal Navy. First it had to evacuate the British Army from the ports and beaches of mainland Europe, notably from Dunkirk. Contrary to popular legend the small craft mobilized by the navy to assist in the evacuation carried very few troops all the way back to Britain. Of the 338,226 troops lifted by Operation Dynamo half were lifted by warships and many other ships and craft were Royal Navy manned. The entire operation was organized by the Flag Officer Dover, Admiral Sir Bertram Ramsay, ably assisted by Rear-Admiral W. F. Wake-Walker and Captain W. G. Tennant.

EFFECTS OF THE GERMAN CONQUEST OF FRANCE

The Royal Navy then had the distasteful task of neutralizing the French fleet. A new Gibraltar-based squadron, Force H, was formed under Vice-Admiral Somerville to take over from the French responsibility for the western Mediterranean, and its first task was to be ordered by Churchill, now prime minister and Minister of Defence, to bombard the French fleet at Mers-el-Kebir near Oran. One French battleship was destroyed and a newer battlecruiser disabled, but another battlecruiser *Strasbourg*, a major target of the operation, got away to Toulon, escaping an attack by *Ark Royal*'s slow Swordfish. At Alexandria the Commander-in-Chief, Mediterranean Fleet, Admiral Sir Andrew Cunningham, incurred considerable Churchillian displeasure by his slow but statesmanlike attitude which obtained the disarming of the French ships there without bloodshed. French feelings were understandably outraged by the Mers-el-Kebir attack and an attempt in September to use most of Force H plus units of the Home Fleet and South Atlantic Station to cover a Free French landing at Dakar was driven off, with damage to the battleship *Resolution* and cruiser *Cumberland*.

September 1940, however, saw the Royal Navy succeed in one of its main and most unsung achievements of the war—the deterrence of a German invasion of the United Kingdom itself. The arrival of the German army in north-western France and the Low Countries led the Admiralty to fill the ports on the east and south-east coasts of England with warships to guard against the threat of invasion.

By early September 1940, Admiral Sir Reginald Plunkett-Ernle-Erle-Drax's Nore Command had 38 destroyers and seven 'Kingfisher'/'Shearwater'-class coastal escorts based at Immingham, Harwich, and Sheerness waiting to fall on any German invasion force ill-advised enough to try a Channel crossing. These were backed up by the cruisers *Manchester*, *Birmingham*, and *Southampton* at Immingham, and *Galatea* and *Aurora* at Sheerness. On the other flank of the invasion, Commander-in-Chief, Portsmouth deployed another nine destroyers and five Free French torpedo boats backed up by the old cruiser *Cardiff* and battleship *Revenge*. No fewer than 700 smaller craft were also deployed, with 200–300 at sea at all times from the Wash to Sussex to provide early warning.

The battered German navy was in no position to provide an effective screen and even the Luftwaffe would have faced serious problems preventing a flotilla of this size from a massacre of the Wehrmacht in its improvised invasion barges. The German air force itself thought it had an alternative to 'Seelowe' ('Sealion'), the invasion plan. This was the 'Adlerangriff' ('Eagle Attack'), an attempt to gain air superiority over southern England and then systematically destroy London in a few days. The Luftwaffe had little interest in merely supporting the other two services. The Adlerangriff, the Germans' only credible strategy against a Britain that still commanded the narrow seas, foundered on the rocks of RAF Fighter Command's highly advanced air defence system, a belated vindication of the creation of a separate air service in 1918.

THE ATLANTIC
CAMPAIGN, 1940–1941

The forced deployment of valuable escort vessels to anti-invasion duties helped tilt the Battle of the Atlantic in Germany's favour. More importantly, Germany's possession of the Biscay ports was a force multiplier of the utmost importance for the U-boats. These were circumstances that no pre-war planner could have contemplated. On 17 August, Hitler declared a total blockade of the British Isles; all shipping of whatever nationality was to be sunk on sight. German U-boat commanders called the period from July to October 1940 the 'Happy Time'. No fewer than 217 merchantmen fell victim to the U-boats, whose captains made their reputations picking off individual, unescorted ships as they passed outward bound beyond the convoy dispersal point, straggled from the convoys, or sailed their lonely independent courses. By September and October the Germans even began to inflict serious attrition on convoyed shipping for the first time using 'wolf pack' night attacks.

The escorts were not able to do more because they were totally unprepared for the scale and nature of the massed night attacks. There were no proper communications between the ships beyond signal lamps and the different escort vessels often arrived unbriefed about tactics or even the nature of the convoys. The commanders did not know each other and operated as uncoordinated individuals. With the invasion threat reduced, more ships could be allocated to escort tasks, and at the end of October the close escort line was moved further westward. Escort groups began to be formed and proper training began to be given to new anti-submarine crews, notably by the terrifying Commodore 'Monkey' Stephen-

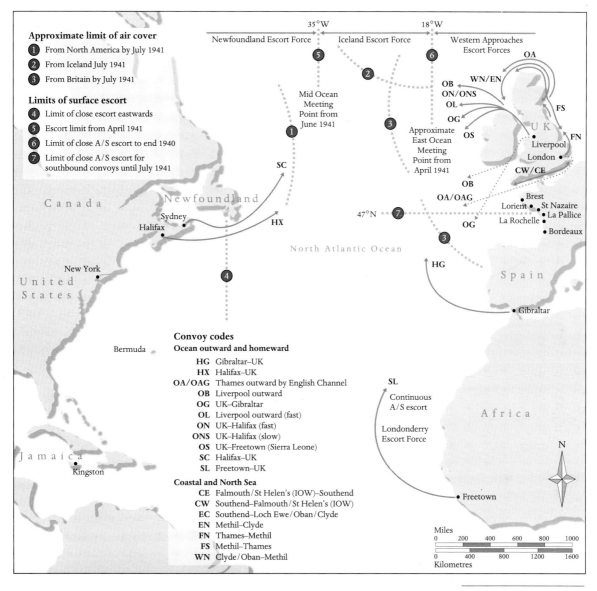

Approximate limit of air cover

1. From North America by July 1941
2. From Iceland July 1941
3. From Britain by July 1941

Limits of surface escort

4. Limit of close escort eastwards
5. Escort limit from April 1941
6. Limit of close A/S escort to end 1940
7. Limit of close A/S escort for southbound convoys until July 1941

Convoy codes

Ocean outward and homeward

HG	Gibraltar–UK
HX	Halifax–UK
OA/OAG	Thames outward by English Channel
OB	Liverpool outward
OG	UK–Gibraltar
OL	Liverpool outward (fast)
ON	UK–Halifax (fast)
ONS	UK–Halifax (slow)
OS	UK–Freetown (Sierra Leone)
SC	Halifax–UK
SL	Freetown–UK

Coastal and North Sea

CE	Falmouth/St Helen's (IOW)–Southend
CW	Southend–Falmouth/St Helen's (IOW)
EC	Southend–Loch Ewe/Oban/Clyde
EN	Methil–Clyde
FN	Thames–Methil
FS	Methil–Thames
WN	Clyde/Oban–Methil

Building the Atlantic Convoy System, 1939–41.

son at HMS *Western Isles* at Tobermory. Radio telephones allowed effective tactical co-operation and the standard depth charge pattern was doubled in size.

Losses to U-boats increased to a new peak of 61 ships in June 1941, but the tide was already turning. The situation improved for the British for four main reasons. First, more escorts became available which allowed convoys to be extended. The 50 old destroyers obtained from the USA came into service in the escort role, later supplemented by former US Coastguard Cutters. The Royal Navy began to get the vital assistance of a fast-growing Royal Canadian Navy and forward basing in Iceland extended reach. From April 1941 convoy anti-submarine surface escort was extended to 35° west. Then in May the first convoys were given escort all the way across the Atlantic, and in mid-July convoy was finally extended to Sierra Leone.

Secondly, better equipment was developed. The problem of U-boat attack on the surface at night was solved by the fitting of radar to escorts, first the metric wave 286 and then the revolutionary centimetric 271 that had sufficient resolution to detect reliably the low silhouette of a surfaced U-boat. When starshell proved ineffective at illumination it was replaced by brighter 'snowflake' illuminants. ASDIC crews also became more adept at exploiting their instruments to follow surfaced submarines passively. Wolf-pack tactics meant much radio indiscretion by U-boats and their commanders, and Royal Navy escorts began to be fitted with High Frequency Direction Finders (HFDF).

Admiral Sir John Tovey, Commander-in-Chief, Home Fleet, 2 December 1940– 14 April 1943. Best known for sinking the *Bismarck*, Tovey was involved in covering the early Arctic convoys to the USSR. Tovey was an officer of much moral courage, trusted by his men who knew he would support their interests, even at the cost of displeasure in Whitehall. He is seen on board his flagship, HMS *King George V*.

Thirdly, the Admiralty developed an efficient submarine tracking system. This was at first based on radio direction finding alone, but in 1941 the coded messages began to reveal their secrets. Between March and June the U-boat code was effectively broken by the government cryptanalysis centre at Bletchley Park. German signals could usually be read within 36 hours, sometimes immediately. Convoys could thus often be routed away from known enemies.

Fourthly, co-operation with RAF Coastal Command improved. It was placed under Admiralty operational control and joint area headquarters were created, notably for Western Approaches Command and 15 Group at Derby House, Liverpool. For all these reasons, U-boat productivity slumped permanently and the 'aces' began to disappear. In March 1941 three notable U-boat captains, Prien, Schepke, and Kretschmer, were all killed or captured in attacks on convoys. The cutting-edge in these significant successes were the First World War 'V/W'-class destroyers, a vindication of the Admiralty's decision to retain sufficient of these valuable ships in reserve with exactly the role of convoy escort in mind.

The Royal Navy continued to contain the surface threat. In October, Captain E. S. Fogarty Fegen of the armed merchant cruiser *Jervis Bay* earned a posthumous Victoria Cross covering the scattering of convoy HX 84 when it was attacked by the pocket battleship *Admiral Scheer*. On Christmas Day 1940, the cruiser *Admiral Hipper* was driven away from the Middle East troop convoy WS 5A by the strong escort, the cruisers *Berwick*, *Bonaventure*, and *Dunedin* supported by the carrier *Furious*. Early in 1941, *Scharnhorst* and *Gneisenau* were able to cruise the North Atlantic for almost two months, but on the two occasions they found convoys they were deterred from attacking by the old 15-inch gun battleships which the British were using as ocean escorts specially with this eventuality in mind. Some of Germany's disguised merchant raiders were also brought to book. In May 1941, the heavy cruiser *Cornwall* destroyed the *Pinguin* in the Indian Ocean, and the specially successful *Atlantis* finally succumbed to the cruiser *Devonshire* in November.

In May 1941 Germany's first full-sized battleship, *Bismarck*, broke out into the Atlantic. In company with the cruiser *Prinz Eugen* she attempted to pass through the Denmark Strait. The Commander-in-Chief, Home Fleet, Admiral Sir John Tovey, who had replaced Forbes in December 1940, was covering this passage with the battlecruiser *Hood* and new battleship *Prince of Wales*. This looked a more powerful pair than it was. The mighty *Hood*, largest capital ship in the fleet and pride of the inter-war navy, was more a fast battleship than a true battlecruiser, but had old-fashioned armour protection that was vulnerable to long-range fire and above-water torpedo tubes that threatened to blow the ship in half if hit. *Prince of Wales* was so new that her crew was not yet worked up; indeed she still had civilian workmen on board to deal with teething troubles. The Germans with typically accurate opening salvoes were able to exploit these weaknesses to annihilate *Hood* and drive off the hapless *Prince of Wales*. Tovey tried to use his available fleet carrier *Victorious* to slow the Germans down but she was as new and ill-prepared as *Prince of Wales* and her strike with Swordfish torpedo bombers was ineffective. Then, to add to Tovey's woes, the Germans were able to shake off their shadowing cruisers. *Bismarck* had suffered one damaging hit from *Prince of Wales* that caused her to leak oil and she was forced to separate from her consort and make for Brest to join *Scharnhorst* and *Gneisenau*. Poor radio discipline, a little cryptanalysis, and an RAF flying boat brought the battleship back into contact and she was attacked by Somerville's Gibraltar-based Force H in the shape of a strike of Swordfish torpedo bombers. Despite an initial mistaken attack on the cruiser *Sheffield*, the Swordfish eventually crippled *Bismarck* with a hit in the stern. Unable

THE *BISMARCK*

to steer, she was closed by Tovey in his flagship *King George V*, supplemented by HMS *Rodney* that had been diverted from a voyage to America to begin a refit. The two British capital ships pounded the one German into a wreck that had to be scuttled by her own crew.

THE MEDITERRANEAN, 1940–1941

The *Bismarck* affair, with the German battle group being faced by a Home Fleet reduced to one fully operational, completely modern capital ship, showed how thinly spread was the Royal Navy by the spring of 1941. The main reason for this was the diversion of strength into the Mediterranean following Italy's entry into the war in June 1940. In 1939 naval planners had toyed with the idea of a forward strategy to knock out Mussolini. The architect of this thinking, the then Deputy Chief of Naval Staff Andrew Cunningham, was sent to the Mediterranean to put it into effect if required. Cunningham had been forced to take a back seat in the opening months of the war, but now he had every intention of handling his reinforced fleet as aggressively as possible.

Malta, Cunningham's main fleet base, was held to be untenable as such because of its proximity to Sicilian air bases and the Commander-in-Chief moved his main base to Alexandria. In July, both the Mediterranean Fleet and the Italian fleet were at sea covering convoys in the central Mediterranean and they came into contact off Calabria. The British had three capital ships to the Italians' two. Cunningham's fleet flagship, the modernized *Warspite*, was able to hit the Italian *Giulio Cesare* at disturbingly long range, causing the Italians to withdraw. Sadly, however, the British carrier, the old *Eagle*, had great difficulty in slowing the enemy down with its small number of Swordfish. The Italians were thus able to get away, but Cunningham had succeeded in asserting an important moral ascendancy.

Both sides increased their strength in the succeeding months. Cunningham received the second modernized 'Queen Elizabeth'-class battleship, *Valiant*, and the first of the new fleet carriers, *Illustrious*. The Italians tripled their battleship strength to six ships, including two new 15-inch gun units, *Littorio* and *Vittorio Veneto*. Cunningham therefore decided to even up the odds once more by a strike on the enemy fleet in its Taranto base. *Illustrious* was used with an air group reinforced from *Eagle*. On the night of 11 November, 21 Swordfish succeeded in torpedoing three of the Italian battleships, including the *Littorio*. At a stroke Italian strength was halved and the surviving members of the fleet withdrew to less exposed bases.

The remarkable success of Taranto allowed the British to begin passing convoys through the Mediterranean once more, covered by Force H in the west and Cunningham in the east. The Italians sent out two of their surviving battleships against the first of these, but *Ark Royal*'s slow Swordfish, with less well-trained crews were less effective against the fast-moving Italians by day than *Illustrious*'s similar aircraft had been against anchored opponents by night. Somerville had to break off the pursuit.

Much worse was to follow, however. As Mussolini faced strategic failure everywhere, Hitler was forced to come to his aid. Fliegerkorps X was deployed to Italy

The greatest Royal Navy commander of the Second World War and one of the greatest leaders in its history, Admiral Sir Andrew B. Cunningham ('ABC') seen during his first period as Commander-in-Chief, Mediterranean Fleet, 1 June 1939–1 April 1942.

at the beginning of 1941 and immediately registered its appearance by disabling
Illustrious in a heavy Stuka attack on 10 January. Fliegerkorps X, which sank the
cruiser *Southampton* the following day, had completely reversed the decision of
Taranto. It also opened the first air siege of Malta that struggled with increasing
difficulty to keep up some pressure with air and submarine forces on Axis supply-
lines to North Africa.

    As Cunningham campaigned for better anti-air warfare capability and
improved shore-based air cover his preoccupations shifted to the deployment of
British troops to Greece that began in March 1941. The Germans prevailed upon
the Italians to use their still-powerful surface fleet to interdict this movement, and
the result was the Battle of Cape Matapan at the end of the month. Cunningham
had been reinforced via the Cape with the carrier *Formidable*, and her Albacores
hit the battleship *Vittorio Veneto* slowing her down for a time. She was eventually
able to get away but the Italian heavy cruiser *Pola* was damaged in a second air
strike. The Italian commander sent back two more similar ships to assist and this

Force H, the Royal Navy's
Gibraltar-based task force, a
key squadron in 1940–1.
Commanded by Admiral Sir
James Somerville and
composed of the modernized
battle cruiser *Renown*, fleet
carrier *Ark Royal*, and light
cruiser *Sheffield*, it fought in
both the Atlantic and
Mediterranean.

sealed the fate of all three, which were sent to the bottom by gun and torpedo fire in a night surface action. The Italians were totally unprepared for such encounters for which the Mediterranean Fleet had been training for a decade under successive Commanders-in-Chief.

This success at sea was soon overshadowed by failure ashore. The Germans swept through Greece and Cunningham was faced with the need to cover evacuation. Over 50,000 men were evacuated for the loss to air attack of four transports and two destroyers. Then Churchill insisted that an emergency convoy, 'Tiger', be fought through the Mediterranean to reinforce both Malta and Egypt. This operation was successfully carried out and Cunningham received a valuable new battleship, the third modernized 'Queen Elizabeth', *QE* herself, but *Formidable* suffered significant losses to her small fighter complement. This was to create problems as the Mediterranean Fleet's greatest travails from the air were about to begin.

Malta provided the base for the famous 10th Flotilla of small U-class submarines that preyed on Axis supply shipping in the central Mediterranean. Submarines played a subordinate role in the Royal Navy's operations in the Second World War being most useful where, as in Malta in 1941–2, the British had lost command of the sea.

The Germans wished to complete their occupation of Greece by taking Crete through combined airborne and seaborne attack. The Royal Navy was able to massacre the seaborne component of this attack but the paratroops, at great loss, forced the British troops—drawn from many parts of the commonwealth and empire—into yet another evacuation. The attack was covered by Fliegerkorps VIII, whose aircraft soon cut their teeth all too effectively in the anti-shipping role. Fliegerkorps X also joined in from North African bases. In the operations around Crete, Cunningham lost two light cruisers, an AA cruiser, and six destroyers. Three battleships (including *Warspite* that had to go to the United States for repairs), five cruisers, another AA cruiser, and 7 more destroyers were damaged, as was *Formidable*, crippled by the same air forces that had smashed her sister. The Mediterranean Fleet approached breaking-point as it repeatedly ran the air gauntlet, but over half of the garrison of 32,000 was transported from Crete to Egypt. The Commander-in-Chief's powers of leadership were strained as tautly as the fleet itself, but Cunningham was able to gain an extraordinary response to his appeal not to let the army down.

Only the diversion of much of the Luftwaffe to the Eastern Front took the pressure off the Mediterranean in the second half of 1941. The Luftwaffe remained strongest in the eastern basin and Malta had to be supplied from the west. Repeated carrier operations flew in fighters while two major convoys were run covered by Force H reinforced from the Home Fleet. More developed techniques of radar fighter control allowed *Ark Royal* to provide effective air cover.

Thus sustained, Malta began to be a more important submarine and air base. The small boats of the U class based there were established formally as the 10th Flotilla in September 1941, but despite much heroism and considerable success, notably by Lieutenant-Commander M. D. Wanklyn in *Upholder*, who sank two Italian liners in a pack attack on a convoy, the pressure was marginal at best. There were more supplies reaching Axis North African harbours than could be transported by truck to the front. Only surface forces could really cut the Axis jugular, and in October Force K was sent there, composed of the small light cruisers *Aurora* and *Penelope* and two destroyers. Force K was able to annihilate supply convoys located by code-breaking and the Axis supply crisis finally became real. Force K was reinforced by Force B of similar size and the Axis disasters continued. A destroyer force on passage to reinforce Cunningham added to Axis woes by sinking two Italian fast light cruisers trying to rush fuel to Tripoli.

The Germans were forced to react. In October Hitler decided to send U-boats to the Mediterranean. This had the useful side-effect of reducing the pressure in the Atlantic, but the increased threat in the Mediterranean soon took its toll. On 13 November, while returning from yet another reinforcement of Malta's air strength, *Ark Royal* was torpedoed and sunk. Eleven days later, U-331 penetrated the screen of the Mediterranean Fleet and sank the battleship *Barham*. British woes continued in December. Although the Italian fleet was kept from the fast supply ship *Breconshire* off Sirte by a cruiser force from both Malta and Alexandria, when the Malta cruisers sailed immediately afterwards to attack another convoy they ran into a minefield losing a cruiser and destroyer sunk, and another cruiser damaged. The Malta striking-force was effectively neutralized. In Alexandria harbour itself the remaining Mediterranean Fleet battleships *Queen Elizabeth* and *Valiant* were sunk at their moorings by Italian 'Maiale' human torpedoes. Within a few days the main Royal Navy striking potential in the Mediterranean had been destroyed. Finally, in December, Fliegerkorps II was deployed to Sicily to reopen the air siege of Malta.

<table>
<tr><td>

JAPANESE ENTRY INTO
THE WAR: THE FAR
EAST, 1941–1942

</td><td>

The expensive Mediterranean campaign that cost the Royal Navy so dear meant that when American pressure looked as if it might drive Japan to war in late 1941 the pre-war concept of an Eastern Fleet based on Singapore would be impossible to maintain. The Admiralty produced cautious plans for a fleet based in the Indian Ocean, but Churchill insisted that a small but fast deterrent force be sent all the way to Singapore. This was to consist of the battleship *Prince of Wales*, battle-cruiser *Repulse*, and carrier *Indomitable*. In late October, the prime minister brow-beat Pound into sending *Prince of Wales* to join *Repulse*, which was already in the Indian Ocean, but *Indomitable* never appeared as she was damaged in a grounding accident. Force Z was nevertheless formed, commanded by Admiral Sir Tom Phillips, a gifted staff officer who had fallen foul of Churchill. A sea command was a good way to get Phillips away from London but he had no recent operational experience.

</td></tr>
</table>

Phillips's force was unbalanced, with no cruisers and only four destroyers of which half were First World War veterans. When the Japanese invaded Malaya on 8 December, Force Z sailed to interfere with the landings at Singora. Phillips seems to have misinterpreted the information that fighter cover was not available in the north as meaning that no fighter cover was available anywhere. He never subsequently asked for air cover, despite the provision of a fighter squadron at Singapore specially for the purpose. Abandoning his original plans because of being spotted by Japanese reconnaissance aircraft, he was misled into closing the coast at Kuantan thinking that landings had taken place there also. Force Z was spotted by a striking force of Japanese navy land-based strike aircraft. Torpedo bombers first disabled the flagship and then sank *Repulse*, despite excellent ship handling by Captain Tennant, the hero of Dunkirk. *Prince of Wales* then succumbed to her extensive flooding. The Singapore base itself, linchpin of inter-war British strategy, fell with the rest of the island two months later.

The remnants of the Royal Navy in the Far East became a contribution to an ill-fated Allied force of cruisers and destroyers which under Dutch command fought vainly to defend the Dutch East Indies. Without the basics of Allied tactical co-operation such as a common signalling system, and outfought by a Japanese navy that had developed surface torpedo warfare to perhaps an all-time peak, the Allies were defeated at the Java Sea and their squadron sunk or dispersed. Among the losses was HMS *Exeter*, hero of the River Plate.

In order to protect India the Admiralty reverted to their policy of building up as powerful an Eastern Fleet in the Indian Ocean as possible. By the end of March,

all four old 'R'-class battleships and the repaired *Warspite* flying the flag of Admiral Sir James Somerville were based in Ceylon and at Addu Atoll in the Maldives. With the battleships were three carriers, the armoured-hangar *Indomitable* and *Formidable* and the old *Hermes*. Between them these ships carried only 90 aircraft, Albacore and Swordfish biplanes and a mixed bag of Martlet, Fulmar, and Sea Hurricane fighters. These were soon to be faced by virtually the entire carrier fleet that had attacked Pearl Harbor—five carriers with 300 modern, high-performance fighters and strike aircraft, screened by four fast battleships and a dozen smaller escorts. Somerville, newly arrived in the theatre, had warning of the Japanese approach and planned to try to manœuvre into a position where he might use his unique ability to fly at night together with a supposed—and misplaced—superiority in night action to bring on a night engagement. He knew he could not fight the Japanese carriers by day.

In the end the two fleets missed each other. The Japanese attacked Colombo and Trincomalee but only sank detached portions of Somerville's forces, notably the cruisers *Cornwall* and *Dorsetshire* and the carrier *Hermes*. The Japanese, on a raiding expedition to protect their western flank before trying conclusions with the Americans in the Pacific, withdrew. Yet Churchill was right to call this 'the most dangerous moment' of the war. If the Japanese had put more effort into an Indian Ocean offensive to cut Allied supply-lines, Somerville's fleet would have been doomed. His 'fleet in being' policy was exactly the right one. A major defeat in the Indian Ocean, with the loss of perhaps all four of the old and vulnerable 'R's, each one a potential *Hood*, would have been a blow from which British prestige,

'The most dangerous moment' of the war was when the full might of the Japanese carrier fleet made a foray into the Indian Ocean at the beginning of April 1942. Isolated members of the Eastern Fleet were sunk, including the heavy cruiser *Cornwall* which suffered nine bomb hits in 12 minutes in a dive-bombing attack.

and the Churchill government, would have found it hard, if not impossible, to recover, especially in the aftermath of the fall of Singapore. All that could be done was to rely on the US fleet to defeat Japan in the Pacific while safeguarding as far as possible communications in the western Indian Ocean. Madagascar was occupied in a well-executed amphibious operation in early May.

HOME WATERS AND ARCTIC CONVOYS, 1941–1942

Early 1942 was not a good time for either the British empire or the overstretched Royal Navy. In February, the battleships *Scharnhorst* and *Gneisenau*, with *Prinz Eugen*, carried out the daring Channel dash to return home from Brest up the Channel. The limited surface and Fleet Air Arm forces available in the area were unable to prevent this débâcle, despite much heroism. Air-laid mines were, however, able to damage the German ships, one of the many successes of the joint Admiralty–Air Ministry air mining campaign. The Royal Navy supplied the weapons and coastal and bomber commands the delivery aircraft.

Minelaying was also a role for coastal forces. The fall of France had given small fast-attack craft a new relevance to the Royal Navy. German motor torpedo boats ('E-boats') preyed on coastal convoys, and motor gunboats (MGBs) were developed as countermeasures. British motor torpedo boats (MTBs) attacked German coastal convoys by night while motor launches (MLs) were used for escort and patrol. By November 1942, 90 MGBs, 101 MTBs, and 263 MLs were deployed round

Motor torpedo boats and other coastal craft came into their own in the operations in the narrow seas caused by enemy occupation of France and the Low countries. German 'E-boats' were larger and more heavily armed and such vessels as these needed the support of motor gunboats (MGBs).

Britain's coasts. In October 1942, they scored a major success when the disguised raider *Komet* was sunk by MTB 236 as it moved down-channel. Coastal forces played a significant role until the end of the war, in the Mediterranean as well as the narrow seas. Over the whole war they sank 40 enemy merchantmen and 70 warships.

Since the late summer of 1941, the maritime supply of the Soviet Union had been a major new commitment for the Home Fleet. The Arctic convoys were as much a political gesture as a real logistical link. Over the war as a whole three-quarters of the Anglo-American supplies to Russia went via the Pacific or the southern route via Iran. Nevertheless, in 1941–2 the Arctic route was dominant in terms of supplies delivered, and at a time when there were few Anglo-American troops in action no better proof of commitment could have been found than the efforts made to fight the merchant ships through to Archangel and Murmansk against some of the bitterest resistance and in the vilest weather conditions faced by seamen in any theatre in any war. The overwhelming burden of the escort and support of the PQ convoys, as they were first code-named, fell on the already heavily laden shoulders of the Royal Navy.

In early 1942, the Germans built up their surface forces in Norway. *Tirpitz*, *Bismarck*'s sister ship, arrived in January and in early March, escorted by three destroyers, she put to sea to intercept PQ 12. Code-breaking allowed Tovey to sail with *King George V*, *Duke of York*, and *Renown* to intercept. All depended on *Victorious*'s ability to slow *Tirpitz* down, but the carrier's Albacores could not score any hits. It was a disappointment but not a complete failure. The Germans decided to be still more cautious, especially if a British carrier was in the vicinity.

Three destroyers were used to attack PQ 13 which sailed at the end of March. The Germans were, however, caught by the covering force, the British cruiser *Trinidad* and the destroyers *Fury* and *Eclipse*. The weather conditions were vile, with a heavy snowstorm blowing and freezing spray. The big, heavily armed but top-heavy German ships were at a disadvantage and Z26 was sunk. Unfortunately, *Trinidad* suffered the indignity of being hit by one of her own torpedoes that reversed course in the extreme cold! She was given temporary repairs in Russia but was bombed and sunk on the way home in May.

As the spring days rapidly lengthened in the far north Pound warned the Cabinet of the dangers of running Arctic convoys in these conditions. U-boat attacks and an increasingly effective Luftwaffe took their toll of merchant ships and escorts alike. A major loss was the large light cruiser HMS *Edinburgh* torpedoed by U-456, again by three German destroyers, and finally given a *coup de grâce* by the British destroyer *Foresight*. The Germans had to pay for their success, however. One German destroyer was sunk by the doomed *Edinburgh*'s extraordinarily accurate fire. Despite such set-backs, however, the convoys were able to get through in both directions with remarkably little loss considering the circumstances.

The Germans were dissatisfied with these results. In June 1942, therefore, they determined to strike the next convoy with their full battle group. The target was PQ 17, a convoy of 35 merchantmen which came together in Hvalfiord in late June.

The close escort was strong: six destroyers, four corvettes, three minesweepers, four armed trawlers, and two AA ships. The Anglo-American covering-force was made up of the heavy cruisers HMS *London*, HMS *Norfolk*, USS *Tuscaloosa*, USS *Wichita*, and three destroyers. In support and hoping for a major fleet action was the main body of Tovey's Home Fleet, the battleships HMS *Duke of York* and USS *Washington*, aircraft carrier HMS *Victorious*, two more cruisers, and eight destroyers. PQ 17 had become the bait in a trial of strength between the Allied and German navies.

As long as the convoy kept together it was safe from the air and submarine threats but then, on 4 July, mistakenly believing the German heavy units to be at sea, Pound personally decided to scatter the convoy. The escort and covering forces concentrated, expecting imminent action. They were not to know that at that time the German Battle Group was still at Altenfjord. There was no need for valuable surface forces to be risked and after a brief foray the German surface ships were recalled. In all, U-boats and aircraft sank 22 merchant ships from PQ 17, taking to the bottom 430 tanks, 210 aircraft, 3,350 other vehicles, and almost 100,000 tons of cargo. Pound had committed one of the worst professional errors of the war against the best intelligence advice. The First Sea Lord's judgement was clouded by the effects of the brain tumour that would kill him a year later.

THE MEDITERRANEAN, 1942

Even if PQ 17 had not been such a disaster the Arctic convoys would have had to stop, as the Mediterranean began to dominate the allocation of main fleet assets. The year 1942 saw Malta, in Correlli Barnett's brilliant phrase, become 'the Verdun of the naval war', an island neutralized by Axis air power, kept more for reasons of morale than anything else and acting as a kind of strategic black hole sucking in aircraft, merchantmen, and warships. The only silver lining to this strategic cloud was the skill and courage shown by the Royal Navy in fighting through the Malta convoys. The year opened with one of the finest actions in the entire history of the service when Rear-Admiral Philip Vian used his three 'Dido'-class light cruisers supplemented by HMS *Penelope* still based at Malta, ten fleet, and six escort destroyers to drive off a powerful Italian surface force composed of a battleship, three cruisers, and ten destroyers. The threat of torpedo attack was skilfully used to keep the enemy at arm's length. Sadly it was all for nothing as Fliegerkorps II sank all four ships of the convoy, including the naval supply ship *Breconshire*. Only 20 per cent of the convoy's supplies were landed.

This was ABC's last major action before he was sent to the United States as Pound's representative on the combined chiefs of staff in Washington. His successor was Harwood, a fine squadron commander and favourite of Churchill, but no Cunningham. Harwood's first major operation was a convoy to Malta from Alexandria, 'Vigorous'. After heavy attacks by Axis aircraft and German E-boats, the convoy, threatened by the Italian Fleet, was ordered to turn back with the loss of a cruiser, three destroyers, and two merchantmen sunk, and two more cruisers damaged. At the same time another convoy, 'Harpoon', approached Malta from the west. The two old carriers *Eagle* and *Argus* had difficulty putting up effective

The Second Battle of Sirte,
22 March 1942. A picture taken
during the engagement from
the 'Dido'-class cruiser
Euryalus with Vian's flagship
Cleopatra steaming ahead
laying a smokescreen. The dual
purpose armament of 5.25 inch
guns is clearly visible.

fighter cover, but only the cruiser *Liverpool* was damaged. This, however, meant
that no cruiser was available to support the close escort after the covering forces
withdrew. An Italian surface force drew off the escort opening the convoy to air
attack. Close to Malta some of the survivors hit mines and only two merchant
ships arrived out of six for the loss of two destroyers and serious damage to a
cruiser, three destroyers, and a minesweeper.

The only solution was the largest convoy yet from the west. Other main fleet
operations ceased as a huge force including two battleships, seven cruisers, and 24
destroyers was put together for 'Pedestal'. *Victorious*, *Indomitable*, and *Eagle*,
equipped with 72 Fulmars, Sea Hurricanes, and Martlets, worked up before the
operation into the most effective carrier fighter force yet deployed by the Royal
Navy. The carrier *Furious* was to fly in more fighters to Malta. *Eagle* was sunk by
U-boat attack, *Indomitable* was put out of action by bombing, and the destroyer
Foresight was sunk by a torpedo bomber. After the main covering force withdrew
the convoy was attacked by mutually supporting submarines, E-boats (both
German and Italian), and aircraft. The AA cruiser *Cairo* and light cruiser *Manchester* were sunk and in the end only five out of 14 merchant ships got through. It was
just enough; Malta, the George Cross Island, did not fall.

The focus now shifted to French North Africa, where Churchill had persuaded
the Americans to mount the first major Allied amphibious landing of the war to
help the Mediterranean sea route and open offensive opportunities against
Europe. The Royal Navy had progressively improved its techniques of combined
operations. A Combined Training Centre was set up at Inveraray at which increasing numbers of crew were trained for the proliferating array of landing craft. In

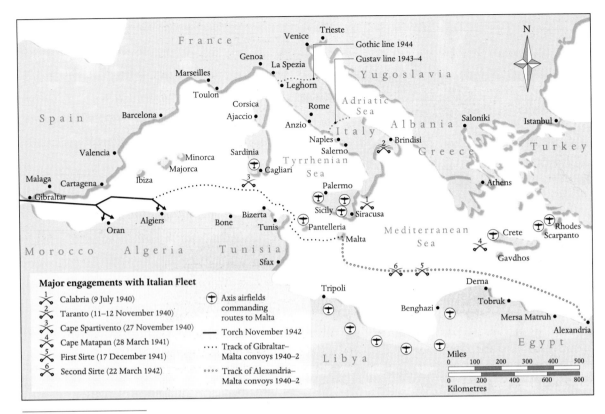

Major engagements with Italian Fleet

1 Calabria (9 July 1940)

2 Taranto (11–12 November 1940)

3 Cape Spartivento (27 November 1940)

4 Cape Matapan (28 March 1941)

5 First Sirte (17 December 1941)

6 Second Sirte (22 March 1942)

(T) Axis airfields commanding routes to Malta

— Torch November 1942

···· Track of Gibraltar–Malta convoys 1940–2

∘∘∘∘ Track of Alexandria–Malta convoys 1940–2

The Mediterranean Theatre, 1940–44.

1942 there were two contrasting major raids across the Channel, the successful destruction of the Normandie dock at St Nazaire, and the costly raid on Dieppe. Dieppe demonstrated the need for proper planning, bombardment, and headquarters ships containing the commanders of specially designated naval assault forces.

The Royal Navy took prime responsibility for ensuring the success of the landings on the Mediterranean shore of North Africa. Cunningham returned as Allied Naval Commander Expeditionary Force. His deputy was Admiral Bertram Ramsay, who had lifted the British Expeditionary Force from Dunkirk and who would become the Royal Navy's greatest exponent of the painstaking and complex staff work required for major amphibious operations. One hundred and sixty Royal Navy warships were assembled with two battleships, a battlecruiser, and seven carriers, including three of the new escort carriers. Direct attempts to capture Algiers and Oran proved costly, with the loss of a destroyer and two former American coastguard cutters, but a counterattack by French destroyers was beaten off by HMS *Aurora*. The Italian fleet did not intervene. Only the Germans' rapid reinforcement of Tunisia prevented the full strategic benefit being reaped. The advance of the Eighth Army opened the road to Malta. The last convoy to be opposed was 'Stoneage' in November and by the end of the year the Malta surface striking force, Force K, had been reconstituted together with a companion force at Bone, Force Q. Together with submarines, aircraft, and later MTBs, they inflicted heavy losses on the Axis forces' supply shipping. In April, the Italians lost over 100 merchantmen. In May, Cunningham, now once more Commander-in-Chief, Mediterranean fleet, mounted Operation 'Retribution' with the memorable order: 'Sink, burn and destroy: Let nothing pass.' The Axis armies had been defeated ashore in Tunisia and evacuation was expected, but in effect the Royal Navy had locked the Axis force in. With their surrender in large numbers the full fruits of the Mediterranean strategy could at last be enjoyed. The first through-Mediterranean convoy made the passage in May. The saving in shipping was considerable.

Spring 1943 also saw the major turning-point in the Atlantic. The heavy shipping losses in the first half of 1942 had been more the responsibility of the US Navy than the Royal Navy as the U-boats were able to slaughter unescorted shipping off the American coast. Few ships were sunk in the convoys in mid-Atlantic as the U-boats found easier unescorted targets elsewhere. The success of the better-organized and equipped Atlantic escorts in 1942 was in spite of the intelligence loss that occurred in February 1942 when the Germans introduced a fourth wheel into their coding machines. Until December the skilled plotters in the Submarine Tracking Room at the Admiralty lost their major insight into Dönitz's activities, but the experience of the preceding months and other intelligence sources combined to mitigate the situation. Only towards the end of 1942, with a full-scale convoy system created in the western Atlantic, were the U-boats regrouped for their climactic assault on the Atlantic convoys.

THE ATLANTIC CAMPAIGN, 1942–1943

There were now many more U-boats available. In January 1942 there were only 91 U-boats operational and fewer than a dozen at any one time were doing serious damage to Atlantic shipping. By July there were 140 operational boats, by October 196, and by April 1943 no fewer than 240. These began to concentrate their efforts where the escorts were at their weakest—the 'gap' or 'black hole' south of Greenland where air cover was still not available. In December, Bletchley Park was able to break the new U-boat code, but decrypts were not always immediately available. More important, there were so many U-boats at sea by March 1943—about 70—that evasive routeing was virtually impossible.

In the first 20 days of March 1943, 54 merchant ships were sunk, two-thirds in convoy (of which 21 were lost from the especially hard-pressed SC 122 and HX 229). The large convoy proportion was partly due to the smaller number of ships being sailed independently, but it led the Admiralty to reconsider its whole convoy strategy. Yet, if the British thought they were losing in March 1943, the Germans did not think they were winning. Every ship sunk had to be paid for in hours of fruitless attacks worsted by the ever more efficient escorts. If Allied aircraft were present the situation for the U-boats became almost impossible. The straw that finally broke Dönitz's back, therefore, was the closing of the black hole in mid-Atlantic where air cover was not available.

The Admiralty had to fight a long and bitter struggle with the RAF, the US

Navy, and Churchill to obtain sufficient aircraft with the range to plug the mid-Atlantic 'gap'. Sufficient Liberators were only made available in the spring of 1943. But escort carriers were equally late in appearing. The first, HMS *Audacity*, was used successfully in late 1941 but was sunk escorting convoy HG 76. American-built replacements were supplied to the Royal Navy in 1942, but they were deployed to higher priority tasks, the North African landings and Russian convoys. Only in April 1943 were the first escort carriers, HMS *Attacker* and HMS *Biter*, deployed in their primary theatre of operations. The following month the first merchant aircraft carriers sailed with convoys. These were cargo ships fitted with flight decks and a small complement of Swordfish Anti-Submarine Warfare (ASW) aircraft.

Another key factor was the formation of Support Groups—extra escort groups added to threatened convoys to hunt attacking submarines to destruction. The ships for these groups came from three sources. First, operational research had demonstrated that bigger convoys were statistically safer than small ones. This economized on escorts. Secondly, there were simply more ships with the coming to fruition of the wartime corvette- and frigate-building programmes in Britain and Canada and the entry into service of mass-production American frigates. Thirdly, Arctic convoys were suspended in March releasing forces to the Atlantic.

In April shipping losses were halved and the escorts claimed a U-boat for every ship sunk in convoy. For the first time U-boats failed to press home attacks; it was simply getting too dangerous. The climax came with Convoy ONS 5 which sailed from Liverpool and the Clyde on 21–2 April. Its close escort was the B7 group led by Commander Peter Gretton, one of the navy's brightest officers. B7 consisted of two destroyers, a frigate, four corvettes, and two rescue trawlers. It was supported

Convoy and land-based maritime patrol aircraft, shown in this view of a convoy being overflown by a Coastal Command Sunderland flying boat.

by two support groups—the First Escort Group composed of a sloop, three frigates, and an ex-US coastguard cutter, and the Third Escort Group with five destroyers transferred from fleet work. Land-based air cover was provided, including Liberators. No fewer than 42 U-boats were deployed against ONS 5, of which six were sunk, five damaged seriously, and 12 more lightly. A dozen merchantmen sunk was poor reward for such a concentration of force by the Germans, especially when combined with such losses. The Germans were now smashing their heads against the brick wall of a convoy escort system of ships and aircraft that for the first time in the war had no loopholes to be exploited.

Admiral Sir Max Horton, Commander-in-Chief, Western Approaches, and an ex-submariner, sensed it was the turning-point. The remaining convoys in May had better weather and more air cover from Liberators and escort carriers. Convoy attacks never again reached ONS 5's level of intensity. SC 129 lost two ships, but its escort sank a U-boat and a pack of ten boats was driven off. In May, the exchange ratio, once running at 100,000 tons per U-boat sunk, came down to 10,000 tons per boat. Forty-one submarines were sunk, 14 by convoy surface escorts and 11 by air escorts. With losses at such 'unbearable heights' Dönitz felt he had little alternative but to call off his wolves to lick their wounds.

The news was little better for the Germans in the north. After the PQ 17 disaster, Tovey had decided to give the next Arctic convoy an exceptionally powerful 'fighting destroyer escort' to help keep German surface ships at bay without either scattering the convoy or risking his heavy ships. To defend against the air threat the new escort carrier *Avenger* was assigned. The convoy sailed in September 1942 and was heavily attacked by aircraft and submarines. It lost 13 ships, but this cost the Germans three U-boats and 22 aircraft.

Apart from one homeward-bound convoy there was now a pause in Arctic convoys while the 'Torch' operations consumed the lion's share of Allied naval forces. When they resumed in December they were re-coded; convoys to Russia were now JW and those in the opposite direction RA. The new convoys were often run in two parts and although JW 51A had a quiet passage JW 51B was blown southwards by a storm. In the Arctic twilight of the last day of 1942 the latter was attacked by *Hipper* and *Lutzow* escorted by six destroyers. Captain Sherbrooke's escort of six destroyers, supported by Rear-Admiral Burnett's covering force of HMS *Sheffield* and *Jamaica* and two more destroyers, sank the destroyer *Friedrich Eckholdt* and kept the rest of the German battle group at bay for the loss of the minesweeper *Bramble* and destroyer *Achates*. Sherbrooke lost an eye when his ship HMS *Onslow* was hit by four 8-inch shells but he won the Victoria Cross for a well-conducted action that became known as the Battle of the Barents Sea.

Hitler was incensed and ordered the decommissioning of the German major surface units. Raeder resigned to be replaced by Dönitz, who was able to obtain a reversal of the scrapping order in order to keep a fleet in being, tying down Royal Navy assets. *Tirpitz* emerged from self-maintenance in January and *Scharnhorst* arrived in March. The Admiralty tried to neutralize the threat by using new

SUCCESS IN THE ARCTIC, 1942–1943

Convoy PQ 18 being fought through to the Soviet Union in September 1942. The destroyer in the foreground is the 'Tribal'- class HMS *Ashanti* which later tried to save her stricken sister *Somali*.

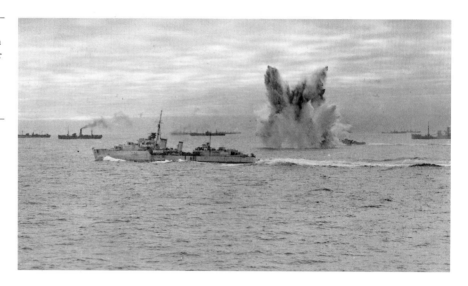

midget submarines called X-Craft in a daring attack on 22 September. The explosion of the side charges laid by these boats caused grievous damage to *Tirpitz*. The ship's armament and equipment was severely shaken and put out of action and she was immovable and unsteerable. Repairs to the ship would take a long time. The two surviving commanders, lieutenants Cameron and Place, were awarded well-deserved Victoria Crosses.

Scharnhorst escaped damage in this attack but met her nemesis in December when she attacked convoy JW 55B (convoys had restarted the previous month). This was supported by Sir Bruce Fraser, the new Commander-in-Chief of the Home Fleet with his flagship the battleship *Duke of York* accompanied by the cruiser *Jamaica* and four destroyers. Rear-Admiral Burnett's cruiser covering force comprised *Sheffield*, *Belfast*, and *Norfolk*. *Scharnhorst* was kept wrong-footed throughout by her signals being read by the British. She was surprised by Burnett's cruisers on the morning of 26 December and hit by two 8-inch shells which knocked out her forward radar. Partially blinded, *Scharnhorst* made off into the Arctic darkness. She tried again to find the convoy but was again caught by the British cruisers. Burnett shadowed on radar and homed in Fraser. When *Duke of York* and *Jamaica* opened fire *Scharnhorst* tried to get away and a chase developed. She might have escaped, but a shell penetrating her engine room almost brought her to a standstill. The delay was enough for Fraser's destroyers to attack with torpedoes. They hit the German ship four times, slowing her down again and allowing Fraser and Burnett to catch up. *Scharnhorst* was finished off by multiple torpedoes fired by *Jamaica*, *Belfast*, and returning convoy RA 55A's fighting destroyer escort. The Battle of the North Cape was the last battle between big-gun battleships in the history of the Royal Navy.

Only *Tirpitz* remained. Informed by Enigma intercepts when she would be ready for trials the British planned a major carrier air strike, Operation Tungsten, for early April 1944. Fraser was well informed enough to make his attack just as the battleship was about to leave her anchorage for sea trials on 3 April. Two fleet carriers were used, *Victorious* and *Furious*, together with an escort carrier force of four ships. These launched 40 Barracuda bombers escorted by 79 fighters. *Tirpitz* was hit or near-missed 16 times and knocked out once more. The bombs were dropped too low to penetrate her main armour but her design made her vulnerable to damage above the armoured deck; also, her engines were once again disabled by shock damage. Attempts to mount subsequent carrier strikes were foiled by smoke screens and AA defences, and the final destruction of *Tirpitz* had to be left to Bomber Command.

Late 1943 saw the U-boats attempt a counter-attack in the Atlantic. They were armed with new homing torpedoes to take on the escorts. In the first of the new convoy battles in September, ON 202 and ONS 18 were attacked. Warned by intelligence derived from code-breaking, the convoys were concentrated with a combined escort; a Canadian support group added five more. A merchant aircraft ship (MAC) was with the convoy and Liberators gave additional air support. Three escorts and six merchant ships were sunk, all when flying was impossible. The Germans lost two U-boats and two more were damaged. This was the best the Germans could do before the escorts acquired 'Foxer' decoys to combat the homing torpedoes. In October, the Germans lost no fewer than 23 U-boats attacking convoys. SC 143 was typical; one of 39 ships in the convoy and one escort sunk in exchange for three U-boats.

In early 1944 the U-boats returned to concentrate in the Western Approaches. Here the support groups such as Captain 'Johnny' Walker's Second Escort group of six sloops were able to hunt them to destruction around the convoys. In February, Walker's group sank six boats out of a total of 11 destroyed in operations to the west of Ireland. Escort groups such as Walker's were by now superbly well-trained

DOMINANCE AT SEA AND SUPPORT OF AMPHIBIOUS LANDINGS, 1943–1944

Captain F. J. ('Johnny') Walker in typical pose leading his Escort Group on VHF inter-ship radio. A long-time Anti-Submarine specialist, Walker was passed over for promotion before the war but rose to prominence as a most effective killer of submarines from 1941 to 1944. The VHF radio telephone was a key factor in commanding effective escort group operations.

teams that could operate with the minimum of signalling and on the basis of mutual confidence. The Germans gave up attacks on convoys in March. U-boats concentrated on covert, submerged patrols, utilizing the schnorkel to stay submerged as long as possible. Until the end of the war British coastal waters witnessed a dangerous game of cat and mouse with submarines stalking their prey and ships equipped with sophisticated anti-submarine weapons such as the Squid ahead-throwing mortar hunting them down. In January 1945, there were about 335 Royal Navy destroyers, frigates, sloops, and corvettes deployed in the anti-U-boat role around Britain's coast escorting convoys and patrolling focal areas.

The U-boats were unable to prevent the great amphibious landings of 1943–4 which re-established Allied military power on the European continent. The Mediterranean offensive continued in 1943 with the invasions of Sicily in July and Italy in September. Ramsay masterminded the 'Husky' plan for Sicily, which was covered by six British battleships and two fleet carriers. The Salerno landings saw another display of the Royal Navy's strength, with fleet and escort carriers providing vital air cover. Italy had capitulated the day before and, in scenes reminiscent of 1918, her fleet surrendered itself at Malta. The strategy conceived in 1939 had finally been consummated and it was a suitable point for Cunningham to return to London to replace Pound. During the Salerno landings the Germans introduced guided anti-ship missiles, one of which damaged *Warspite*. When the Anzio landings took place in January the cruiser *Spartan* was sunk by missile attack, the last major Royal Navy warship loss of the war.

The demand for a mass production ocean escort was met by the 'frigate', a larger twin-screw development of the earlier, smaller corvette. This is one of the original 'River'-class ships of the 1940 Programme, HMS *Test*, completed in 1942.

HMS *Warspite*—one of the most important ships in the history of the Royal Navy. She fought in many key engagements in both world wars between which she was given a thorough modernization.

The greatest amphibious operation of all was, however, Normandy, in June 1944—Operation Neptune. This was fundamentally a Royal Navy affair which provided the Naval Commander, Ramsay, and the lion's share of naval assets, from bombardment, covering, and escort forces to landing ships and craft. Of the 2,468 major landing vessels in the two Task Forces deployed on 6 June, only 346 were American. Of the 23 cruisers covering the landings 17 belonged to the Royal Navy. With Rear-Admiral Tennant in command the Royal Navy also developed the innovative prefabricated Mulberry harbours to be built off the beaches. In all, Neptune was a 'never surpassed masterpiece of planning' which demonstrated that the Royal Navy's capacity for detailed staff work was greater than often estimated. The landings saw considerable naval activity beyond shore bombardment, extensive and crucial though that was. Clandestine surveys by Combined Operations Pilotage Parties, and a huge minesweeping operation using 255 minesweepers and danlayers, prepared the way for the invasion armada. Meanwhile, escort groups acted as an effective U-boat barrier on both flanks. On 8 June, a sharp destroyer action disposed of the main surface threat when four British and two Canadian destroyers with two Polish destroyers sank two German destroyers, *ZH1* and *Z32*, and heavily damaged *Z24*. A campaign then had to be waged against manned torpedoes, E-boats, and explosive-laden motor boats. Once the Allied armies began their advance, naval parties who had honed their skills in the Mediterranean grappled with the difficult task of clearing captured ports.

In the north the Home Fleet covered the Arctic convoys, which continued to run with little opposition. Also using both surface action groups and escort carrier forces, forward offensive operations were carried out against German sea communications and bases along the Norwegian coast. These culminated in a successful sortie in May 1945 by a force of three escort carriers, two cruisers, and seven destroyers commanded by Rear-Admiral Rhoderick McGrigor to destroy the German U-boat infrastructure in Vestfiord.

The end of the Italian and German fleets allowed the Royal Navy to begin recovering its position in the Far East. Much planning went into major offensives by South-east Asia Command, but resources were never available. By the spring of 1944, however, Somerville's Eastern Fleet was built up sufficiently to go over to the offensive. In April, the carrier *Illustrious* together with the USS *Saratoga* struck Sabang and Soerabaya. Two more British carriers arrived in July and Somerville mounted another, all-British, attack on Sabang supplemented by shore bombardment and a destroyer torpedo raid.

It had been decided at the Cairo summit in November 1943 that Britain would send a fleet to the Pacific. In this theatre the key weapons had been carriers and their aircraft, and with the help of the Americans in terms of aircraft, flying training, and techniques of carrier operation, the Royal Navy was able to cram larger air groups of 50–70 aircraft on to their fleet carriers. Commanded by Sir Bruce Fraser the British Pacific Fleet, the carriers *Indomitable*, *Victorious*, *Indefatigable*, and *Illustrious*, together with the battleships *King George V* and *Howe*, five cruisers and 14 destroyers, duly arrived on station in March 1945. Two more carriers, *Formidable* and *Implacable*, joined later. It was a brave show but only a single Task Force's worth in terms of the gigantic fleet deployed by the Americans. Much had to be learned about distant fleet support, fleet train, and replenishment at sea, lessons that would stand the navy in good stead in the post-war era. The British Pacific Fleet saw considerable action, not least off Okinawa, where its armoured hangar carriers proved resilient under Kamikaze suicide attack, but the Americans made sure that it was never in the forefront of the final defeat of Japan. The fleet did, however, participate in the final bombardment of the Japanese home islands and its battleships were the last British gun-armed capital ships to fire their guns in anger.

Admiral Sir Bruce Fraser (*left*) arrived in Ceylon in August 1944 to succeed Admiral Sir James Somerville (*right*) as Commander-in-Chief, Eastern Fleet.

In the Indian Ocean, although the amphibious assault on Rangoon was pre-empted by Japanese withdrawal and the long-awaited landing in Malaya did not take place until after Hiroshima, surface, escort carrier, and submarine forces kept up a maritime offensive. In May, in an operation that combined the traditions of surface torpedo attack with the new action information organization necessitated by three-dimensional sensors, the Twenty-sixth Destroyer Flotilla sank the Japanese cruiser *Haguro*. The following month another cruiser, *Ashigara*, was sunk by the submarine *Trenchant*, and at the end of July, in a daring midget submarine attack, *Takao* was sunk by charges laid by *XE3* and *XE1*.

On 2 September 1945 Admiral Sir Bruce Fraser signed the Japanese surrender on behalf of the United Kingdom in Tokyo Bay. On the following day Penang was reoccupied by Royal Marines from HMS *London*. On the 5th, the destroyer *Rotherham* began the reoccupation of the Singapore naval base. On 9 September, the main landings to reoccupy Malaya took place. Reasserting control was even more important given the massive responsibilities faced by the South-east Asia Command's forces in maintaining order in the former French and Dutch empires as well as Britain's own.

The Second World War had cost the Royal Navy dear. Some 224 major surface ships of corvette size and above had been sunk. One hundred and thirty-nine destroyers were lost, 31 more than had been in service in 1939. A total of 1,525 British warships of all types were lost, of over two million tons. Over 50,000 naval personnel lost their lives, many of them members of the Royal Naval Reserve and Royal Naval Volunteer Reserve. The service had fought well, better than in 1914–18, and if not all of its initial capabilities had been proved adequate its capacity to learn from experience and apply new technology had been second to none. The doubts of 1918 had been dispelled; the service had been vindicated.

CONCLUSION

The Royal Navy had also grown enormously both in size and scope. The strength of the fleet had increased from almost 400 major combatants (of which over one-third were in reserve) to almost 900. In terms of personnel the growth was even greater, from a pre-war strength of 129,000 to 863,500 by mid-1944. Of this last figure, 72,000 were members of the Women's Royal Naval Service (WRNS) who carried out a wide range of shore roles from mine modification to general administration. As the official historian put it: 'no naval establishment was without its complement of "Wrens" and their conduct, courage and capacity won the affection and admiration of the whole service.' Women in uniform had become an essential feature of naval life.

The increased rate of growth of personnel numbers compared to ships reflected the extent that much of the wartime navy was based ashore. Developments in technology required training, research, and support facilities that had never before been necessary. The Fleet Air Arm required a major shore organization as well as aircrew serving at sea. New types of naval forces such as landing craft and coastal forces also added to the personnel demands. The Admiralty itself had grown into a huge organization, with major outstations at Bath and Taunton.

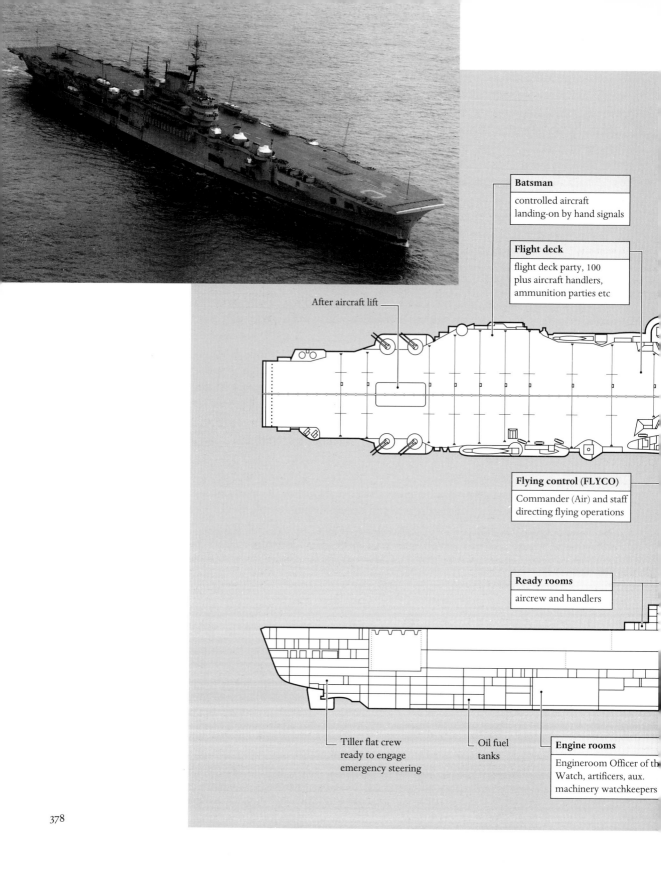

Batsman

controlled aircraft
landing-on by hand signals

Flight deck

flight deck party, 100
plus aircraft handlers,
ammunition parties etc

After aircraft lift

Flying control (FLYCO)

Commander (Air) and staff
directing flying operations

Ready rooms

aircrew and handlers

Tiller flat crew
ready to engage
emergency steering

Oil fuel
tanks

Engine rooms

Engineroom Officer of the
Watch, artificers, aux.
machinery watchkeepers

'IMPLACABLE'-CLASS CARRIER (1945)

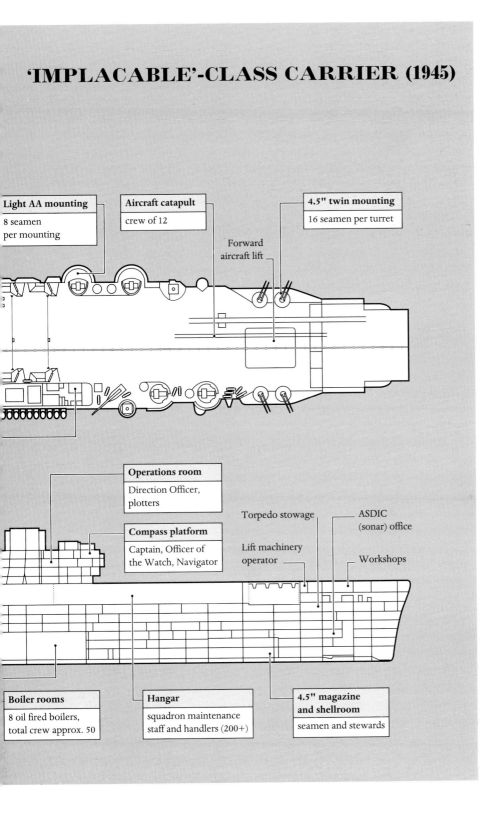

Light AA mounting

8 seamen
per mounting

Aircraft catapult

crew of 12

4.5" twin mounting

16 seamen per turret

Forward
aircraft lift

Operations room

Direction Officer,
plotters

Compass platform

Captain, Officer of
the Watch, Navigator

Torpedo stowage

Lift machinery
operator

ASDIC
(sonar) office

Workshops

Boiler rooms

8 oil fired boilers,
total crew approx. 50

Hangar

squadron maintenance
staff and handlers (200+)

**4.5" magazine
and shellroom**

seamen and stewards

This increased the problems of those who planned demobilization and the size of the post-war fleet. Like their predecessors before the war they were forced into difficult choices. Before 1945 was out massive cuts had taken place in the wartime building programmes, and an argument was in full swing with the new Labour government on the size and shape of the post-war fleet. In these discussions the government made two fundamental points: Britain would have to accept that the United States was now the dominant naval power and that the coat of British defence policy would have to be cut from the cloth of a British economy in severely straitened circumstances. So was the scene set for the future.

The core of the British Pacific Fleet on arrival were the four fleet carriers *Indomitable*, *Indefatigable*, *Illustrious*, and *Victorious*, seen here at Manus before the Okinawa operation in company with the aircraft maintenance ship *Unicorn*. The presence of *Unicorn*, and in the background a fleet tanker, is a reminder of the vital importance of the fleet train to the reach required for Pacific operations.

13

**THE
REALITIES
OF MEDIUM
POWER, 1946
TO THE
PRESENT**

J. R. HILL

*T*he Royal Navy ended the Second World War as a very successful fighting service. It had fought and won a vital campaign in the Atlantic, and played a leading part in countless other parts of the conflict. It had demonstrated its ability to absorb a tenfold expansion of personnel and an unprecedentedly high degree of technical change, and had coped adequately with the logistic problems involved. Most of all, it had proved over and over again its courage and tenacity. Those qualities would be no less in demand in a turbulent, straitened future.

The Second World War had been directed by the prime minister, as Minister of Defence, and the three Service Chiefs of Staff operating from the power bases of their single-service ministries. Thereafter, the higher organization of defence moved steadily towards centralized control.

THE HIGHER ORGANIZATION OF DEFENCE, 1946–1994

There were several driving factors in this trend. First, the need for economy that has always applied to defence provision in time of peace (or what passes for peace) was sharpened by Britain's disappointing economic performance. Secondly, war experience suggested that operations were likely to involve more than one fighting arm, and joint planning was needed. Finally, technology offered so many new opportunities and options that critical scrutiny was needed to develop those that would best serve the overall defence interest.

That, at any rate, was the rationale for the centralizers. The tendency of the single services was to oppose rapid change and centralization, on the grounds that professional management, advice, and expertise must not be lost. The resulting tensions will be apparent in the rest of this chapter.

Two major watersheds can be identified in the centralization process. The first occurred in the late 1950s and early 1960s, when under Defence Ministers Harold Macmillan and Duncan Sandys, advised by Earl Mountbatten, Lord Ismay, and Sir Ian Jacob, the office of Chief of Defence Staff (CDS) was instituted, some staffs were integrated, and many others collocated in a single Ministry of Defence under a full Secretary of State. The Permanent Under-Secretary (the senior civil servant in the Defence Department) became the accounting officer for the whole defence budget. Scientific advice and procurement also became institutionally more centralized. Nevertheless, the single services retained much autonomy, and single-service ministers continued until the early 1980s.

The second radical movement to central control began at that time, under

Defence Secretary John Nott and CDS Sir Terence Lewin, and was given added impetus by a later Defence Secretary, Michael Heseltine. In these reforms the central staffs under CDS were much strengthened, single-service staffs correspondingly reduced and high-level financial planning and control tightened under the Office of Management and Budgets. At the same time, under the New Management Strategy (NMS), day-to-day control of expenditure was delegated to budget holders at much lower levels than previously. This reform was implemented on 1 April 1991; it is rather early to try to gauge its results.

LESSONS OF THE
AFTERMATH, 1946–1957

Plans and Policy From a perspective of 50 years, it is hard to view the policies and plans of the Royal Navy immediately after the end of the Second World War as anything but grossly unrealistic. Regularly, in both official and unofficial documents, there appeared proposals for force levels and independent strategies that in hindsight are absurdly ambitious.

Nevertheless, they are understandable. The country had just fought a massive war, and had vast overseas commitments. The American example encouraged any defence planner to think on a large scale. Moreover, a great deal of material existed, in ships and aircraft and in shore bases. It is therefore unsurprising that in 1947, in support of the first attempt at a post-war strategy (defence of the UK, defence of sea communications, security of the Middle East) the navy bid for manpower totalling well over 150,000 and money amounting to two-thirds of the total available for defence. The other services were no less prodigal in their demands.

The first realistic attempt to match strategy, money, and force levels was a study by Admiral Harwood in 1949 which proposed an 'insurance' force based mainly in home waters on a manpower base of under 100,000. This was rejected as too sweeping, and the less radical plan of a balanced 'core force' adopted to cover peacetime commitments and provide a basis for mobilization. This was overtaken by the more rapid implementation of existing plans as a result of the Korean War.

The preponderance of American power was apparent, and not only off Korea, for in the Western European and Atlantic theatre it was plain that the Soviet threat could not be contained without the engagement of the United States. The conclusion of the North Atlantic Treaty (1949) achieved that engagement. Its effects on the Royal Navy will be apparent throughout this chapter, and here only one element need be highlighted: the NATO sea area extends only to the Mediterranean and the Atlantic south to the Tropic of Cancer.

In the early 1950s, in spite of the Korean War, exercises in future planning—and retrenchment—continued. The Chiefs of Staff produced in 1952 a 'Global Strategy' paper taking into account the Soviet threat, NATO, and nuclear weapons—which had not been fully studied before. The navy, secure in limited-war roles, had more difficulty with the notion of a devastating nuclear conflict, but relied on its perceived utility in a period subsequent to a nuclear exchange, dubbed 'broken-backed' warfare.

Inevitably, the services' plans were challenged by the Treasury, and 1953–4 was dominated by a process called the 'Radical Review', as the Chancellor tried to prune still further the services' plans. Pressed by influential politicians, notably Duncan Sandys (then Minister of Supply) who regarded aircraft carriers as expensive luxuries, the Admiralty argued skilfully to justify its case. What Chatfield in the 1930s had called 'the power of the defensive in Whitehall' worked in the navy's favour this time: no immediate savings were to be had from scrapping carriers, and the failing Churchill was unable to force the Sandys line on a Cabinet that knew a change of prime minister and an election were pending. Moreover, the Admiralty used the Strategic Review of 1954, prompted by the advent of the hydrogen bomb, to produce a case that would stand it in excellent stead for the next few years. The H-bomb, the Chiefs of Staff believed, made general war less likely: thermonuclear deterrence was held to be secure, so limited aggression was the only realistic option for the Cold-War opponent, with the West's response correspondingly limited. In this situation the utility of maritime forces was apparent. Such was the doctrine inherited by Mountbatten when he became First Sea Lord in 1955.

Operations Between 1946 and 1950 the Royal Navy was greatly involved in regrouping and reverting to an organization more appropriate to an uneasy peace. The home and overseas stations settled down once more to something like their pre-war condition, and many organizations formed for war purposes only were closed down.

However, these years were not short of incident. A considerable shock was administered to the Mediterranean Fleet in 1946 when a passage through the Corfu Channel, intended as a warning to Albania not to flex its muscles in sight of the Royal Navy, ended in the mining of the destroyers *Saumarez* and *Volage* with the loss of 44 lives. The subsequent case before the International Court of Justice established the right of innocent passage for warships through international straits, and arguably wars have been fought for less; but it was a straw in a strengthening wind.

Again in the Mediterranean, the formation of the state of Israel, out of the British mandated territory of Palestine, with the associated illegal immigration of groups of desperate Jews displaced from Europe, involved the navy in a messy, distasteful, and often violent series of operations throughout 1947. The brunt fell on the destroyers, whose boarding parties were issued with cricketers' boxes as essential items of equipment—hat-pins being a weapon of choice for the opposition.

It was, however, in the Far East that the most headline-catching incident occurred. In April 1949, HMS *Amethyst*, a small frigate, came under fire from Communist Chinese forces some way up the River Yangtse, was damaged, and suffered casualties. The political and military situation was extremely confused, the respective banks of the Yangtse being occupied by communist and nationalist forces, but after an extended lull while the *Amethyst* remained at anchor, evacuating casualties and recouping materially but running out of supplies, she made a dash down-river by night on 30 July 1949. She rejoined the fleet the next day to general acclamation for her replacement Commanding Officer, Lieutenant-Commander Kerans, and her ship's company.

In spite of its public impact the Yangtse Incident could be regarded as something of an anachronism: what was a gunboat doing in the middle of a civil war in China anyway? The Korean War which broke out in 1950 raised few such questions, since to most observers it was a clear case of cross-border aggression and the response, somewhat fortuitously, had full United Nations backing. The Royal Navy's part in this major war was consistent and considerable. From September 1950 to July 1953 (with one 'spell' from the Royal Australian Navy) a light fleet aircraft carrier was on station, generally off the west coast of Korea, with its aircraft operating in support of the armies ashore, particularly against enemy communications. At least one British cruiser and three or four destroyers or frigates were generally on station also, employed on a variety of tasks from 'train-busting' bombardments to escort. There was also a considerable and necessary minesweeping effort. There is no doubt that the effectiveness of the Royal Navy's effort impressed the US Navy and strengthened the case for the Royal Navy's carrier force.

During the early 1950s it was not only Korea that figured in the navy's operations. In 1951 destroyers and cruisers, mostly from the Mediterranean Fleet, were deployed to the Persian Gulf in reaction to the Iranian nationalization of the Anglo-Iranian Oil Company; a projected operation to evacuate British nationals and retrieve equipment was not implemented, and the problem was solved politically. Further east, the Chinese complications continued; although Britain was actively engaged against Chinese Communist forces in Korea, trade from Hong Kong with the Chinese mainland continued and was protected by British naval forces on an *ad hoc* basis at the entrances to Foochow and Amoy, where it was under threat from nationalist Chinese ships based on the offshore islands.

The heavy forces of the navy remained based in home waters, where, supplemented by the still-numerous ships of the flotillas and training squadrons, they played a major part in the first two large-scale NATO exercises, Mainbrace in 1952 and Mariner in 1953. These demonstrated naval forces' ability to strike at the Soviet Union from the far north and challenge the Soviet navy there, and many lessons were learnt.

However, it was in the Middle East that trouble was most likely to erupt. The problems of Egypt, poor, ill-governed, heavily populated, and humiliated by failure in its 1948 war against Israel, had long been apparent. A strike of local labour in the strategically sensitive Suez Canal was made ineffective in February 1952 by the simple expedient of the Royal Navy's running all the vital canal services. The evacuation of the Canal Zone by British forces was negotiated by Anthony Eden as Foreign Secretary in 1954, but in 1956, Colonel Nasser, the new and popular ruler of Egypt, precipitated a crisis by nationalizing the Suez Canal Company.

The British and French reaction was sharpened by a perception of Egyptian aims that owed much to previous experience of dictators, and to evidence that

Soviet arms were flooding into the country. Discounting the possibility of a *coup de main*, Britain and France embarked on a build-up of amphibious forces, and Israel was involved at an early stage of the planning. It was too cumbersome. By the time, three months after nationalization, the allied assault on the Canal began, world opinion had mobilized against military action if not actually in favour of Egypt. It hardened further after clear evidence of collusion with Israel's invasion of Sinai and a week's preliminary bombardment from British and French forces (including three RN aircraft carriers). Eventually, although the landings were militarily successful—and, incidentally, featured the first helicopter-borne landing of troops on a large scale—superpower pressure, much of it economic, forced the withdrawal of the allied forces by the end of the year with none of the aims achieved. Indeed, it is not certain what the political aim might have been at any stage: the question most often asked, even at the time, was 'What do you do with it when you have got it?'

The first helicopter-borne amphibious landing was made from HMS *Ocean* at Port Said in 1956. Twenty years later the technique had developed into the highly sophisticated organization seen here in HMS *Hermes* in 1978.

Plans and Policy The Suez débâcle demonstrated that Britain, and France for that matter, were no longer great powers with full freedom of action on the world scene. In the following ten years came the slow realization of that fact; its profound effect on the navy; and the almost inevitable overswing of the pendulum.

Almost immediately after Suez Harold Macmillan became prime minister in place of the ailing Anthony Eden, and Duncan Sandys became Minister of Defence. By April 1957, Sandys published his Defence White Paper, the most radical such document since the Second World War. It abolished conscription; placed great reliance on nuclear weapons (the British thermonuclear weapon was about to be tested); and contained the ominous words 'the role of naval forces in total war is uncertain'.

Mountbatten, the First Sea Lord, responded in various ways. He built on the inheritance from his predecessors by seeking to justify balanced naval forces, including carriers and amphibious forces, primarily in limited war. He also emphasized the political crisis that would be caused in the NATO alliance by abandoning the British contribution to its Atlantic command. He sought support from the other services who had their own fish to fry, and he worked on Sandys at the personal level. This multi-faceted approach by a consummate political operator succeeded in preserving the fleet which would deploy more widely as perceptions increased of Soviet and Chinese initiatives East of Suez.

At the same time, Mountbatten did not neglect the capacity of the navy at the higher levels of warfare. He had long been a proponent of nuclear power for submarines and, by forming a good working relationship with Admiral Rickover, the father of the American nuclear submarine project, managed to get the first British nuclear-powered boat, *Dreadnought*, into the water by the end of 1962.

It was in good time. By the middle of 1962 the American Skybolt system, which the Royal Air Force planned to fit into its V-bombers as the next generation British deterrent, was on the verge of cancellation. At the Nassau meeting in November, Macmillan and President Kennedy agreed that Britain should be supplied with the submarine-based Polaris system instead. This could not have occurred without evidence of British expertise in nuclear-powered submarines. Thus, in one of those mixtures of coincidence and foresight that are so often the ingredients of history, the navy found itself with a new role, about which it was ambivalent. Mountbatten had encouraged contingency planning for Polaris but did not want to provoke Air Force hostility by going for the missile too enthusiastically. Once the system was forced upon the navy, however, the implementation of the remit followed in a brilliantly driven seven-year development and construction programme and the subsequent continuous deployment of the British deterrent, which is one of the great success stories of the post-war Royal Navy. But the cost elsewhere proved as heavy as expected.

The Royal Navy had a strong case that its carriers were the most flexible means of providing air power. But new hulls would be needed to sustain the policy and even within the Naval Staff there were doubts whether they could be afforded.

The problem sharpened when a Labour government came into power in February 1964. Denis Healey, the new Secretary of State for Defence, instituted a thorough defence review to stop the rise in the defence budget. The Royal Air Force, chagrined by the loss of its primary deterrent role and of its advanced aircraft the TSR2, argued skilfully that aircraft working from island bases round the world could do everything that carrier aircraft could, with the possible exception of supporting an opposed amphibious landing. The naval case was weakened by the Naval Staff's refusal to accept less costly ships than the proposed 50,000 ton, fixed-wing carriers.

The outcome became increasingly inevitable and was announced in February 1966: Britain would not embark on major operations without allies; the new carriers would not be built, but the existing force would run on to the early 1970s. The Minister for the Navy, Christopher Mayhew, and the First Sea Lord, Sir David Luce, both resigned; the trauma throughout the navy was intense.

The new First Sea Lord, Sir Varyl Begg, who had been sceptical of the value of large carriers, set himself the task of reproviding, so far as necessary, the capabilities that would be lost, and of restoring naval morale. He set up a Future Fleet Working Party and this, by the middle of 1967, produced a blueprint for a naval force—including 'cruisers' with significant air assets—that would meet the commitments envisaged by the Healey defence review.

Indeed, the Supplementary Defence Statement of 1967 continued to stress the East of Suez role, and it was that role that the Future Fleet Working Party had concentrated upon, though Captain David Williams, the far-sighted Director of Plans, had ensured that NATO was not neglected in the text. The wisdom of this was demonstrated in January 1968, when in the face of an economic crisis the government imposed further heavy cuts in defence expenditure plans. The remainder of the East of Suez policy was abandoned; the Bahrain and Singapore bases would be evacuated by 1971; the carriers would go at the same time; the F-111, foundation of the RAF's island-base strategy, would not be bought; NATO was henceforth to be the justification of the British defence effort.

As a shift in strategic priorities, the decision of mid-January 1968 stands far and away ahead of any comparable post-war British decision before or since. It was highly palatable to the Treasury, who now had a simple yardstick by which defence expenditure could be measured (or, bluntly, beaten). It was less so to the Foreign Office, though the more European wing would see advantage in it. It was not mortal to the army and Royal Air Force, who had large forces already dedicated to the NATO role. But it posed grave risks to the Royal Navy as a balanced force capable of exerting sea power in the way it had been accustomed to doing.

Operations Defence policy in the late 1950s and for most of the 1960s has sometimes been given the title 'East of Suez'. As has been suggested above, that is an over-simplification. However, so far as operations actually performed were concerned the description is more precise.

It was indeed almost inevitable that this should be the focus of the navy's efforts, since the country was going through the most rapid phase of divestment from empire. Macmillan's speech in 1960 ('The wind of change is blowing through this continent') recognized a truth reaching wider than Africa alone; and retreat from empires, however orderly its conduct, leads to turbulence.

It was first felt in the Middle East, where in 1958 the Iraqi monarchy had fallen in bloodshed, bringing to power an authoritarian leader in Colonel Kassem. In June 1962 Kassem threatened to annex Kuwait. British forces were fortunately placed and naval forces, particularly, redeployed with prescience. In an extraordinarily short space of time a three-service force with amphibious and armoured elements was assembled which, although its organization left much for the purist to deplore, fulfilled its deterrent purpose.

A more active intervention occurred in the fledgling state of Tanganyika in

Multinational naval forces began to evolve in the mid-1960s, under the aegis of NATO. This picture shows the nineteenth activation of the Naval On-Call Force Mediterranean in 1979.

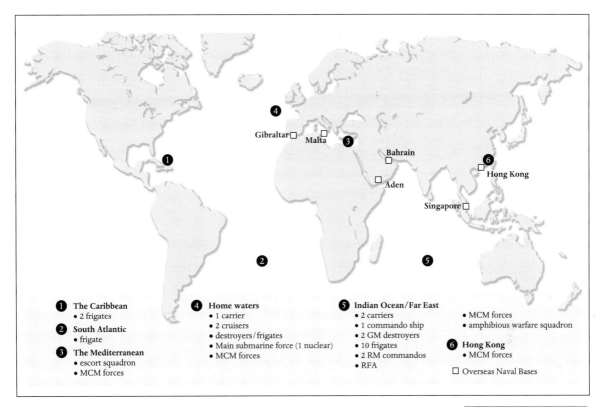

Gibraltar
Malta
Bahrain
Aden
Singapore
Hong Kong

❶ The Caribbean
• 2 frigates

❷ South Atlantic
• frigate

❸ The Mediterranean
• escort squadron
• MCM forces

❹ Home waters
• 1 carrier
• 2 cruisers
• destroyers/frigates
• Main submarine force (1 nuclear)
• MCM forces

❺ Indian Ocean/Far East
• 2 carriers
• 1 commando ship
• 2 GM destroyers
• 10 frigates
• 2 RM commandos
• RFA

• MCM forces
• amphibious warfare squadron

❻ Hong Kong
• MCM forces

☐ Overseas Naval Bases

Deployment of the Fleet, 1965.

1964, when Royal Marines from HMS *Centaur* landed and suppressed an army mutiny, thereby saving the regime of Julius Nyerere.

The sternest test of all was to come in 'Confrontation' in South-east Asia in the mid-1960s. The state of Malaysia—at that time consisting of Malaya, Singapore, and Sabah, all former colonies of Britain—had come into being with the blessing of the United Nations, some doubt about the level of indigenous support and strong opposition from the Indonesian regime headed by General Soekarno. From 1964 to 1966, British arms, with increasing support from the young Malaysian forces and steadily strengthening will among the civilian population, fought and won one of the strangest, most economical, and most brilliant counter-guerrilla wars in history, which ended in the internal overthrow of the Soekarno regime and of the Indonesian Communist Party with which he had increasingly allied himself. The navy's part in Confrontation was twofold: to provide the deterrent backing to the whole operation with the striking power (never used but known to have exerted profound influence) of the fleet aircraft carriers; and to provide appropriate active support ranging from disembarked helicopter squadrons, which operated in the jungle to great effect and with envied flexibility, to patrols in coastal waters and the Malacca Strait where nerves, and rules of engagement, were fully tested and Indonesian probes and attempted landings regularly blunted.

389

An operation of a different sort resulted from the Unilateral Declaration of Independence by the white Rhodesian regime in 1965. This was ruled illegal by the British government and a patrol to block oil supplies through the port of Beira was instituted with full United Nations backing. The Beira Patrol, though initially involving the fleet carrier *Eagle*, soon became a humdrum job for frigates supported by the Royal Fleet Auxiliary, working to strict and, after inevitable initial flurries, well-understood rules of engagement. It went on, steadily attenuating, until the formation of Zimbabwe in 1980.

Some limited confrontations did occur west of Suez: they were seldom against the principal perceived threat from the Soviet navy, though brushes with Soviet units did occur particularly in the Baltic and during major exercises, where territorial sensitivity and intelligence-gathering imperatives sometimes combined to produce touchy situations.

But, west of Suez, the more organized and politically charged events concerned allies or at least countries that were regarded as basically friendly. Iceland, a NATO member whose economy depended on fish and particularly cod stocks, declared a fishery zone in 1958 which, judging by the standards of later international conventions, looks extremely modest; but at the time the British government took the view that it infringed the freedom of the seas and embarked upon the protection of British trawlers fishing in the area. This 'first cod war' lasted for over a year and ended with an exchange of letters that conceded a good deal of the Icelandic case.

Other ventures in the western hemisphere were more successful. Spain's claims over Gibraltar and its waters and airspace were complex and many-layered; the manœuvres of the British and Spanish guardships resembled a courtly dance with infinite variations. It was a civilized demonstration of expressive sea power. In the same way, the rounding-up of a private-venture invasion of the Falkland Islands by 30 Argentine nationals was conducted by the Royal Marine garrison without bloodshed, but with the offshore presence of a British frigate.

Finally, though naval forces were as has been seen deployed on numerous low-intensity operations during this period, more formal exercises and deployments were not neglected. It was possible for whole squadrons of ships to go through a two-year commission without an 'angry' deployment or diversion from the planned programme. Major exercises more often than not included allies, with NATO most often involved but with CENTO (Turkey, Iran, and Pakistan) and the South East Asia Treaty Organization (SEATO) sometimes the focus, to say nothing of bilateral exercises. Towards the end of the period, also, multinational standing forces began to make their appearance: the Standing Naval Force Atlantic averaging five frigates or destroyers was permanently assigned to NATO, always with British participation. Little difficulty was found in getting national units to work together, aided by common signal books and tactical doctrines (one of the great unsung achievements of NATO); the experiment of mixed-nationality manning within a single ship, instituted to support the short-lived and politically motivated Multilateral Nuclear Force concept, was, however, quickly discarded.

Plans and Policy From 1968 to 1982 the Naval Staff had the task of justifying the Royal Navy's continued existence as a major and balanced maritime force in terms of the NATO alliance. The task was difficult for several reasons. First, the NATO sea area extended at most 3,000 miles from British shores. If taken strictly, this would mean a restriction of reach that had not been imposed on the navy since early Tudor times. Second, the principal ally, the USA, was extremely powerful at sea, capable in its own view of taking on the Soviet navy—certainly the strongest, and to NATO the only credible, threat at sea—single-handed. Third, the other services were firmly entrenched on the mainland of Europe with, in the case of the army, a clearly defined section of front to defend and in the case of the Royal Air Force a well-established Tactical Air Force commitment and command. Finally, money was tight.

It would be wrong to say that those in charge of the navy did not believe in the NATO commitment. On the contrary, they made a good job of convincing themselves that the navy's role, particularly in the eastern Atlantic NATO sea area, was critical in seeing through reinforcement shipping from the United States, without which it was inconceivable that any land campaign in Europe could be brought to a successful conclusion, and highly important in other ways such as anti-submarine support for the US Striking Fleet. But they were never convinced that it was the only role for the navy. Several hundred years of the exercise of sea power in all its forms had taught that the unexpected operation in the unplanned place was quite as likely to happen as the set scenario, and that it was the navy's duty—whether so instructed or not—to be ready for such contingencies.

In consequence, the navy in Whitehall led to some extent a double life. It was not difficult to justify high-quality forces because the Soviet navy represented a sophisticated threat. What was more difficult to justify was reach, the capacity to operate at long distances from the home base, particularly at the higher levels of conflict and unsupported by shore-based aircraft. Three particular aspects of capability stand out as signal successes for the Naval Staff's arts of persuasion: the Sea Harrier short take-off/vertical landing (STOVL) aircraft, justified on the grounds of 'seeing off' Soviet shadowers, probing Soviet surface forces, and attacking such forces in the last resort; the nuclear-powered fleet submarine force, justified on the ground that they were the best anti-submarine vehicles available; and the Royal Fleet Auxiliary, justified by its contribution to NATO's overall logistic effort and Britain's acknowledged lead amongst European navies.

A Soviet 'Oscar'-class submarine: archetype of the massive perceived threat against which the NATO-orientated navy of the 1970s was conceived.

A further exercise in justification by the Royal Marines was equally successful: only a few days after the policy decision of January 1968 was known, the Corps reconstituted itself as a force primarily for Arctic warfare, building on a capability that was indeed already there in part. This move—it could almost be called a coup—had far-reaching effects apart from the preservation of the Royal Marines: it gave great comfort to the Northern Flank of NATO; it added a further task for the Royal Navy in the support of amphibious forces in the Norwegian Sea; it solidified the existing entente between the British and Dutch Marine corps; and, most of all, it gave focus to a Corps that had been subject to a diffusion of roles.

In its efforts to remain a balanced, high-capability force of long reach, the navy was helped by some remissions in the pressures towards decline. During the period of Conservative government of 1970–4, there was some extension of commitments beyond the 'irreducible' NATO contribution; the Five Power Defence Agreement in the Far East was honoured; and some extension of life was granted to the old *Ark Royal*. The decision was also made to proceed with the new carriers (still, cosmetically, called 'cruisers'); but it was a Labour government in 1975 which, confirming its predecessor's NATO emphasis, gave final approval to the Sea Harrier.

Again, in the late 1970s, the Labour administration under James Callaghan, alarmed by signs of Soviet expansion, set in hand studies on capability outside the NATO area. Even though these were 'no-cost or low-cost' items, they effectively checked any possible cut. Stemming from the same alarm, the 1977 decision by NATO as a whole to increase defence spending by 3 per cent annually in real terms eased financial pressure on all the services.

Nevertheless the navy's position remained precarious. It could not bid for

Justified on a NATO scenario and seen here exercising in a simulated nuclear-war environment, the 'Invincible'-class carriers were versatile in roles at many levels of conflict.

resources on anything but a NATO ticket; an attempt by the Naval Staff in 1975 to give an honest, published assessment of what the navy thought it was for—accepting that there were national as well as NATO tasks and suggesting that both must be planned for—was ruled out of order. Moreover, in 1979, on the return of a Conservative government, the replacement of the Polaris force by Trident was approved—and the navy was to foot the whole bill.

Financial pressures now appeared from all sides, and in 1981 John Nott was appointed as Secretary of State for Defence with the task of rationalizing defence policy and spending. He took a firmly minimalist NATO line as to strategy. He had no difficulty in accepting that the long-standing and well-defined commitments of the army and Royal Air Force should remain, but was unconvinced by the Royal Navy's necessarily more convoluted reasoning. His scientific advice was that the emphasis of anti-submarine effort, on which the navy set much store, should shift to nuclear-powered submarines and shore-based aircraft. He appeared, moreover, to discount the importance of the amphibious role. Keith Speed, the Minister for the Navy, resigned in May 1981 in protest at the decisions he saw coming.

The ensuing document, entitled *Defence Policy: The Way Forward* (Cmnd 8288), imposed a severe planned cut in the Royal Navy's force levels. Only two carriers were to remain; specialized amphibious shipping was to be much reduced; destroyer and frigate numbers were to be heavily cut; and manpower was to be set at a lower level than it had been for over a century. Effectively, the Royal Navy's general purpose forces were to be reduced to what was indeed a 'contribution to NATO'; and no one knew how low such a contribution might subsequently sink. That was the prospect on 31 March 1982.

Operations and Exercises The 1970s were a less busy time operationally for the navy than were the 1960s. The aftermath of empire had passed its peak of activity, and East of Suez was less pressing in its demands, which was just as well since there were not enough ships to sustain operations in that area.

Nevertheless, calls continued, both inside and beyond the NATO sea area. In 1974, in the aftermath of the Yom Kippur War, British forces played a leading part in clearing the Suez Canal of mines and obstructions. In the same year, in the turmoil following the Turkish invasion of Northern Cyprus, HMS *Bulwark* was involved in evacuation of civilians and casualties.

Fishery disputes continued with Iceland. The second and third 'cod wars' erupted in 1973 and 1975–6. Icelandic gunboats, designed for the work, confronted British frigates which decidedly were not. Every device and manœuvre, short of the use of firearms, was employed. Casualties were minimal, damage was not. Though a decision was reached in 1974 by the International Court of Justice that might have been thought favourable to Britain, her fishing rights in effect continued to erode.

On the other side of the Atlantic, two examples occurred in 1977 of the low-level deterrence that could so clearly be exerted by the navy. In response to threats from

Top and bottom Low intensity and presence operations continued to occupy the navy in the 1970s. The evacuation of casualties and civilians from Cyprus after the Turkish invasion in 1974 and the first good will visit by British warships to the People's Republic of China in 1980 were typical.

Guatemala against neighbouring Belize, still then a colony, the fixed-wing carrier *Ark Royal* on one of her final sorties defused the situation. Similarly, a build-up of tension over the Falkland Islands was countered by the deployment of a nuclear-powered fleet submarine with distant support from two frigates. It has never been admitted that the Argentine government was aware of this deployment, but in the nature of things it would be most surprising if they had not some wind of it.

Closer to home, the navy was finding a novel role in what became known as the 'offshore tapestry'. The regulation of good order in the waters round the United Kingdom had become ever more complicated, with sophisticated fishing effort by many nations; exploration, and the beginning of exploitation, of undersea oil and gas resources; the control of pollution from serious tanker accidents, of which the *Torrey Canyon* (1967) had been the first; and counter-gunrunning requirements in the approaches to Northern Ireland. These occupied much of the time of the navy's smaller vessels and the misemployment of some large ones.

Towards the end of the decade two more commitments East of Suez emerged. In Hong Kong, a surge of illegal immigration from China threatened the stability of an already overcrowded colony. The small Hong Kong squadron, operating in sometimes uneasy co-operation with the Hong Kong Marine, was fully employed in attempting to counter the fast 'snake boats' that carried the immigrants. The problem was solved, though, far more by a tightening of the civil power over immigrants than by naval or military measures. More international in character was the Iran–Iraq War which broke out in the Gulf in 1980. The Armilla patrol of two destroyers or frigates with RFA support was instituted almost at once, but for many years its activity was confined to a watching brief; in spite of numerous missile attacks on merchant ships, close escort by British warships was not allowed, to the frustration of their commanders and crews and to the fury of the merchant marine.

In a decade with so much emphasis on NATO, exercises with that alliance necessarily occupied much of the navy's time. The Teamwork series, watched always with interest by the Soviets (as indeed their exercises were by the West) concentrated on the Norwegian Sea and demonstrated British commitment to the full. National exercises were more specific in their objectives, and, for example, Exercise Highwood in 1971 exposed difficulties in supporting naval forces by combat aircraft working from shore.

Operations On 2 April 1982 Argentine amphibious forces landed in the Falkland Islands. After several hours' intensive fighting with, surprisingly, only three fatal casualties on the Argentine side and none on the British, the Royal Marines garrison, by now heavily outnumbered, outgunned, and surrounded, surrendered. The Argentine Commander issued proclamations assuming the Governorship of the Islands, to be renamed the Malvinas.

London had had some 48 hours' firm intelligence of Argentine intentions. Admiral Sir Henry Leach, the First Sea Lord, had used the time to some advantage. He had convinced the Prime Minister personally, in the face of more pessimistic advice from officials and ministers, that a naval force could be assembled capable of convincing Argentina of Britain's intention to regain the Islands; and, if necessary, of recovering them by force. Mrs Thatcher agreed that the task group should be formed and sail as soon as practicable.

Some naval forces were fortunately placed. Exercise Springtrain was in progress off Casablanca, and under Rear Admiral 'Sandy' Woodward nine warships from that exercise, with one fleet auxiliary, sailed south. Meantime in the home bases the rest of the Task Force was hurriedly prepared. The first ships included two carriers—the old *Hermes* and the nearly new *Invincible*, both sentenced by the cuts announced by Mr Nott some months before—and there followed a steady stream of warships, RFAs, and STUFT—the inelegant acronym for Ships Taken Up From Trade. The final tally amounted to some 40 warships, over 20 RFA, and 45 STUFT. Carried in the Task Force were ten Naval Air Squadrons and one RAF Harrier Squadron; three Royal Marine Commandos; two Parachute, two Guards, and one Gurkha battalions; and artillery and supporting arms.

The force assembled at Ascension Island—a fortunately placed airhead. There, faced with increasing evidence that attempts at a diplomatic solution would fail, plans were made in concert with the Fleet (and Task Force) Commander, Admiral Sir John Fieldhouse at Northwood, for the recovery of the Islands. Woodward had severe problems of force defence: he must be close enough to the Islands to protect his landing force with Harriers against the air threat from the Argentine mainland, but must not unduly expose his carriers on which all depended. He had to contend with submarine and surface threats too. He was without direct operational control over British submarines. A 200-mile Exclusion Zone around the Islands complicated his rules of engagement.

On 1 May the Royal Air Force mounted a bombing raid against Stanley Airfield involving a single Vulcan bomber supported throughout most of its long flight from Ascension by no fewer than 12, mostly tanker, aircraft. Three hours later Sea Harriers from the Task Force attacked targets in East Falkland, and shore bombardments were carried out by destroyers. The Argentine air force reacted throughout the day, 56 sorties being flown against the Task Force; four Argentine aircraft were destroyed.

There is thus no doubt that the engaged parties regarded themselves as being at war. The following day Woodward, with intelligence of Argentine surface forces to the south-west and north-west, just outside the Exclusion Zone, obtained approval for an attack on the cruiser *General Belgrano* to the south-west. The attack was carried out by the nuclear-powered submarine *Conqueror* on the evening of 2 May; the *Belgrano* was sunk with the loss of over 300 lives. Remaining Argentine surface forces, including the carrier *25 de Mayo*, returned to harbour and took no further part in the campaign.

On 4 May the Task Force suffered its first significant loss, the destroyer *Sheffield* being set on fire by an Exocet missile fired from Argentine shore-based aircraft. She later sank. On 21 May, British amphibious forces landed in strength from San

The Falklands conflict required almost every aspect of the modern naval art. A Sea Harrier takes off from a carrier to give air support with a RFA supply ship in the background (*below*); Royal Marines prepare to land at San Carlos from the liner *Canberra*, a ship taken up from trade, on 21 May 1982 (*far left*). One element of necessary capability was dangerously, and potentially fatally, absent: Airborne Early Warning helicopters (*left*) were not developed in time to take part in the operation. It had not been possible to justify them to the Treasury on the restricted NATO scenario.

Carlos Water some 50 miles from Stanley. Argentine aircraft from the mainland provided fierce opposition; the frigates *Ardent* and *Antelope* were sunk, and there would have been more casualties had Argentine aircraft not been forced to attack from very low level, causing fusing problems for their bombs. Nevertheless, the essential amphibious shipping escaped and the build-up on shore continued.

Further naval casualties occurred on 25 May, when the destroyer *Coventry* was sunk by bombing and the STUFT *Atlantic Conveyor*, carrying important heavy-lift helicopters, by an Exocet; and on 8 June, in circumstances that had some of the elements of Greek tragedy, the landing ships *Sir Galahad* and *Sir Tristram* were set on fire in Bluff Cove with heavy loss of life to their RFA crews and the Welsh Guards embarked. But these reverses did not seriously impede the successful land campaign, which with continuing naval support achieved the repossession of the Islands on 14 June.

Given that the Falklands were 8,000 miles from Britain and only 400 from Argentina, and that the Argentine forces were numerous and well-armed, the South Atlantic campaign was an extraordinary feat. It was inevitably, as Woodward echoing Wellington said, a near-run thing; it could not have been achieved without extremely high states of training and organization, a flair for improvization, and rapid action. Nor could it have been achieved at all by the forces envisaged in the Nott decisions, made nine months earlier. In this, as in many other ways, the navy was lucky. So was the government, which avoided humiliation; the country, which recovered some pride; and the international community, where the principles of deterrence and the rule of international law were reinforced.

The navy did not rest on its laurels after the Falklands. The Armilla patrol in the Gulf continued, and in 1988 permission was at last given for British-flag ships to be escorted. Deployments East of Suez by forces of a carrier and some escorts and RFAs occurred at least once every two years, giving opportunities for diplomatic and naval contacts with a variety of nations. NATO exercises continued unabated, the Teamwork 1988 exercise being the apotheosis of the American Maritime Strategy (see below). In 1991, as part of the successful US-led Desert Storm operation to liberate Kuwait (which had been overrun by Iraq the preceding year), substantial British naval forces operated in the Gulf, with notable success in air defence and mine clearance.

Table 13.1 Strength of the fleet, 1950–90

Year	Aircraft carriers	Large amphibious ships	Cruisers	Destroyers/ Frigates	Ballistic missile submarines	Nuclear-powered submarines	Conventional submarines	Mine and coastal craft
1950	12	—	29	280	—	—	66	66
1960	8	2	14	156	—	—	54	207
1970	3	4	3	81	4	4	31	54
1980	2	2	—	64	4	12	28	52
1990	3	2	—	51	4	16	9	54

Source: Statements on Naval/Defence Estimates (HM Stationery Office).

In 1993, the complex civil war in former Yugoslavia required the deployment of naval forces in the Adriatic to provide 'poised' support for humanitarian effort on the ground and to enforce sanctions particularly against the warring parties. Substantial British naval forces, led by a carrier, operated in the area throughout the year.

Plans and Policy A looseleaf Foreword to the 1982 Defence White Paper, over the signature of John Nott and dated 18 June, said (after a brief tribute to the Armed Forces): 'The events of recent weeks must not, however, obscure the fact that the main threat to the security of the United Kingdom is from the nuclear and conventional forces of the Soviet Union and her Warsaw Pact allies. It was to meet this threat that the defence programme described in Command 8288 was designed. The framework of that programme remains appropriate.'

Though Nott left office soon afterwards, the civil servants who had supported his proposals remained largely in place, and although the Royal Navy was able as a result of the Falklands experience to add back certain assets, notably amphibious and carrier capabilities, and also to replace war losses, it continued throughout the 1980s to struggle to maintain its general position.

It was helped by two factors. The first was national; under two Chiefs of Defence Staff of notable vision, Field Marshal Sir Edwin Bramall and Admiral of the Fleet Sir John Fieldhouse, Whitehall had been (in the former's words) nudged into a more open acceptance of the out-of-area role. The second, which was external, was the emergence of an American 'Maritime Strategy' based on forward operations, particularly in the Norwegian Sea and north-west Pacific if deterrence appeared to be about to fail. In the late 1980s the Naval Staff linked itself strongly to this policy as a justification for high-quality forces. While this appealed to the more mechanistic school of planners, and was useful in approaches to the Treasury, it had flaws. It supported what was arguably an over-aggressive policy with strong elements of bluff and, perhaps more seriously, it was vulnerable to changes in the overall strategic situation.

These duly occurred in 1989–90 when the Soviet empire and the Warsaw Pact crumbled with a suddenness that surprised the world. Whatever other results it had, and as this history is written they are still developing rapidly and unpredictably, one was certain: a massive assault against western Europe with little warning was no longer conceivable. The NATO scenario was undermined.

NATO responded with new strategies and with an outreach to the countries of the former Warsaw Pact. Britain, after some temporary policy statements emphasizing a new need for flexibility and mobility in the armed forces, came out with a more precise statement of strategic aims in the 1992 Statement on the Defence Estimates: 'To ensure the protection and security of the United Kingdom and our dependent territories, even when there is no major external threat; to insure against any major external threat to the United Kingdom and our allies; to contribute to promoting the United Kingdom's wider security interests through the maintenance of international peace and stability.'

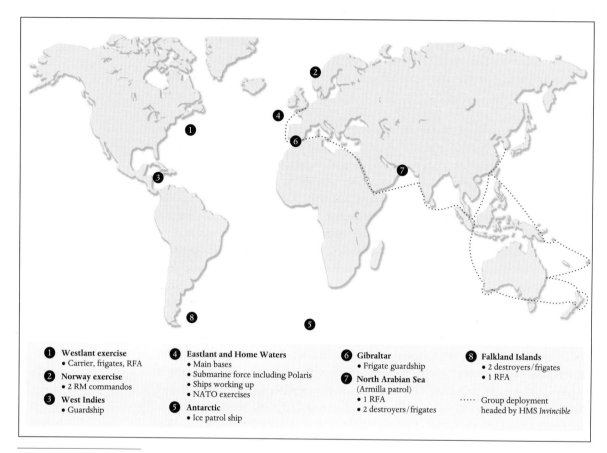

1 Westlant exercise
• Carrier, frigates, RFA

2 Norway exercise
• 2 RM commandos

3 West Indies
• Guardship

4 Eastlant and Home Waters
• Main bases
• Submarine force including Polaris
• Ships working up
• NATO exercises

5 Antarctic
• Ice patrol ship

6 Gibraltar
• Frigate guardship

7 North Arabian Sea
(Armilla patrol)
• 1 RFA
• 2 destroyers/frigates

8 Falkland Islands
• 2 destroyers/frigates
• 1 RFA

····· Group deployment
headed by HMS *Invincible*

Activities of the Fleet,
early 1984.

This put the Royal Navy in a better philosophical position than at any time since 1968. In practical terms, however, financial pressures were again being felt; an 'Options for Change' exercise in the Ministry of Defence had already reduced force levels though no essential arm of the naval service looked like being severed completely.

The Treasury tourniquet tightened throughout 1993 and a Defence Costs Study was begun. The navy restated its core capabilities as: three carriers; amphibious forces with associated specialist shipping; and the nuclear-powered submarine force. Destroyers, frigates, constabulary craft, and afloat support, while accepted as essential to back up the core forces, were accorded lower priority for procurement. Straitened though it was, the navy insisted on maintaining standards of material and training that would keep it in the front rank, and able to expand effectively if major crises threatened.

THE NAVY'S PEOPLE,
1945–1994

In the early years after the Second World War, naval personnel matters were dominated by the needs of demobilization. This was handled in a much more sensitive way than in the corresponding period after the First World War. At the same time attention was paid to continuity; entry to the service, at officer and rating level, never ceased, though the flow of recruiting slackened. As the post-war structure evolved, the navy was helped by a government ruling on conscription: only some 11,000 conscripts were allowed in the navy, a much smaller proportion than in either of the other services, and motivation was assured by stipulated previous membership of the Volunteer Reserve.

In 1946 the age of entry to the Royal Naval College at Dartmouth was set at 16, having previously been 13. This was intended to eliminate class barriers and widen the officer entry, but it satisfied hardly anyone because no school wanted to prepare a candidate for a special examination at that age. However, given the comprehensive selection techniques of the Admiralty Interview Board, the continuance of the Special Entry at 18, and a flow of 'Upper Yardman' candidates from the lower deck, an adequate intake was sustained for the next ten years.

Also in 1946 the Electrical Branch was formed, initially from Torpedo Officers and Ratings with a technical bent and those Special Branch officers who had served, often as Reserves, in the war. A swift influx of graduates helped to turn the 'greenies', as they were called due to the colour of the officers' distinguishing cloth, into an influential body.

Rating structure generally went through more evolutionary change. The lower deck hierarchy was still dominated by the requirements of experience and professional qualification, save that Artificers held rank appropriate to their accelerated assumption of responsibilities. Names changed (Stokers became Stoker Mechanics and subsequently Marine Engineering Mechanics) and these reflected new emphases in the jobs ratings did. Educational qualifications for entry varied according to the recruiting climate, sometimes being set so low that the navy's level of professionalism was jeopardized, though the more rigorous standard for the Artificer entry was maintained.

During the late 1940s and 1950s initial shore training was prolonged, both for officers and ratings, reflecting the early age of entry (boy ratings were often no more than 15). Once personnel were at sea, the pattern of ships' commissions did not always help training. It was common for a ship to remain on a foreign station for 18 months, sometimes without much base or staff support, and standards varied. Towards the end of the decade, with more emphasis on structured commissions with a 'home leg' and pre-commissioning training in the establishments, the situation improved.

By the mid-1950s, the need for change in officer structure and training was apparent, and a comprehensive review resulted in the issue of Admiralty Fleet Order 1/56. Not long before, the distinguishing cloth on the sleeve, denoting an officer's specialization, had been abolished; this was a symbolic precursor. Now 'military command', for long the prerogative of the seaman officer, devolved by seniority; only 'sea command' remained reserved. Technical and Supply specialists could now command shore establishments, and the number of appointments open to officers of any specialization was increased. The shortage of sea commands compelled the authorities to split the lists of seaman commanders and

captains into 'Post' and 'General' (colloquially, 'Wet' and 'Dry') and only post list officers could normally expect sea command.

The structure of AFO 1/56 went hand in hand with changes in officer entry and training. The entry age was standardized at 18 or thereabouts, to be followed by a seven-term course at Dartmouth, of which the equivalent of two terms was spent at sea in the training squadron. There were modifications for technical and supply specialists, Upper Yardmen, and the growing cohort of limited-career officers, particularly aircrew. Finally, AFO 1/56 envisaged that Special Duties officers (previously called Branch Officers and before that Warrant Officers), promoted from the lower deck at a later stage than the Upper Yardmen, should form a substantial part of the officer corps. They were at that time separately trained, and it was not until the 1980s that all officers' initial training was concentrated at Dartmouth.

Conscription ended, as a result of the Sandys reforms, in 1961. The navy had never found it easy to fit conscripts into its highly professional training pattern and though there was regret at losing the leaven of the better-motivated Volunteer Reserve conscripts, it was mingled with relief.

Possibly the most marked development of the 1960s in the training field was the increasing merger of training for sea with training at sea. For ratings, this was seen in an increasing tendency to quite short (well under a year in the case of adults) initial shore training followed by a sea draft. This did not apply to Artificers who still needed considerable shore training in their trades. For officers, the educational standard demanded by the AFO 1/56 reforms quickly proved to be too low. The 'Murray Scheme' which succeeded it in the early 1960s required advanced-level attainment for full career officers, who then had a shorter time at Dartmouth before going to sea in the Fleet as Midshipmen, thus restoring an old and tried formula. Engineer officers went on to be trained to degree level at the Royal Naval Engineering College (with application courses in the sub-specializations of Marine, Weapon, or Air Engineering) and the opportunities for other officers to gain university degrees were expanded.

However, it was whole-ship training at sea that was the hallmark of the 1960s. Portland became the working-up base for ships at appropriate stages of their commissioning cycle. A Flag Officer Sea Training, with a full staff of 'Sea Riders', supervised and exercised every detail of a ship's operational efficiency and readiness. Several weeks' intensive training, with interim and final examination by 'Hallmark' and, later, 'Thursday War' exercises, resulted in graded departmental and overall assessments. 'Good' was the best, rarely awarded, epithet. Ships from other NATO navies, particularly those from northern Europe, regularly appeared to be put through the process.

1970 saw the end of an institution once whimsically characterised by Churchill (along with sodomy and the lash) as one of the navy's abiding traditions: the daily rum ration. Though accompanied by predictable funerary rites, the passing of the 'tot' was accepted by the lower deck, who well understood that a highly technical fighting force, depending for its survival in war on quick and accurate reactions,

Facing: Training for all ranks and ratings concentrated on the assumption of individual responsibility in carrying out professional tasks. An initiative test in river-crossing at Dartmouth (*below left*), a damage control exercise in realistic conditions (*below right*), a classical unarmed-combat shot from the Royal Marines (*above right*), and operations room personnel staffing consoles that direct weapons and co-ordinate information (*above left*).

HMS *Cumberland*, 1991.

BATCH 3 TYPE 22 FRIGATE (1993)

HMS *Chatham*, 1993.

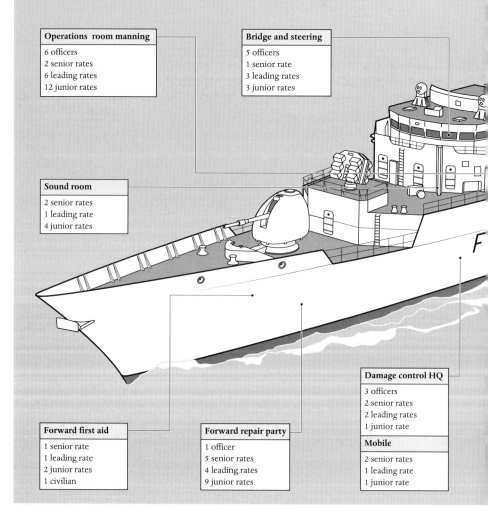

Close range

2 leading rates
10 junior rates

Resupply

1 leading rate
3 junior rates

Operations room manning

6 officers
2 senior rates
6 leading rates
12 junior rates

Bridge and steering

5 officers
1 senior rate
3 leading rates
3 junior rates

Sound room

2 senior rates
1 leading rate
4 junior rates

Damage control HQ

3 officers
2 senior rates
2 leading rates
1 junior rate

Mobile

2 senior rates
1 leading rate
1 junior rate

Forward first aid

1 senior rate
1 leading rate
2 junior rates
1 civilian

Forward repair party

1 officer
5 senior rates
4 leading rates
9 junior rates

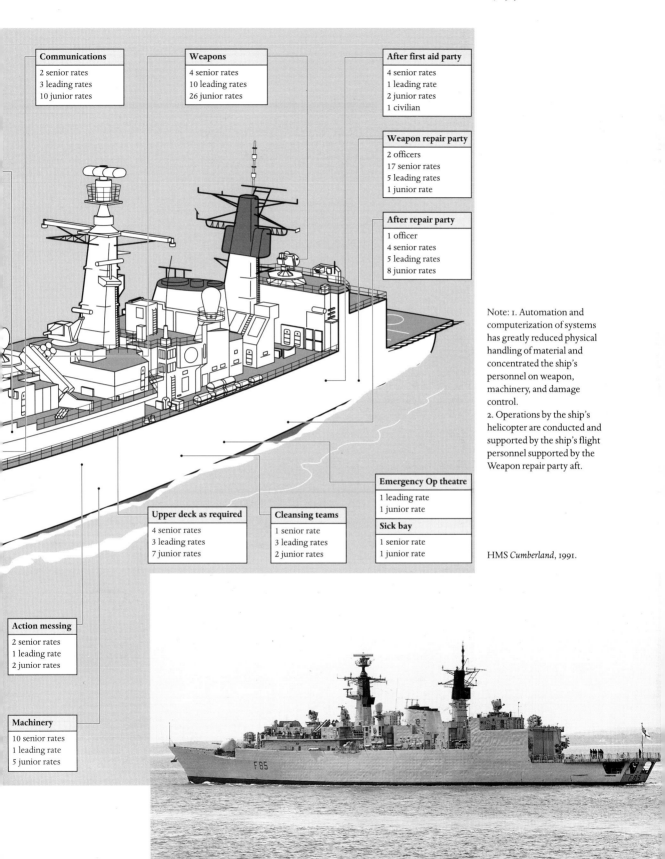

Communications

2 senior rates
3 leading rates
10 junior rates

Weapons

4 senior rates
10 leading rates
26 junior rates

After first aid party

4 senior rates
1 leading rate
2 junior rates
1 civilian

Weapon repair party

2 officers
17 senior rates
5 leading rates
1 junior rate

After repair party

1 officer
4 senior rates
5 leading rates
8 junior rates

Note: 1. Automation and computerization of systems has greatly reduced physical handling of material and concentrated the ship's personnel on weapon, machinery, and damage control.
2. Operations by the ship's helicopter are conducted and supported by the ship's flight personnel supported by the Weapon repair party aft.

Upper deck as required

4 senior rates
3 leading rates
7 junior rates

Cleansing teams

1 senior rate
3 leading rates
2 junior rates

Emergency Op theatre

1 leading rate
1 junior rate

Sick bay

1 senior rate
1 junior rate

HMS *Cumberland*, 1991.

Action messing

2 senior rates
1 leading rate
2 junior rates

Machinery

10 senior rates
1 leading rate
5 junior rates

The Women's Royal Naval Service went to sea in the complement of combatant ships for the first time in 1990, and was fully absorbed into the Royal Navy in 1994.

was not well served by the issue of one eighth of a pint of raw spirit per man per day. The money saved by the Crown was applied to a new Sailors' Fund, the income from which has financed many desirable welfare projects afloat and ashore.

The navy now believed it had got its sea-training pattern right, but the pace of modern and particularly missile warfare dictated further change in the organization of seaman officers. Up to 1971, such officers had sub-specialized as they had since the late nineteenth century, in Gunnery, Torpedo and Anti-Submarine, Navigation, or Communications; or had remained non-specialist 'salt horses'; or had been aviators or submariners. Now the old tribal sub-specializations were swept away and non-aviation and non-submarine officers were to be Principal and later Advanced Warfare Officers (PWOs and AWOs), trained in the arts of the Operations Room across the whole spectrum of warfare. While the PWO system went through modifications and some sub-specialization itself, the underlying concept remained sound and was proved in action in 1982.

Rating structure reflected these changes in a reconstituted Operations Branch. More dramatic, however, was the introduction of the rank of Fleet Chief Petty Officer, bringing the navy into line with the other services at the highest non-commissioned level. In this way the responsibilities of about 1,100 top ratings were rewarded with enhanced pay and status.

Pay was a recurring problem in the 1970s. The decade began with the introduction of the Military Salary, abolishing many old allowances and establishing comparable rates of pay with civilian equivalents, augmented by an 'X' factor for the rigours and dangers of service life. Charges were exacted for accommodation and food, but for naval personnel at sea these were ameliorated.

Recruiting and retention temporarily improved. However, from 1973 onwards the imposition of successive pay policies nation-wide had a disastrous effect on service pay. Soldiers and sailors could not move from job to job, which was the accepted way in civilian life of beating a pay freeze. In consequence, by 1978 the comparability gap between service and civilian pay had reached 37%. Recruiting and retention suffered in all the services; there was insufficient manpower for the available arms.

When the Thatcher government came to power in 1979 it quickly rectified the situation. The cost, however, was apparent in pressures on the rest of the defence budget; and the 'Way Forward' review in 1981 laid down particularly heavy cuts in naval personnel. These were slowed down, but not reversed, in the aftermath of the South Atlantic campaign. They called for extremely tight planning of drafting, training programmes, and the ratio of sea to shore service. Successive reviews in the 1980s did nothing to lighten the situation, and by 1992 a programme of compulsory redundancies, the third since the Second World War, was being imposed.

A decision was taken in 1990 to send personnel of the Women's Royal Naval Service to sea in combatant vessels for the first time. The WRNS had been an important part of the service since 1917, with a temporary closure between the wars; it was of high quality with an array of talents. There was some misgiving about sending women to sea particularly in small ships' companies where every member had literally to pull their weight in activities such as replenishment. However, initial reports from sea, including operational deployments, were encouraging and the navy rode out a few inevitable scandalous situations without much difficulty. The continuation of the experiment led to the integration of women into the Royal Navy, and abolition of the WRNS as a separate entity, in 1994.

The Royal Naval Reserve and Royal Naval Auxiliary Service were by the early 1990s under great pressure. Both had clear and indeed vital roles in large-scale conventional war, in Naval Control of Shipping and Mine Countermeasures, but planning for war of this sort was unfashionable and financial pressures were strong. In 1993 the abolition of the RNXS was announced, as were deep cuts in the RNR. The savings associated with this risky development turned out to be pitifully small. The same could be said of an equally controversial contraction, the closure of the Royal Naval Engineering College at Manadon, Plymouth, a centre of technical excellence that had served the navy outstandingly well and whose loss could have very serious effects.

In its Pacific campaign the US Navy had learnt the importance of afloat logistic support. The British Pacific Fleet's performance in this field lagged well behind. It was a decade and a half before the British caught up; but by the early 1960s they were equal to the US Navy, ahead of other NATO navies, and even further ahead of the Soviet navy in their ability to replenish at sea with fuel, stores, and ammunition. That ability has been retained since.

SUPPORT FOR THE FLEET, 1945–1994

This could not have been achieved without a well-endowed Royal Fleet Auxiliary making use of modern techniques, backed by a shore organization that provided the right stores in the right ships. This essential work was done by the Royal Naval Supply and Transport Service in spite of steady constriction in its depots.

Support for the material state of the fleet was more subject to criticism. The general perception in the service was that ships came out of dockyard hands late and in an unsatisfactory condition. Blame was attributed to a variety of factors, from restrictive labour practices, through poor interdepartmental organization, to lack of leadership; ships were, often justly, accused of overbidding for facilities and unscheduled improvements. When operational, ships' maintenance was in the hands mainly of their technical staffs with the help of shore-based maintenance teams who were deployed (by air to foreign parts) for assisted maintenance periods.

The contraction of the fleet and its concentration in home waters meant that permanent repair and maintenance facilities abroad steadily diminished. Simonstown, Malta, Singapore, Gibraltar, and Hong Kong dockyards passed into other hands along with many lesser facilities; although visiting naval forces could call upon most of them, it would be on commercial terms. But even at home the basis

Development and mainten-
ance of the ability to replenish
fuel and stores at sea, including
the retention of a modern Fleet
Auxiliary, was an essential
element of the fleet's mobility
in the post-1945 period.

on which the dockyards did their business was altered in the 1980s. Chatham was closed in 1984 and Portsmouth became a naval base, no longer of dockyard status, at the same time. The Devonport and Rosyth yards, in parallel with other developments in British industry, came under the management of privately funded companies. Work-forces were heavily cut and more flexible working practices introduced.

Delegation of financial management in many other areas continued. By the end of 1991 the New Management System was fully in place, and naval Budget Holders at several levels wrestled with financial intricacies to which they had not been used. The flexibility this conferred was generally welcomed. However, it extended only so far: in the higher reaches of defence finance, Treasury control remained as powerful as ever. The navy had every reason to suppose that, as always in peacetime, further demands would be made upon its ability to make do with resources it considered inadequate, and that politicians and officials would impose ever more stringent tests of its ingenuity and tenacity.

14

COLD AND HOT WARS

*T*n 1945 the Royal Navy found itself in a naval world radically different from that of 1939, largely due to such wartime technical innovations as radar, effective sonar, jet aircraft, missiles, fast submarines, influence mines, and atomic weapons. Over the following decades, it adapted to its new technological—and economic—situation. To an important extent British innovations developed in this adaptation were adopted by the world's other navies.

Strategically, the post-1945 situation resembled that in Europe before 1939. Relatively strong Western navies faced a hostile navy—in this case the Soviet navy—designed to deny use of the sea (and to defend the Soviet coastline), rather than to contest (as in the Japanese wartime case) sea control. The Royal Navy also retained its pre-war role of policing the empire and commonwealth. The war had released enormous nationalist tensions in much of the empire. The Soviet Union was perceived as capable of exploiting those tensions. Particularly after about 1948, the policing role was often called the Cold War role. The Royal Navy, then, had to prepare to fight a Hot War (against the Soviets) while quite actively conducting a series of Cold War campaigns. Major Cold War operations included the suppression of Malaysian Communist terrorists (1948–60), protection of Malaysia and Brunei against Indonesia, and police duties in East Africa. The day-to-day Cold War role required continuous naval presence in areas such as the Persian Gulf for a variety of anti-piracy and even anti-slavery missions. The Royal Navy also participated in the Korean War, which was a Cold War operation of a different sort.

The Hot War issue was always whether the Russians could or could not be precluded from invading western Europe. Until the mid-1950s it seemed unlikely that any Western nuclear threat could actually deter the Russians, and Britain, like the United States, built towards a posture suited to prolonged war. That entailed maintaining large numbers of convoy escorts and building defensive minecraft, as well as keeping some coastal forces to help dominate the Channel and the North Sea should the Russians repeat the German successes of 1940. Most of the surviving Second World War escort fleet was maintained in reserve against the requirements of a prolonged Third World War.

Less-expensive alternatives were eagerly sought. The most attractive was to rely heavily on nuclear (later thermonuclear) deterrence. The 1957 Defence

HMS *Ocean en route* to the Korean War, 21 July 1952. She symbolizes the early post-war Royal Navy, with its reliance on its wartime legacy and its vital Cold War role. Her aircraft were all Second World War piston types.

Review, with its wholehearted emphasis on this strategy, has been described in Chapter 13. Ultimately, the British element of nuclear deterrence became the Royal Navy's Polaris submarine force. Prolonged-war naval forces, such as large numbers of Second World War-built escorts and coastal forces, were discarded. The Cold War force was, however, maintained because no degree of thermonuclear deterrence could affect nationalist tensions or Communist subversion. Keeping the submarine deterrent viable entailed maintaining local Anti-Submarine Warfare (ASW) and mine countermeasure forces to protect the vital ballistic missile submarine base at Faslane.

ECONOMIC
CONSTRAINTS

Economically, the Royal Navy's position in 1945 was far weaker than it had been in 1939. The Second World War had effectively bankrupted the country. Resources, perhaps particularly scarce engineers and scientists, had to be shifted into exports and hard-currency earners such as British-flag merchant ships. For example, work on major warships under construction in 1945 was greatly slowed because electrical draughtsmen had to be shifted to merchant ships and exportable equipment. Moreover, many in the government imagined that the war had demonstrated that air power alone might suffice to protect the country in the future. The navy's share of the drastically reduced defence budget was itself cut to pay for an increased share for the Royal Air Force, and also for an air-based British nuclear deterrent.

In 1948 the First Sea Lord had to admit that the Royal Navy could no longer, by itself, guarantee protection of vital seaborne supplies. In any future war it would have to be allied with the US Navy. That alliance enormously influenced the development of British naval technology. For example, it was assumed that in wartime the United States would supply most British naval aircraft. Carrier hangars had, therefore, to be large enough to accommodate US aircraft. The Royal Navy planned to adopt the 3-inch/70 heavy anti-aircraft gun, using the same barrel and ammunition (but fortunately not the same flawed mechanism) as the US Navy. Similarly, it was assumed that in wartime British submarines would have to fire some US homing torpedoes, so British submarine fire control systems were adapted to handle them (as well as British weapons). Oddly enough, British surface-to-air missile launchers seem never to have been designed to accommodate US weapons. The relationship also had profound effects on the US Navy, most of which will be described below.

On the other hand, the Royal Navy emerged in 1945 with a valuable legacy, in the form of existing warships, both completed and under construction. This capital lasted for about two decades, allowing the British Government to put off painful decisions to cut British naval commitments and the forces to fulfil them. Once the cost of replacement did have to be faced, the weakness of the British economy made drastic cuts inevitable.

The question was always whether some major slice of British naval (or other defence) roles could be abandoned to alleviate the pain. For example, in 1945, a large standing garrison conscript army, carrying heavy expenses, had to be maintained to police the empire; costs were actually higher than before the war

HMS *Daring*, the leader of the last class of British destroyers ordered during the Second World War.

because of the loss of the Indian army, which had provided much of the garrison of the Middle East and East Africa. When national service was finally eliminated (as announced in the 1957 Defence White Paper), much of the Cold War role 'East of Suez' was taken up by mobile carrier task forces. Then they, too, became unaffordable. In 1967 the British government formally announced that it was withdrawing from East of Suez in the hope of eliminating the Cold War role almost entirely. The fleet would operate wholly in the north-east Atlantic, within range of shore-based aircraft. That made the Falklands War, the clearest echo of the old Cold War mission, a great shock.

The amphibious ship (LPD) *Fearless* was symbolic of the East of Suez mission. She could launch landing craft to carry heavy loads to a beach, or helicopters to carry them inland.

MODERNIZATION

It was clear in 1945 that the fleet would have to be modernized to deal with the new technology. The Cold War role made it impossible to finance modernization entirely by drastic shrinkage in the active fleet, although numerous ships were laid up or scrapped (for example, the older battleships were quickly discarded, where the US Navy retained their contemporaries in reserve). By 1949 the British Government had found an alternative. Surely the new technology would quickly pass through a series of intermediate stages before attaining its ultimate stable form. Buying short-lived systems could be quite expensive. Ideally, then, the services would retain their 1945 equipment until the 'ultimate' systems were available. That seemed practicable because the Soviet Union had been so badly crippled by the Second World War that it would be some years before she was ready for new adventures, by which time 'ultimate' systems might be available. In particular, the British (and almost certainly their US allies) estimated that Stalin would not take his chances until he had amassed about a hundred atomic bombs, about five years after the first test, which no one expected before 1952. That made 1957 the 'year of maximum danger' (a phrase first used in 1934, presciently referring to the possibility of war with Germany in 1939) and hence the target date for modernization.

That did not quite work out. Stalin exploded his first bomb in 1949, so the US Government advanced the calendar to 1954. Then Stalin showed that he was much readier than had been imagined: his North Korean satellite started a war in the Far East in June 1950. Many imagined that it was really the opening of the Third World War, and the West began to rearm. For a Britain barely recovering from the Second World War, the Korean War was disastrous. The feared intermediate technology had indeed to be adopted, often at the cost of the hoped-for 'ultimate' (1957) systems.

The Royal Navy surface force of the first post-war decade might be characterized as a battle fleet, built around one or more strike carriers, plus convoy escort or ASW support forces (including ASW carriers), plus mine countermeasures (mainly home defence) and some coastal and amphibious forces. After 1952, the battle fleet was intended both to participate in NATO strikes on North Russia (as part of the NATO strike fleet) and to protect NATO convoy escort forces (which were primarily intended to deal with submarines) from the Soviet main fleet and land-based naval air arm. The carrier offered a combination of long-range air defence (using her fighters) and long-range striking power (supported by her own radar aircraft). The carrier's escorts provided back-up anti-air and anti-submarine firepower; they benefited enormously from the carrier's superior command system and radar. In the ASW role she contributed invaluable command facilities as well as aircraft and, later, helicopters.

Cold War missions could easily become hot. HMS *Ardent*, a Type 21 frigate, burns in Falkland Sound after an Argentine air attack. She had been hit, not by a modern stand-off weapon, but by unguided bombs dropped by Argentine jets.

In the police role the carrier's striking power was essential to support troops or marines ashore. Amphibious assault was less important as long as British garrisons were often already on the ground. Later it became much more important. At Suez in 1956, the British were the first to mount a helicopter assault from a carrier, HMS *Ocean*. During the confrontation with Indonesia over Malaysia a decade later, the Royal Navy mounted a major night amphibious assault. Amphibious assault was the centrepiece of the Falklands operation in 1982.

The reasoning of the late 1940s was repeated in 1957. Surely missiles would soon replace manned aircraft. The more sophisticated manned aircraft in development might be characterized as interim systems. The political deadline would probably be set by the pace of Russian ballistic missile development; the Russians would hardly fight until they had a reasonable chance of disarming Western bomber forces by a first strike. The Royal Navy argued that land bases, whether for aircraft or missiles, were inherently vulnerable; an influential part of it wanted to move the country's deterrent to submarines, using the US Polaris. The British Government adopted a RAF proposal to carry a US-developed air-launched missile, Skybolt, aboard bombers on airborne alert. When the United States killed the Skybolt programme in 1962, Britain bought Polaris.

THE NAVY OF THE SEVENTIES DEVELOPED

The high cost of the nuclear-powered Polaris submarines had to be borne just as the carrier replacement issue had to be faced. The Royal Navy argued that Soviet missiles would soon make fixed tactical air bases untenable. Moving carriers would have a far better chance. Ballistic missiles could only attack fixed points; it took aircraft to deal with targets moving unpredictably. Unlike ballistic missiles, those aircraft could be shot down. Had the Royal Navy prevailed, it would have consumed by far the bulk of the British defence budget. The Royal Air Force did not find the idea of its effective demise attractive; it won the fight to retain its customary fraction of defence spending. Britain could not afford both Polaris and new large carriers. The death of the carrier programme was announced in 1966 (the last British full-size carrier, HMS *Ark Royal*, left service in 1978).

Even with this economy, fleet running costs (meaning manpower) had to be drastically reduced. A new generation of warships was conceived, all powered by gas turbines, which required far smaller crews than the earlier steam warships.

Heavy ASW helicopters and their command systems were still essential, so a command/helicopter cruiser (later designated a carrier), HMS *Invincible*, was developed. The carrier and her surface consorts were vital for a NATO role, support of the NATO strike fleet which was intended to head north early in any war with the Soviets (to neutralize Soviet heavy naval forces, including anti-ship bombers, threatening ocean shipping in the North Atlantic). The carrier could double as an assault (troop-carrying) ship. With the specialist fighters gone, the 'Invincible'-class ship would have to depend more heavily on anti-aircraft missiles carried by her consorts. A minimum-cost missile destroyer, Type 42 ('Sheffield' class) was designed. To make up for the loss of anti-ship firepower, the Royal Navy bought and widely deployed the new French Exocet anti-ship missile (it was later

HMS *Resolution* was the first British Polaris submarine. Her appearance changed the balance of roles between the Royal Navy and its chief rival, the Royal Air Force, since in the past the air force had justified itself largely on the basis of its strategic role.

largely superseded by the US Harpoon, which has about twice the range). There was also a new ASW frigate (Type 22) and a new general-purpose 'gunboat' frigate (Type 21).

An internal Royal Navy study made at this time showed that the saving due to eliminating the carrier might well be illusory. It was tempting to imagine that the important anti-ship strike mission could be taken over by ship-launched missiles like those recently introduced by the Soviet navy. However, the missiles would represent only a fraction of the total cost of attacking enemy surface ships. Some means would still be required to detect and localize those targets in the first place. That seemed to mean that some sort of sea-based aircraft would still be needed; the likely alternative of a ship-detection satellite system was impossibly expensive. Moreover, the anti-aircraft missile ships lacked the range needed for fleet air defence. The Royal Air Force claimed that it could provide fighter cover in north European waters, but exercises showed that was an illusion.

Nor did the move away from East of Suez Cold War eliminate the need to land and support troops. Just as the large carriers were being abandoned, NATO strategy moved away from depending on pure deterrence to preclude the Hot War. Even though the Soviets might not welcome nuclear catastrophe, their nuclear weapons would deter NATO from using its own. In this stalemated situation, both sides might well fight a non-nuclear war, at least for a time. In a way, that returned

415

to the ideas of the early 1950s. Roles such as wartime convoy reinforcement of
Europe were once more quite important. In the 1970s, for example, a major Royal
Navy mission was to transport amphibious troops to Norway, to prevent the
Soviets from operating freely on the northern flank of NATO. The troops might
well need naval support.

By this time, the resources of the early 1950s were gone, particularly the carriers, the battleships, and cruisers with their gun firepower (so useful for supporting amphibious operations), and the mass of war-built convoy escorts. Many of
those ships could not have functioned effectively against modern threats, but even
so the contrast with the Royal Navy of two decades before cannot have been a
happy one to contemplate.

Fortunately, it turned out that a new kind of sea-based aircraft, the short take-
off/vertical landing (STOVL) Harrier, could operate from the new helicopter
carrier. It could carry effective anti-ship missiles and it provided a very useful
degree of fleet air defence in the Falklands, and it also supported troops ashore.
However, the new 'Invincibles' were clearly too small; they could not support
anything like enough Harriers. Moreover, the Harriers could not provide surface
search like the earlier carrier radar aircraft. A partial solution, applicable in NATO
situations, was to fuse all available intelligence on surface ship location in a headquarters ashore, at Northwood. This OPCON surface picture was provided to
deployed ships via a dedicated satellite link. The parallel US system, comprising a
shore headquarters (FCC), a shipboard terminal, and a satellite link (OTCIXS),
was conceived to support the Tomahawk anti-ship missile, equivalent to the long-
range anti-ship weapon considered in the British studies of the late 1960s. In effect,
a Harrier armed with anti-ship missiles functioned as a manned Tomahawk.

The carriers and their escorts still cost a great deal. In the late 1970s it was sometimes argued that costs could be cut drastically by restricting the wartime role of
the Royal Navy to anti-submarine warfare, prosecuted almost entirely by
submarines. For example, diesel submarines in the Greenland–Iceland UK Gap
could detect Soviet submarines en route to their patrol areas. Although they
might be unable to intercept the Soviet units, they could call in shore-based RAF
Nimrods to attack them. In 1981, Minister of Defence John Nott decided greatly to
weaken the surface ASW capability (which was justified partly to support the
NATO Strike Fleet) and the ability to land troops in Norway. The Falklands War
the following year showed that these capabilities, however marginal in a NATO
context, were still extremely important because there was a world beyond NATO.

SIZE AND SHAPE IN A
NEW STRATEGIC
SITUATION, 1990

Now the Soviet threat has collapsed, at least for the time being. The post-Cold
War world is likely to be extremely turbulent, and the navy is the most flexible
means of projecting British power where it is needed. Recent examples are the
Gulf War of 1991 and the continuing presence in the Adriatic (the lessons of which
caused the British Government to order a helicopter assault carrier in 1993). There
are still also remnants of empire, for example in the Caribbean, where Royal Navy
warships help try to interdict the drug trade. What is gone is the old need to

prepare for the Hot War against the Soviets, particularly for large-scale North Atlantic anti-submarine warfare. What used to be called Cold War capability is, if anything, far more important.

Thus, the key 1966–7 decision, to abandon out-of-area capability (as provided by strike carriers), no longer seems nearly as wise. Some influential opinion seems to be shifting towards a larger next-generation carrier, still operating STOVL aircraft. She would be supported by a relatively small number of very capable escorts, more like the cruisers of the past than like the frigates and small destroyers conceived in the 1960s. To balance their cost, the number of lesser ships, designed primarily for convoy escort, would be drawn down. Yet the drawdown will have a limit, because the Royal Navy will still have to cover foreign stations, such as the Persian Gulf. Other economies will come from drastic reductions in submarine strength; much of the submarine force had been justified by its ability to cover the massive Soviet submarine fleet. The survivors are likely to be nuclear-powered only, since only nuclear power offers the mobility to work with long-range surface forces. The prototype of such operations will surely be the successful attack on the Argentine cruiser *Belgrano* (by HMS *Conqueror*) during the Falklands War.

Admiral of the Fleet Lord Fraser, then First Sea Lord, characterized the new technological challenges in 1949: fast submarines (such as the German Type XXI which appeared at the end of the Second World War) armed with long-range homing torpedoes, jet aircraft armed with guided missiles, pressure mines, and nuclear weapons. Four and a half decades later, the Royal Navy could be reasonably sure that it had solved the problem of jet aircraft (but that solution in turn had dissolved with the loss of the carrier fleet). The Type XXI problem had been largely solved, only to be superseded by the far worse problem of the nuclear submarine. Nuclear weapons seem to have been neutralized by world politics, at least until they become common in the Third World. The pressure mine problem (indeed, that of all modern mines) has never really been solved.

THE TECHNOLOGICAL CHALLENGES

Moreover, the character of each problem changes depending on the political environment in which the navy finds itself. For example, the Soviet submarine problem seemed solvable largely because in the Third World War submarines could have been attacked almost anywhere in the world ocean, without much regard to niceties such as avoiding neutrals. In a future limited conflict, as in the Falklands, the ASW battle area is very limited; offensive action (for example, attacks on the submarines' bases) may be precluded altogether. Instead of concentrating on enemy submarines, it may be more important to protect ships against torpedoes fired, by surprise, before the outbreak of war.

Type XXI was the first major submarine designed primarily to operate submerged, where it was faster than on the surface. Much of the success of wartime anti-submarine warfare could be traced to the U-boats' need to spend much of their time on the surface, where they could be detected by airborne and shipborne radars. In 1944 the Germans fitted their submarines with snorkels, by means of which they could run their diesels underwater. However, these U-boats were still relatively slow submerged, and thus could not, for example, intercept convoys unless they spent considerable time running on the surface.

Type XXI, however, was faster submerged than many existing escorts, particularly if the latter were running into a sea. High submerged speed had been bought partly by streamlining, which also drastically reduced the effective sonar cross-section of the submarine, making it more difficult to detect. There had been some hope that the submarine would be noisy, hence detectable, at high underwater speed, but that turned out not to be the case; Type XXI was actually very quiet. Beyond Type XXI lay the even more threatening Type XXVI peroxide (Walter) submarine. Type XXI could sustain its 16 knots underwater for an hour or two. Using a closed-cycle hydrogen peroxide powerplant, Type XXVI was designed to maintain 24 knots for as much as six hours, outrunning any escort chasing it. It did not yet exist (although its powerplant had been built in shore-based form), but the Germans had completed several slower Type XVII closed-cycle U-boats. Most observers thought that anything the Germans had promised, they could build. In 1945 many also considered nuclear submarine propulsion a medium-term proposition. It offered essentially unlimited endurance at Type XXVI speed.

The Germans had, moreover, developed a variety of homing and pattern-running torpedoes for their submarines. U-boats had low-frequency passive sonars, which could detect convoys from beyond the detection range of the escorts' higher-frequency active sets. Thus a fast submarine could shadow a convoy from beyond the escorts' sonar range. From that safe position it could fire long-range self-guided torpedoes into the convoy without risking effective counterattack.

A new kind of anti-submarine warfare was needed.

Jet aircraft offered similar challenges. The Royal Navy was already painfully aware that its standard anti-aircraft fire control systems could not cope with high-performance piston aircraft; it had bought some US systems. The higher speeds of jets made matters worse; they also greatly complicated the work of the Action Information Organization (AIO) used to co-ordinate fleet fighter aircraft and anti-aircraft guns. The AIO also found it difficult to cope with the enormous increase in the number of targets. Most wartime bombers had approached a fleet *en masse*, in hopes of crashing through anti-aircraft defences. Although there might be many bombers, the AIO saw them as only a limited number of distinct raids. However, a single aircraft armed with a guided missile (or, worse, an atomic bomb, which could be tossed from a distance) could expect to sink a target without approaching close enough to have to worry about anti-aircraft fire. There was no point in massed raids.

A 1948 experiment on board the carrier *Illustrious* showed that large numbers of individual attackers could overload an AIO. The AIO was built around a plot, typically on a perspex screen, showing all air targets, depicting the tactical situation on the basis of which decisions were made. Each time a radar operator detected an aircraft or raid, that information was passed verbally to a plotter who entered it in chinagraph pencil. Each stage of the process imposed a delay, so that the perspex screen always showed a delayed version of the true situation: the operator took time to make a report, and the plotter took a bit longer. Each time-lag cost a little accuracy. As the number of plots built up, the entire system began to slow down imperceptibly. It turned out that no manual plot could accommodate more than about twelve raids per hour.

Both to deal with enemy aircraft and to strike targets, the fleet had to operate its own jets. Their engines did not accelerate rapidly enough to allow unassisted (rolling) take-offs. Unfortunately, existing hydraulic catapults lacked sufficient power. Moreover, jets approached the carrier at speeds too high for the existing 'batsmen' to cope. The 'batsman' controlled the landing aircraft by observing its approach and signalling corrections to the pilot. No matter how quick his reactions, the sequence of observation, decision, and signal cost time. Signals might well be inappropriate by the time they were seen. Consequences often included deck crashes.

Jet engines had first appeared in Britain, so it was natural for an interesting offshoot to become available in 1945: the marine gas turbine, the naval equivalent of the new turboprop. It offered high power on dramatically less weight than a conventional steam plant and, as it turned out, with drastically fewer personnel. As it happened, gas turbines appeared at about the same time the Admiralty had to finance a new generation of steam plants to supersede its obsolete existing types, which operated at relatively low pressure and temperature and therefore at low efficiency. Gas turbines were not yet mature enough to place in widespread service, but they were clearly the ideal powerplant of the future. They first entered British service in the early 1960s, combined with new-generation steam plants.

Missiles, used in wartime by the Germans (and, in effect, by the Japanese kamikaze pilots) presented new problems. For example, radar-controlled missiles

Modern mines were, and are, a very difficult problem. British post-war rearmament plans always included a large mine-sweeper programme. HMS *Coniston*, one of the new generation, is shown on 19 June 1958.

could be fired in weather too grim for fleet fighters. Work on the obvious counter-measure, fleet anti-aircraft missiles to replace existing heavy anti-aircraft guns, began as early as 1944. At that time the Royal Navy also envisaged anti-ship and anti-submarine missiles, the former to replace existing heavy naval guns.

The wartime pressure mines, laid on the sea-bed, responded to the drop in water pressure due to a ship passing overhead. Earlier acoustic and magnetic mines could be swept because the signatures setting them off could be simulated relatively simply. Unfortunately, only a ship could provide the pressure signature: the mine was, in effect, unsweepable. That was potentially disastrous. Britain depended heavily on coastal traffic, for example, to provide coal for London. During the war, it had taken an enormous fleet of sweepers to keep the east coast ports open in the face of heavy German mining, mainly by aircraft and by fast attack boats; this operation was practicable only because existing mines could be swept. It was affordable only because small sweepers tended not to set off the rela-tively crude magnetic mines. By 1945, much more sensitive magnetic mines, well suited to attacking sweepers, could be made relatively cheaply. Thus the large fleet of war-built sweepers might well be unusable in a future war, quite aside from their inability to deal with pressure mines.

Almost half a century later, the only counter to pressure and other sophisti-cated influence mines is still the one developed late in the Second World War: hunting. The sea floor must be searched for possible mines, and those objects identified and neutralized one by one. Natural debris around an industrial area would greatly complicate (and slow) any such search. Modern mine counter-measures ships must be carefully (and expensively) silenced to avoid acoustic and magnetic mines. They generally also have complex navigational and combat systems (for example, to register where mines have and have not been found, so that follow-up ships will not repeat their work). The combination makes such ships quite expensive, so they cannot be built or operated in anything like the numbers of the past. Since a hunter covers far less ground per hour than a sweeper, and since there are now many fewer hunters than there were sweepers in the past, the situation can hardly be considered promising. In the recent past the situation has been further complicated by the appearance of rising mines, which are very suitable for planting in deep water, and which can sometimes attack ships approaching at a considerable distance. Mines can be cleared *after* the end of hostilities, as in the Persian Gulf, but it is by no means obvious that they can be cleared quickly enough to permit a big amphibious operation, for example.

Finally, in 1945 there was the larger question of whether nuclear weapons had made surface navies obsolete. They did offer enormous firepower in a very small package suitable, by the early 1950s, for carriage by carrier aircraft and then even by helicopters. The Royal Navy sometimes saw them as a way of making up for drastic reductions in the number of available aircraft and weapons. For example, the first British carrier-borne atomic bombs were justified as a means of dealing with Soviet cruiser raiders, at a time when the Royal Navy could muster very few anti-ship attack aircraft. Much later, nuclear bombs made it possible (in theory) for even the few Harriers on board an 'Invincible' to deliver a useful degree of support to British troops ashore in Norway. The Royal Navy also considered nuclear depth bombs the best counter to such fast deep-diving Soviet submarines as the 'Alfa' class of the 1970s. Thus it was a very mixed blessing for the Royal Navy when such weapons became effectively unusable owing to the nuclear stalemate.

Anti-submarine warfare was the first post-war priority. The Soviets had captured examples of the new German technology during their advance to Berlin, including twenty partly-built Type XXIs. The peace settlement allocated them four completed ones (the US and Royal Navies each got two). They also obtained the Walter factory, with its closed-cycle power-plants. Although British submarine operations against surface ships might no longer be terribly important (the Soviets had few merchant ships, for example), modern British submarines were needed in order to learn about likely Soviet developments and tactics, and to test countermeasures. Initially, the Royal Navy decided to concentrate on closed-cycle submarines, on the theory that they represented the most important future technology. For example, in June 1945 the British occupation authorities tried to order Type XXVIs from the Blohm & Voss yard in Hamburg, only to discover that the submarine was much further from operational status than had been advertised. Vickers received a contract to restore a Type XVII to serviceability, and to develop a Walter powerplant suited to an operational British submarine.

The British rejected the diesel-electric Type XXI as an irrelevant intermediate step; they did not realize until about 1948 that the Walter project was so difficult that this interim step might be the more important one. They did, however, make a very important choice as early as 1945, to make special efforts to silence submarine machinery, by sound-mounting diesel engines and by adopting diesel-electric rather than direct drive. An unexpected consequence was to encourage work in passive sonar for long-range submarine detection. The post-war 'Porpoise' and 'Oberon' classes were significantly quieter than their American equivalents.

The silencing technology developed for submarines was applied to post-war minesweepers, to protect them against acoustic mines. It evolved into the idea that noisy machinery should be mounted on a rigid platform or 'raft', which could be sound-insulated from the hull proper. Rafting was included in the first British nuclear submarine designs (which were, however, abortive because the engineers involved had to be transferred to the more urgent civilian nuclear power industry). The main source of noise in submarine nuclear power-plants was gear whine: the plant drove propellers through reduction gearing, and it produced electricity through geared turbo-generators. Turbines, generators, and gearing could all be placed on a raft. In 1956, when the US Navy determined to silence its new nuclear submarines, it adopted the British rafting technique. Although the first few British nuclear submarines used un-rafted powerplants similar to early US installations, later ones were rafted for silencing. Later the Royal Navy introduced a second silencing device, a pumpjet to replace the conventional propeller, a secondary source of noise.

The 'Oberons' were the ultimate development of early British post-war submarine design, quieter than any of their foreign contemporaries. Ships of this type were sold to Australia, Brazil, Canada, and Chile.

421

The British became interested in intercepting enemy submarines before they could reach their patrol areas. Their own submarines could lurk near enemy bases. In order to make good progress out to distant patrol areas, enemy submarines would have to proceed either on the surface or continuously (and loudly) snorkelling. In either case, they could be attacked using conventional straight-running torpedoes. However, homing torpedoes would clearly be better, and work began on them.

These submarines could not replace classical convoy escorts. Existing escorts were too slow; it was assumed that any escort had to be 10 knots faster than its quarry. In the approaches to Europe, moreover, a convoy would necessarily face massive air attack as well as submarines. Among its escorts would have to be specialized anti-aircraft and fighter control ships. All the escorts would have to be mass-produced; the hope in 1945 was that a single standard hull, with alternative outfits, could handle all three missions. To be able to cross the Atlantic unre-fuelled while escorting a convoy, it would be diesel-powered. Unfortunately, no existing or planned large-ship diesel offered sufficient power, so the ASW version, the Type 12 frigate, was split off and a new steam plant designed.

HMS *Undaunted*, one of 23 war-built destroyers rebuilt as interim ASW frigates (Type 15). Note the experimental helicopter platform right aft.

HMS *Arethusa*, one of the very successful 'Leander' class, symbolizes the opposite pulls of Cold and Hot War requirements. She was based on a Hot War (ASW) frigate (Type 12), but modified for general purpose duties, with the helicopter (visible aft), and with a good enough air search radar for independent operations.

The new ASW ship was deemed too sophisticated to be built in sufficient numbers. In 1949, the Controller (responsible for warship design), Rear-Admiral R. S. Edwards, suggested a ship carrying half the ASW battery of Type 12 (and lacking its 4.5-inch gun) at half the cost. It became the 'Blackwood' class (Type 14), the 'Third World War corvette'. Unfortunately, the Third World War did not come as advertised, and Type 14 lacked the medium-calibre gun needed for the peacetime policing and presence roles. The specialist ASW frigate developed into the very successful multi-purpose 'Leander', the last British examples of which have only recently been retired (1994).

Several new sonars and ASW weapons were developed. The new submarines could flash in and out of the relatively narrow beam of a conventional sonar. The closer the submarine, the less distance the beam width covered, and thus the shorter the time the submarine was in the beam. The solution, both in Britain and in the United States, was to increase effective range (by reducing frequency, and therefore by accepting a larger sonar) and to project multiple beams simultaneously, to cover a large area. The first important British post-war sonar, Type 177, covered an 80-degree sector out to about 8,000 yds, which was considered enough to deal with a 15-knot submarine. It was superseded by Type 184, whose main new feature was full 360-degree coverage to deal with faster submarines.

Weapon development did not quite match this potential. At the end of the war the Admiralty was working on a new version of the wartime ahead-throwing Squid mortar, which emerged in 1952 as Limbo (maximum range 1,000 yds). An escort version of the submarine-homing torpedo, which would have come closer

423

to Type 177 range, was less than successful. In much the same situation, the US Navy developed a torpedo-carrying missile, Asroc. The Royal Navy did eventually buy the comparable Australian-developed Ikara missile, but most ships were fitted instead with flight decks for light helicopters.

The frigates the Royal Navy could afford could not carry both a helicopter and a specialized ASW missile. The helicopter offered a solitary frigate (for example, on Cold War duty) more options. For example, it could scout or direct gunfire. In the ASW role, the ship could command the MATCH (*MAnned Torpedo-Carrying Helicopter*) into position to drop its weapon. In the 1960s, the Soviets supplied numerous torpedo- and missile-attack craft to their Third World allies; preparation for defence against them became an important Cold War issue. The MATCH helicopter offered a solution: it could carry a small guided missile (initially the French SS-12, later the British Sea Skua) with which to attack torpedo or missile boats. This concept was demonstrated dramatically during the Gulf War (1991), when Royal Navy Lynx helicopters successfully attacked Iraqi fast attack craft with Sea Skua.

Passive towed-array sonars, now fitted in all modern Royal Navy frigates, reach well beyond Type 177/184 range, to the point where they cannot locate submarines precisely enough for missile or directed-helicopter attack. Whatever platform reaches the area of the submarine must search that area to re-acquire it before attacking. Hence the importance of the big ASW helicopters on board the carriers, with their own dipping sonars. The new Merlin helicopter, to be carried also by the current Type 23 frigates, will have similar capability.

Long-range passive sonars had another significance for the Royal Navy. Frigates equipped with them were planned to patrol in the Greenland–Iceland–UK (GIUK) Gap, through which Soviet submarines would have had to come to reach the North Atlantic in wartime. They did not carry dipping-sonar helicopters of their own, but could call in land-based Royal Air Force Nimrod patrol planes, which in turn would have re-acquired the submarines (using sonobuoys) and attacked them with torpedoes and, if authorized, nuclear depth bombs.

This technique of ASW, in which submarines were detected in the open sea and then sunk by aircraft, had first developed during the Second World War. At that time submarines were generally detected either by their own radio emissions (HF/DF), or by breaking the codes ordering them into position. In 1945 it seemed unlikely that Russian codes would be broken as easily as those of the Germans had been, or that the Russians would be foolish enough to transmit widely on submarine radios. The only remaining techniques were submarine blockade of Soviet base areas and escort of convoys; aircraft might be useful, however, in searching the areas around sinking merchant ships ('flaming datum' searches). During the 1950s, however, it became possible to use large low-frequency fixed arrays (the British Corsair, the US SOSUS) to detect submarine noises at long range. Such operation offered an alternative to the contemporary concept of convoy escort supported by short-range carrier-based aircraft. In 1954, the RAF seized on the new technology (which was not yet in service) to suggest that the surface fleet, and particularly its carriers, could be replaced by shore-based RAF aircraft. In 1981, John Nott in effect revived the idea, and very nearly had it adopted.

The 1945 generation of escorts, armed only with guns, stood little chance of surviving attacks by jet aircraft. The navy was already developing a heavy anti-aircraft missile, Seaslug, but it was far too massive to fit in an escort. In 1952, the Royal Navy formally proposed that the United States develop a small-ship anti-aircraft missile, small enough directly to replace a 4.5- or 5-inch gun mount. It became the US Tartar (later Standard Missile), but ironically it was not adopted by the Royal Navy. Instead, from the late 1950s on the Royal Navy developed its own frigate missile, Sea Dart, which eventually appeared on board its second-generation missile destroyers.

Like the US Navy, the post-1945 Royal Navy invested heavily in carrier-based ASW aircraft. Indeed, at times its carriers were justified primarily for their ability to support such aircraft. Even when that was not the case, the service suffered because ASW aircraft consumed a large fraction of limited aircraft capacity. Helicopters solved the problem, since they took up less space and could be shifted from the carrier on to major surface ships. When the new carrier (CVA 01) was conceived in about 1960, one projected way of increasing her striking power was to build a smaller carrier, called the escort cruiser, which would combine the necessary ASW helicopter deck with defensive anti-aircraft missiles. The Admiralty realized it could not hope to buy both escort cruisers and a new carrier, but the escort cruiser concept evolved into the 'Invincible'-class light carrier—albeit without the strike carrier with which it was intended to work. Two existing 'Tiger'-class cruisers were converted into limited escort cruisers, their after turrets replaced by helicopter decks.

An RAF Nimrod maritime patrol aircraft flies over a Russian 'Kotlin'-class destroyer. Several times after 1945 the RAF argued that such aircraft could replace most British warships, at least for maritime operations near the United Kingdom.

The Admiralty pursued one other important line of ASW development. The new submarines might be able to avoid detection before firing their torpedoes; that inspired interest in an anti-torpedo weapon. Ideas included anti-torpedo torpedoes and special rockets. All foundered on the problem of classification: there were too many false alarms, so a ship would have to carry too many weapons in order to have enough available when they were needed. The only survivor of this effort was the towed Type 182 decoy, which is designed to attract acoustic homing weapons. The anti-torpedo problem was never really solved, and as this is written the US and Royal Navies are engaged in collaborative work on a new generation of devices.

JET AIRCRAFT
OPERATION

As early as December 1945 a de Havilland Vampire landed on board the light fleet-carrier HMS *Ocean*. The Royal Navy ordered new jets such as the Supermarine Scimitar. To operate them effectively, it had to modernize its carriers. The Royal Navy could not then expect to build new carriers. However, existing ships were

The post-war British carrier fleet at its peak, with HMS *Hermes* leading *Ark Royal* and *Victorious*. Both *Hermes* and *Victorious* show the big nacelle of the Type 984 3-dimensional radar.

relatively small. Only one of six armoured fleet carriers, HMS *Victorious*, was modernized, largely because so much effort was required to rebuild the heavy flight deck integral with her hull. The existing light fleet carriers proved too small for efficient jet operation. That left two heavy 'Ark Royals' and four smaller 'Hermes'-class carriers, ordered during the war but left at so early a stage of construction that major changes could be made before completion.

For jets, the most important innovation was the steam catapult, which offered far more power than a conventional hydraulic unit. First suggested even before the Second World War, it was perfected in the late 1940s. The US Navy gratefully adopted it, having failed to develop an alternative explosive-driven catapult.

The second important innovation was the angled deck. In 1945, carrier aircraft were pulled forward into a deck park after they landed. The park was protected by a wire barrier against aircraft which jumped the arrester wires, but it was not unknown for an aircraft to jump the barrier and crash (and destroy) the deck park. The solution was reached accidentally. Once propellers had been abandoned, naval aircraft could, in theory, land on their bellies on a flexible deck (in effect, a big inflated rubber bladder). They then had to be pulled off to be catapulted off again. One proposal was to cant the flexible deck at an angle, so that aircraft could be pulled off to one side rather than directly ahead. Seeing a sketch of this arrangement, the prospective commanding officer of HMS *Ark Royal* suggested that a similar arrangement would solve the barrier problem. Aircraft landing at an angle to the ship's centreline would not risk crashing into the deck park. If they missed the arresting wires, they could simply apply more power and fly off again. This configuration, which is now practically universal, made carrier jet operation really practical.

The batsman problem was solved by the 'mirror sight'. The approaching pilot saw his position as the apparent position of a light reflected in a mirror alongside the flight deck, and compensated accordingly. The US Navy adopted the mirror, and eventually both navies replaced it with a Fresnel lens system, which operates on the same principle.

These innovations helped make it possible to operate ships of modified Second World War design through the 1960s. They could not, however, shrink the size of carrier required to operate high-performance jets.

It seemed that a further British innovation, the vertical take-off (Harrier) carrier, could solve even that difficult problem. It did not have the universal popularity of the other three. Carrier operation seems to demand numbers, not merely the ability to put a few aircraft into the air. It is true that catapults and arresting gear impose minimum size requirements on a carrier, but so do fuel and ammunition for a useful number of aircraft. A Harrier ship could easily be too small for real efficiency. On the other hand, it offered some air capability to a navy which could not afford a much larger conventional carrier. Harrier ships have since been built for Spain and Italy; one is now being built for Thailand. The US Marines bought Harriers, and operate them from large amphibious carriers not fitted for conventional carrier jets. Proponents of the Harrier point out that the aircraft can often outmanœuvre nominally higher-performance jets, but the issue of sheer numbers remains. For example, the two British carriers in the South Atlantic carried a total of 20 aircraft. When two Harriers collided in bad weather, the force suffered a sudden 10 per cent drop in capability.

A desirable future combination for the post-2000 Royal Navy would be a higher-performance descendant of the Harrier in a hull large enough to accommodate a more useful number of aircraft.

ACTION DATA SYSTEMS

The new digital computers offered a natural solution to the plotting problem revealed in 1948. Radar operators could enter targets directly into a computer rather than reporting them to plotters. Ideally, the computer could estimate target movement (on the basis of several observed positions). It could continuously display all the known targets, with little or no plotting delay. The Admiralty developed a simple digital computer, with one plotting channel per target, called the Comprehensive Display System (CDS). A 48-channel CDS was teamed with a new 3-dimensional Type 984 radar in the carrier *Victorious*. The computer-compiled plot could be transmitted digitally to the carrier's consorts by a system called Digital Plot Transmission (DPT), the first operational digital data link to enter service. CDS was shown to the US Navy, which briefly used an equivalent as EDS, the electronic data system. CDS also inspired the big computer plotting system then being developed for the US SAGE (Semi-Automatic Ground Environment) air defence command system for North America.

Some argued that ASW was already so complex that it, too, needed a computer-based tactical system. Through the early 1960s the Admiralty worked on a computerized 'Cambria' ASW plot, but to little avail. This role was eventually filled by a derivative of the evolving air defence system.

CDS was a pure display system. It could not compare data on different targets; in effect, each channel was a separate computer. Nor could it use target data for further computation, for example for fighter direction. The step beyond was a general-purpose digital computer, which could carry all the targets in a single memory. There they could be compared, and data could be extracted for further use. The first British version was Action Data Automation (ADA), used in HMS *Eagle* in about 1962. The step beyond was to use the same computer for weapon calculations, a system called Action Data Automation Weapon System (ADAWS), first used in the second series of 'County'-class missile destroyers. The US equivalent, developed in parallel, is NTDS. ADAWS probably inspired the Dutch SEWACO combat direction systems, which have been widely exported. A simpler version of ADAWS, CAAIS, was used by the Royal Navy and was also widely exported in various forms. Unlike ADAWS, it was a federated system using one or more separate fire control computers.

AIR DEFENCE

The Royal Navy also needed new anti-aircraft weapons. Its dilemma was that with very limited carrier resources its surface escorts might well also have to deal with enemy surface raiders. Wartime experience with German battleships and cruisers was not soon forgotten; the Soviets were building large numbers of new 'Sverdlov'-class cruisers. Thus it never seemed altogether practical to build single-purpose anti-aircraft weapons.

At the end of the Second World War the main medium-calibre gun was a new twin semi-automatic 4.5-inch, mounted in destroyers and frigates. Work was proceeding on automatic dual-purpose 6-inch and 3-inch/70 guns, which eventually were mounted on board the three 'Tiger'-class cruisers. The 3-inch/70 was planned for the new frigates, but it was not ready in time: they had to be armed with 4.5-inch guns. However, the Royal Canadian Navy did adopt the weapon for its new frigates (destroyer escorts). The ideal weapon was a more powerful automatic gun somewhat smaller than the cruiser's 6-inch. For a time it appeared that a 5-inch/70 would enter service in the mid-1950s. It inspired a project for a large destroyer or 'cruiser destroyer' which might be powerful enough to deal with a full-sized cruiser (banking on excellent fire control and a high rate of fire), yet small enough to be built in numbers, as a convoy escort.

It took a missile to cope reliably with the coming generation of jet bombers and guided missiles. Like the contemporary US Terrier, the first-generation Seaslug rode a radar beam towards its target. Initially it was hoped that it could replace existing 4.5-inch mounts on a one-for-one basis, but that proved impossible. By 1948 it was clear that a single Seaslug mount would be about as large as a cruiser turret. After several false starts, a new 17,000-ton cruiser was designed, to carry Seaslug aft and a pair of 6-inch turrets forward. It perished in the 1956–7 defence cuts.

The navy still badly needed the missile. A class of large destroyers, the 'Counties', was designed to carry it (with a pair of twin 4.5-inch mounts forward). Compared to the cruiser, they lacked flag (command) facilities; they also lacked a long-range 3-dimensional radar. They were likely to be effective only to the extent that they worked with a carrier, which could transmit data from its 3-D radar via DPT. A second series of 'Counties' had some of the earliest British computer command systems (ADAWS), but still lacked any 3-D radar equivalent to that on board foreign contemporaries, and thus still depended on a carrier for long-range radar information.

Seaslug had another peculiarity. Its designers chose to wrap its booster rockets around its body rather than to place them in the more conventional tandem position. Unlike the contemporary US Terrier, then, Seaslug could not be stowed on a short rotating ring. Instead, it had to be carried, like a small aircraft, in what amounted to a carrier hangar running much of the length of the ship. The most serious was the failure of the design of the original bi-fuel liquid sustainer motor. It was replaced by a solid-fuel motor. The system's performance against low level targets was limited.

The affordable missile ship, the 'County', could not stand up to a surface ship such as a 'Sverdlov'. The Royal Navy therefore made special efforts to adapt Seaslug as an anti-ship weapon. Its beam-riding guidance system could not fly it into a target on the horizon, so a special 'dump mode' was developed: the missile

The 'County'-class guided-missile destroyer *Antrim* displays her chief weapons, the big launcher for her Seaslug missile and a Wessex anti-submarine helicopter.

could be made to dive when it arrived over the estimated position of the target. No such attack could be really precise, so another possibility, fitting the missile with a nuclear warhead, received considerable attention.

Beam-riding ultimately limited missile range, partly because the beam spread out as range increased. The US solution, ironically first used in Tartar, the missile conceived by the British, was semi-active homing. Since the missile homes on radiation reflected by the target, accuracy improves as it approaches. There seems to have been no attempt to adapt Seaslug to semi-active operation (as was done with its US contemporary, Terrier), but the next British missile, Sea Dart, operated in this fashion. Semi-active homing provided Sea Dart with a direct capability against surface targets, out to the horizon.

Both Seaslug and Sea Dart were area defence weapons, carried on board a few specialist destroyers responsible for defending a group of other ships. Fleet air defence doctrine envisaged fighters operating beyond missile range, then a missile defence zone (extending out to perhaps 20–40 n.m.). Inside that range, ships would have to fend for themselves.

Much work went into electronic defences to work with the missiles. By 1960 it was clear that attacks, at least by the Soviets, would generally be by missiles, locked on to their targets before they were launched. A powerful jammer could make that very difficult, so that the attacking bomber would have to come close before firing. The closer it had to come, the better the chance that the defending fighters and missiles would destroy it in time. A very powerful jammer on board a destroyer or frigate might also so increase the apparent size of that ship that the bomber would be unable to decide that it was not the carrier.

Ironically, on the one occasion the fleet had to defend itself against large-scale air attack, in the Falklands, the attackers were generally armed with bombs. They had to overfly the fleet in order to hit ships. On the other hand, the fleet's radars were ill-adapted to distinguish aircraft flying low overland; there had not been enough money to modernize or replace the main search sets. The few missile-armed aircraft succeeded largely because airborne early warning (AEW), and the fighters which were supposed to kill the bombers before they could attack, were gone, victims of the 1966 carrier decision. Moreover, they were able to fire from fairly long range because the Royal Navy had abandoned the planned

HMS *Exeter* is a stretched version of the Type 42 anti-aircraft missile destroyer conceived after the 1966 policy shift. Note how much smaller her Sea Dart launcher is than the big Seaslug unit aboard HMS *Antrim*. The Type 909 missile director radar is in the big radome atop her bridge, with another aft.

jamming/deception policy of the early 1960s (although the Argentine pilot who hit HMS *Sheffield* apparently thought he was attacking the nearby carrier *Hermes*). Apparently there was some fear that turning on a very powerful jammer would give away the fleet's location.

In the aftermath of the Falklands, the Royal Navy adopted three kinds of electronic defence. One was a revival of the earlier long-range jamming, probably including measures to create false targets in enemy radars. The other two were self-defensive: deception jammers and decoy (chaff) launchers. In theory a deception jammer could draw an incoming missile into an attack on a cloud of chaff (aluminium foil) created by the decoy. The next step is to place a deception device in the decoy proper, to convince the radar of the incoming missile that the decoy, hanging from a parachute, *is* the ship. That becomes somewhat easier when, as now, the ship is coated partly with radar-absorbing material, which reduces the radar reflector the missile sees.

Many ships obviously could not accommodate any massive missile system, with its big magazine and radar fire control. In the late 1950s an alternative appeared. An aircraft company, Short Brothers, was already making a command-guided anti-tank missile, Malkara. It proposed an anti-aircraft version, which could replace the existing 40 mm. gun on a one-for-one basis. The Royal Navy (and then numerous foreign navies) adopted the missile as Sea Cat. It was considered effective against subsonic jet attackers, though not against the new supersonic anti-ship missiles, many of them bomber-launched, then entering Soviet service.

Then a new kind of threat appeared. From about 1968, Soviet 'Charlie'-class submarines carried SS-N-7 missiles to sea. These weapons could be fired underwater, like torpedoes, albeit from greater ranges. It seemed likely that a group of Soviet submarines attacking a convoy would use SS-N-7s to destroy or neutralize the escorts, then close in to sink the ships with torpedoes.

No area air defence system could deal with such a threat. Even moving at subsonic speed, an SS-N-7 might take no more than three minutes to fly from submarine to target, and it might cross the ship's horizon no more than a minute before it hit. The ship therefore needed a new automatic self-defence missile system. A British Aerospace/Marconi/Vickers Consortium developed a supersonic missile, Seawolf, and the associated radars and fire controls. It was first installed in the new ASW frigate, Type 22, conceived after the 1966 cuts.

Seawolf was the only quick-reacting short-range missile available to the Royal Navy when the Falklands War began. Although it was intended only to defend the ship mounting it, Seawolf-equipped frigates were assigned to protect aircraft carriers against Argentine air-launched missiles, running so close alongside that any missile fired at the carrier would also be, in effect, fired at the frigate. Although the Seawolf never had to intercept an Argentine missile, it very successfully dealt with several Argentine aircraft.

For the future, the old kind of area air defence will continue to be a useful counter to aircraft. Increasingly, however, missiles will fly at very high speeds,

HMS *Boxer*, an early Type 22, symbolizes the attempted shift away from Cold War duties after 1966. Excellently equipped for ASW, she lacked the one most important Cold War weapon, a gun which could be fired across another ship's bow.

offering very short warning times. Thus the successor to Sea Dart will probably work at relatively short range, covering only two or three ships. The Royal Navy calls it the Local Area (defence) Missile System (LAMS), and expects it to equip the next-generation destroyer, whose tripartite staff requirement was signed in December 1992 by France, Italy, and the United Kingdom. Ships will generally also have a very short-range air defence system, probably a small missile, and a variety of decoys.

POSTSCRIPT

J. R. HILL

*I*t is a favourite joke of historians nowadays to say they can teach us nothing. The pace of change, they say, is so rapid that history—even quite recent history—is irrelevant when considering current and future problems.

I first used those words nearly thirty years ago, at an Oxford University seminar in fact, and whatever has happened to History in the meantime it has not changed the Historians' Joke much. I did not agree with it then and do not now; but, keeping a weather eye lifting for historians, this postscript will be relatively modest in the lessons it draws from the material that has gone before.

It will not, that is to say, offer new or old theories of sea power. Nor even will it suggest a new order of sea-based economy, coercion, and conflict, tempting though it is to peddle useful tools for the strategist and planner such as levels of conflict (normal conditions, low intensity and higher level operations, general war) and reach (the distance from the home base at which operations can be sustained). They can be found elsewhere.

But certain threads, simpler in their composition and starker in their colour than any theory or system, stand out in the narrative of this book and all the evidence suggests they are enduring. They may be flagged by three words: Communication, Autonomy, Institution.

COMMUNICATION

Communication across the sea has been the principal concern of the Royal Navy for nearly 1,000 years. One of the earliest uses of Royal ships was to 'waft' other craft, trade, or soldiery; and through all the centuries that followed the protection of that traffic, in convoy or by more dubious means, was a very high priority. Denial of the use of the sea to opponents was its obverse. The interest remains; in spite of every effort, political and technical, to lick geography, well over 90 per cent by bulk of British trade travels by sea, and so in time of crisis does the same proportion of its military effort deployed overseas.

In another sense of the word, communication of thought and information was and is a vital element of the navy's business. At its highest level it operates from government to command and has often in that sense proved the navy a uniquely sensitive politico-military instrument, provided—and there are salutary lessons both ancient and modern—there is sufficient delegation to the commander on the spot. At the planning level, as many examples in the book show, communication was equally necessary but not always forthcoming; the navy was frequently none too good at explaining to the financiers what it was for.

Space has allowed all too little specific mention in this book of communication between units, either of instructions or information. That is not a criticism of the authors, for the subject is immensely complex and in naval affairs pervasive. Brian Tunstall wrote a book (since ably edited by Nicholas Tracy as *Naval Warfare in the Age of Sail* (Annapolis, 1990)) on eighteenth-century naval warfare that was in fact a book about signalling systems: the two were inseparable. It is no coincidence that the book as originally written was so complicated and opaque it was almost incomprehensible to the modern reader. Perhaps all we can do is acknowledge that there is scarcely any naval activity that does not require accurate communication between units and commands: and that when such communication is lacking disaster looms.

It is nowadays a statement of the obvious that communication between individuals, in such a straitened environment as a ship of war, is of the first importance. Yet in earlier times it was a lesson learned by good officers and not learned by bad ones; here surely history has taught us much already, but the lessons are to be relearned by each generation.

Training is a form of communication: and it is easy to trace in this book the increasing formalization of training as the centuries moved towards the twentieth. Yet there was always, at sea, room for initiative: and since the Second World War that scope has increased until the modern sailor operates from a basis of knowledge as a highly responsible, trained individual. In battle, communication reaches its apotheosis: well-trained, well-informed, well-motivated crews, provided they are left to get on with the job and not mucked about, are very likely to win. One could have said the same about an eighteenth-century warship in action.

One final point must be made about communication and it is one where previous centuries have less to teach, at least in degree. The thirst for news that is now satisfied by the media certainly existed in history, and news often got around quicker than you would think; nevertheless, the speed of modern media communication is a new phenomenon and one with which any navy has to come to terms. Recent wars, from Vietnam onwards, have much to teach, and navies as politico-military instruments need to learn how to handle themselves in this environment. It is not enough to be good, one must be clever too, without ever being bogus.

AUTONOMY

There is no doubt that for a substantial part of its history the Royal Navy was a largely autonomous institution. In the eighteenth century it was, as Professor Baugh reminds us, the biggest centrally directed organization in the world. In the nineteenth century it was unthinkable that it should depend on any base, industrial or operational, other than those its own nation could provide; the phrase 'splendid isolation' was coined then, not in hindsight. Dependence on alliances was, in the naval sphere, no less anathema. They were tolerated, certainly, at Navarino and in the Russian War; but they were scarcely to be thought necessary.

How then can this historical autonomy be reconciled with the recent situation where a 'contribution' (how big is a contribution?) to the NATO alliance was the

435

only official justification for outlay on British maritime forces? A facile answer, and not a bad one at the time, was 'The Falklands'. But a more complete rebuttal rested on the innate limitations of the 1968–88 strategy particularly as applied to sea power. The artificial boundary of the NATO sea area at the Tropic of Cancer was bad enough, but the single scenario on which NATO strategy based itself was worse; and whatever efforts were made by NATO to introduce variations, through doctrines of flexible response and maritime contingency plans, these made all too little use of the strategic options conferred by maritime forces and gave far too little recognition to the out-of-area interests and vulnerabilities of member states and to threats other than the Soviet.

The case for maritime autonomy in Britain, an individual power with unique interests, was always there. It was as well that the Royal Navy recognized it and—circumspectly, not always completely, and often without admitting it to itself—planned for it. The difference between the autonomy of earlier times and that of the past few decades was one of degree. The relative decline in the economy saw to that; it was clearly a far cry from any two-power or one-power standard. Independent operations could be contemplated only at certain levels of conflict and reach, against certain degrees of threat. In fact, as one looks back through history, it has always been so; but nowadays the limitation is sharper than it has been for centuries. This does not invalidate the need for autonomy in a medium-power navy, which Britain's now is, for two reasons. First, the nation's interests *at* sea, and its interests that can be promoted or protected *by* sea, are still extensive and are unique. Second, in coalition operations an autonomous force is a more flexible and welcome addition than one requiring big-brother support.

All this of course presupposes that a world system subsists in which nation-states are the principal units of account, and that Britain goes on being such a nation-state. All history can teach us here is that whatever other systems rise and fall—imperial, theocratic, ideological, and federal groupings—the nation is the most common unit, and there is no sign of its losing this position. Indeed, the trend in the last decade of the twentieth century is toward smaller, more intense national units. One consequence is increased international instability; not, it may be thought, a situation in which a flexible instrument like a navy should be forgone.

INSTITUTION

The third key word in this analysis is at least as battered by recent history as the other two. It is clear from this book that for centuries the Royal Navy was a unique institution in the national consciousness, yet recently in an underrated book (*Ruling the Waves* (London, 1986)) the journalist Dennis Barker wrote that 'The Royal Navy and most of the British public have been mutually invisible' for nearly 40 years. It was an overstatement but not much of an overstatement; and since Barker wrote, there may well have been further erosion of consciousness as there has been erosion of the navy itself, and further contraction of its bases and shore-side presence.

One may be forgiven for wondering whether the consciousness is simply latent.

The upsurge of popular feeling at the time of the Falklands—accompanied though it was by evidence of widespread ignorance about the functioning of sea power and even of the facts of geography—and the interest generated by television programmes on naval matters indicate that some feeling for the navy as an institution lurks beneath the surface. The vogue for naval novels of the Napoleonic Wars may be nothing but nostalgia, combined with respect for a very well-researched and now well-understood genre, but it is indicative.

The Royal Navy has, after all, been through many doldrum belts before. Ships were taken out of commission, men paid off, officers put on half pay, at irregular intervals with a ruthlessness that modern redundancy schemes scarcely match, and 'peace dividends' sought for much less laudable purposes than diminishing the Public Sector Borrowing Requirement. Naval voices at such times warned of the difficulty of recovery in emergency, both as to trained manpower and as to the industrial base necessary to rebuild or modernize. It was unusual for public opinion to back them, whatever the national consciousness of the navy as an institution.

What may be different this time is the feeling of inevitability. In the eighteenth century there was always, in the words of the old toast, a 'bloody war or a sickly season' (with its prospect of renewed naval activity and chances of personal advancement) round the corner, in the nineteenth a scare from some European threat or a perceived vulnerability in the trade. It was assumed and accepted that the navy would be reconstituted in time of need. In the view of this writer, there is no such perception now. In so far as people think about it at all, they are dubious about the navy's capacity to respond.

Governments would do well to consider this point and view it in the light of history: the narrow margin by which necessary naval forces were reconstituted in crisis, and the penalty when recovery was not achieved in time, are points which emerge from these pages again and again.

Naval provision, however, is a matter of political will; and in a democracy that largely depends on public opinion. It is up to the navy and students of naval affairs to educate the public in the realities of the ways in which the nation may be served by maritime forces. It is hoped this book may aid that process. No doubt a country will eventually get the navy it deserves; the Legions, however they preserve the ancient disciplines, cannot sustain a failing and confused nation indefinitely. But Britain has rallied often enough before.

In July 1993 the writer had the privilege of embarking in a frigate off Portland to witness a 'Thursday War' conducted by the Flag Officer Sea Training. The ship was modern, one of the Royal Navy's crack vessels, and was in tactical command of several other units, opposed by multiple air, surface, and submarine threats. In spite of weather limitations there was a very high level of realism in the exercise.

The professionalism on board this ship of the modern navy was of the highest order. It was clear that every person in the ship's company was fully trained in their task and capable of taking over related tasks in emergency. Delegation

A DAY OFF PORTLAND

437

ensured that initiatives could be taken at the lowest appropriate level. The atmosphere was akin to that in a really well-operated submarine a generation ago; there was that level of individual responsibility. Communication (that word again) was concise, intense, and articulate at all levels, and the 'sea riding' staff not only planned and conducted the exercise but gave advice on how to enhance performance even more.

Even as a display this would have been heartening. As an exercise conducted by a ship working-up, in no way conceived as a demonstration, it was vastly impressive. If the public could see how its navy performed under action conditions there would probably be no need for any other sort of promotion of its value as a national institution.

It would be both fanciful and pretentious to suggest that such professionalism springs from the lessons of history. The process has been much more subtle: tradition, training, technical challenge, and the sea itself have all played their part. But the result is something of which the nation can be proud, and a sure foundation on which to build for the future.

J. R. Hill

CHRONOLOGY

700

793 First recorded raid by Norsemen on England

800

897 Alfred orders the building of a squadron of specially designed ships

1000

1014–16 Conquest of England by Swein and Cnut
1066 Conquest of England by William of Normandy

1100

1155–6 Earliest surviving charters of the Cinque Port towns

1200

1205 First reference to William de Wrotham as clerk in charge of the royal galleys
1212 Building of the galley base in Portsmouth
1213 Battle of Damme
1217 Battle of Dover
1294 Edward I orders the building of a force of galleys

1300

1337–40 French raids on the English coast
1340 Battle of Sluys
1346–7 Crécy campaign and siege of Calais
1350 Battle of *Les Espagnols sur Mer*
1372 English defeat off La Rochelle
1372–8 Edward III's campaign to build barges and balingers

1400

1413–20 Henry V's great ships built
1416 Relief of Harfleur
1417 Capture of French and Spanish ships in the Channel

1422	Sale of Henry V's ships commenced
1442	Soper retired as Clerk of the King's Ships
1452	Clerkship left empty by the Crown
1480	Rogers appointed Clerk of the King's Ships by Edward IV
1487	*Regent* laid down at Reding Creek
1495	Robert Brygandyne becomes Clerk of the King's Ships
1495–6	Dry-dock built for the royal ships at Portsmouth

1500

1509–10	Building of the *Mary Rose*: first use of gunports
1512	Loss of the *Regent* in action off Brest
1513	Death of Sir Edward Howard, Lord Admiral, in action; building of the *Henry Grace à Dieu*
1524	Sir Thomas Spert appointed to the changed office of Clerk Controller
1539	Invasion scare: south coast fortified
1545	Loss of the *Mary Rose*; creation of the Council for Marine Causes
1547	Benjamin Gonson becomes Treasurer of the Navy
1550	Edward Baeshe appointed to the new office of Surveyor General of the Victuals
1553–4	Sir Hugh Willoughby and Richard Chancellor seek the north-east passage
1555	Muscovy Company formed
1557	'Ordinary' established; Lord Treasurer given overall responsibility
1577	Sir John Hawkins becomes Treasurer of the Navy
1579	Hawkins contracts for maintenance
1577–80	Drake's circumnavigation
1585	Treaty of Nonsuch leads to war with Spain; Drake raids Cartagena
1587	Drake raids Cadiz
1588	The Armada Campaign
1589	Unsuccessful expedition to Lisbon
1596	Successful expedition to Cadiz under Essex

1600

1604	Treaty of London concludes peace with Spain
1608	Commission of Inquiry into naval affairs
1618	Second Commission of Inquiry; Admiralty put into commission
1619	Earl of Buckingham becomes Lord Admiral
1625	Unsuccessful expedition to Cadiz
1627	Unsuccessful expedition to Ile de Ré
1628	Assassination of Buckingham; Admiralty in commission

1634–40	Ship money collected
1642	Navy declares for parliament
1649	Execution of Charles I; Commonwealth establishes an Admiralty Commission and begins a building programme
1651	First Navigation Act passed
1652	Beginning of First Dutch War; Battles of Kentish Knock and Dungeness
1653	Battles of Portland, Gabbard, Scheveningen; Cromwell becomes Lord Protector
1654	Treaty of Westminster ends First Dutch War; 'Western Design' begins war with Spain; Blake in the Mediterranean
1655	Capture of Jamaica
1657	Blake destroys Spanish plate fleet at Santa Cruz
1660	Restoration of monarchy; James, duke of York as Lord High Admiral; Samuel Pepys as Clerk of the Acts
1661	England acquires Tangier; Articles of War first issued
1664	Beginning of Second Dutch War
1665	Battles of Lowestoft and Bergen
1666	Four Days' Fight; St James's Day Fight
1667	Dutch attack the Medway and tow away the *Royal Charles*; Treaty of Breda ends Second Dutch War
1670	Treaty of Dover
1672	Beginning of Third Dutch War as consequence of secret Treaty of Dover; Battle of Sole Bay
1673	Battles of Schooneveld, the Texel; duke of York resigns as Lord High Admiral; Pepys becomes Secretary to the Admiralty
1674	Treaty of Westminster ends Third Dutch War
1679	Pepys resigns; new Admiralty Commission formed
1683	Evacuation of Tangier
1684	Admiralty Commission dismissed; Pepys returns as Secretary of the Admiralty
1685	Death of Charles II; duke of York becomes King James II
1688	'Glorious Revolution': fleet fails to intercept William of Orange's invasion
1689	William becomes King William III; Pepys resigns from Admiralty; declaration of war on France
1690	Battle of Beachy Head
1691	Issue of Russell's *Sailing and Fighting Instructions*
1692	Battles of Barfleur and La Hougue
1693	Destruction of British Smyrna Convoy
1694	Greenwich Hospital founded
1695	Establishment of dockyard at Plymouth
1696	Attempt to create a Register of Seamen fails
1697	Peace of Rijswick

1700

1701	New General Instructions for Victualling Office
1702	Beginning of the War of the Spanish Succession; Rooke's attack on Vigo
1704	Capture of Gibraltar; Battle of Malaga
1707	Unsuccessful assault on Toulon; grain blockade enforced on France
1713	Peace of Utrecht
1718	Battle of Cape Passaro
1725	Admiralty Building constructed
1726	Admiral Hosier blockades the Spanish Main: 4,000 die from tropical disease
1728	Birth of James Cook
1731	First edition of 'Regulations and Instructions Relating to His Majesty's Service at Sea'
1733	Royal Naval Academy opened in Portsmouth; construction of Dockyard in Jamaica
1739	Outbreak of War of Jenkins's Ear; typhus epidemic on mobilization; capture of Puerto Bello
1740	Outbreak of War of the Austrian Succession; unsuccessful assault on Cartagena
1740–4	Anson's circumnavigation
1744	Battle of Toulon
1745	Second Jacobite Rising
1746	Foundation of Naval Hospital at Haslar
1747	Western Squadron established to watch and control exit to Brest; First Battle of Finisterre; Lind's investigation into remedy for scurvy
1748	Peace of Aix-la-Chapelle
1749	Establishment of Halifax naval base; Articles of War revised
1755	Thomas Slade made Surveyor of the Navy; Royal Marines come under Admiralty control
1756	Outbreak of the Seven Years War; loss of Minorca
1757	After initial reverses, concentration of strategy on N. America and the watch on Brest
1758	Hawke attacks Basque Roads
1759	Battle of Lagos; Battle of Quiberon Bay; capture of Guadeloupe; capture of Quebec
1760	Capture of Belle Isle
1762	Capture of Martinique
1763	Peace of Paris
1768–73	Cook's first voyage
1772–5	Cook's second voyage
1775	Outbreak of rebellion in American Colonies
1776	Cook sails for third voyage

1777	Surrender of British army at Saratoga; France enters War of American Independence
1778	Battle of Ushant
1779	Copper sheathing introduced to combat shipworm; Cook killed at Kealakekua Bay; Battle of Grenada; French capture Grenada and St Vincent
1780	Formation of the League of Armed Neutrality against Britain
1781	Battle of Chesapeake Bay; surrender of British army at Yorktown
1782	Battle of Negapatam
1783	Battle of Cuddalore; Peace of Versailles
1790	Nootka Sound incident
1791–5	Vancouver's survey of west coast of North America
1793	Outbreak of the War of the First Coalition against Revolutionary France
1794	Capture of French West Indian colonies; Battle of the Glorious First of June
1795	Order in Council sets up office of Hydrographer; lemon juice issued on regular basis as an antiscorbutic
1796	Hilltop telegraphs constructed to home bases
1797	Battle of St Vincent; Mutinies at Spithead and the Nore; Able Seamen's pay raised for the first time since 1686
1798	Battle of the Nile; capture of Minorca

1800

1800	Capture of Malta
1801	British landing in Egypt; Peace of Amiens
1801–3	Flinders surveying Australian coasts
1803	Renewal of the war
1805	Battle of Trafalgar
1806	Royal Naval Academy renamed Royal Naval College
1807	Act abolishing the slave trade becomes law
1808–13	Support of British forces in the Peninsula; offensive operations in the Mediterranean and Baltic
1809	Walcheren expedition
1810–12	Beaufort in *Fredericksteen* on coast of Karamania
1811	School of Naval Architecture established in Portsmouth
1812	Outbreak of war with United States of America
1815	End of the Napoleonic Wars
1816	Bombardment of Algiers
1819–20	Parry first overwinters in Arctic with *Hecla* and *Griper*
1821	*Comet*, first Admiralty steamship, in service; *Aaron Manby*, first iron merchant ship, in service
1821–6	Owen's great survey of Africa

1858	Royal Commission on Manning
1859	*Warrior* laid down; Statute authorizing Royal Naval Reserve; HMS *Britannia* commissioned
1860	Regular leave authorized
1861	First trials of turret in *Trusty*
1863	Purple stripe for engineers introduced
1866	Effective end of the French naval challenge
1868	Engine Room Artificers introduced
1869	Childers's Admiralty reform; *Devastation*, first battleship without sails, building
1870	*Captain* capsizes; reappraisal of designs
1871	Flogging suspended in peacetime
1872	HMS *Vernon*, torpedo school, opened
1872–6	Oceanographic voyage of the *Challenger*
1873	Royal Naval College, Greenwich opened; introduction of compound engines
1875	*Iris*, first steel hull for the RN, ordered; Cooper Key Committee
1876	*Lightning*, first torpedo boat, ordered; Agnes Weston's first Sailors' Rest opened in Devonport
1877	HMS *Marlborough* commissioned as engineers' training ship
1878	Russian war scare; Fleet to Constantinople; loss of training ship *Eurydice*
1879	Flogging suspended
1880	Royal Naval Engineering College, Keyham, opened
1882	Bombardment of Alexandria
1885	Triple expansion engines introduced in HMS *Victoria*; second Russian war scare
1889	Naval Defence Act; introduction of *Royal Sovereign* design, a major advance in battleship size and capability
1890	Publication of A. T. Mahan's book *The Influence of Sea Power upon History 1660–1789*
1891	Royal Naval Exhibition opened
1892	Belleville watertube boilers introduced
1893	Loss of the *Victoria*; Spencer programme
1897	Diamond Jubilee Review at Spithead; introduction of cemented armour
1898	Passage of first German Navy Law; turbine destroyer *Viper* in service

1900

1899–1902	South African War
1902	Anglo-Japanese Naval Treaty; first RN submarine in service
1903	*Entente cordiale* with France; Fisher–Selborne scheme introduced; Royal Naval College, Osborne opened
1904	Fisher becomes First Sea Lord; creation of Home Fleet marks concentration against Germany

1905	Royal Naval College, Dartmouth opened;
	HMS *Dreadnought*, the first turbine and all-big-gun battleship, ordered;
	Battle of Tsushima allows further concentration of RN in home waters
1908	Commissioning of battle cruiser *Invincible*
1909	Declaration of London; concept of Dominion navies approved
1911	Agadir crisis
1912	First flight of an aircraft from a ship's deck (HMS *Africa*)
1914	*July*: Test Mobilization Fleet Review; *August*: outbreak of the First World War. Battles of the Heligoland Bight, Coronel, Falklands; sinking of three British cruisers by submarine U-9; raids by German ships on east coast towns; raid on Cuxhaven: shipborne aircraft attack shore targets for the first time
1915	Battle of the Dogger Bank; first unrestricted U-boat campaign begins (February) and ends (September); the Dardanelles campaign
1916	Second unrestricted U-boat campaign begins (February) and ends (April); Battle of Jutland
1917	Last unrestricted U-boat campaign begins; USA enters war; Admiralty begins large-scale convoy operations, providing solution to U-boat threat; Women's Royal Naval Service (WRNS) instituted
1918	Creation of Royal Air Force and dissolution of Royal Naval Air Service; Zeebrugge raid; armistice with Germany and internment of German High Seas Fleet at Scapa Flow
1919	Adoption of Ten Years' Rule; RN intervention in Russian Civil War
1921	Work begun on Singapore naval base
1921–2	The Geddes Axe
1922	Washington Naval Treaty; Chanak crisis
1924	Trenchard–Keyes agreement on naval aviation
1925	The 'Great Betrayal' of the Fisher–Selborne scheme; unrest in China brings minor RN intervention
1931	Invergordon mutiny
1933	Abandonment of Ten Years' Rule
1935	Mediterranean Fleet prepares for possible involvement in Abyssinian and Spanish Civil Wars; Anglo-German Naval Treaty
1937	Fleet Air Arm comes under naval control; Nyon Arrangement to counter submarine activities in Spanish Civil War; full-scale rearmament programme under way
1939	Outbreak of Second World War; introduction of convoy system; *Courageous* and *Royal Oak* sunk; Battle of the River Plate
1940	German invasion of Norway; Battles of Narvik; *Glorious* sunk; Dunkirk evacuation; in Mediterranean, Mers-el-Kebir, Taranto, Battle of Cape Spartivento
1941	*Mediterranean*: Battle of Cape Matapan, evacuations from Greece and Crete; *Ark Royal* and *Barham* sunk; First Battle of Sirte. *Atlantic*: *Hood* and *Bismarck* sunk; Arctic convoys to Russia begun. *Far East*: Japan enters war; *Prince of Wales* and *Repulse* sunk

1942   *Atlantic*: *Scharnhorst* and *Gneisenau* Channel dash; St Nazaire raid; Arctic convoys including PQ 17 and PQ 18; Battle of the Barents Sea. *Mediterranean*: Second Battle of Sirte; 'Pedestal' convoy; North Africa landings. *Far East*: Japanese carrier raid into Indian Ocean; *Hermes*, *Cornwall*, and *Dorsetshire* sunk; landing in Madagascar

1943   *Atlantic*: climax of the Atlantic convoy campaign and defeat of the U-boats; *Tirpitz* crippled by midget submarines; Battle of the North Cape; *Scharnhorst* sunk. *Mediterranean*: invasion of Sicily; landings on Italian mainland

1944   *Atlantic*: support groups' increasing success against U-boats; Normandy landings. *Far East*: Eastern Fleet strikes Sabang and Soerabaya; formation of British Pacific Fleet

1945   End of the war in Europe; BPF joins US fleet for Okinawa operation; dropping of atomic bombs on Japanese cities; Japan surrenders; election of Labour government in Britain

1946   Mining of *Saumarez* and *Volage* in Corfu Channel

1947   Counter-immigrant operations off Palestine

1948   Counter-terrorism operations off Malaya

1949   Conclusion of NATO Treaty; *Amethyst* incident in the Yangtse

1950   Outbreak of the Korean War; one light fleet carrier, one cruiser, several destroyers and frigates operate continuously off Korea for three years

1951   Deployment to the Persian Gulf on AIOC nationalization; Conservatives take power in Britain

1952   Suez Canal strike: Mediterranean Fleet operates the Canal; Exercise Mainbrace, first demonstration of NATO strike fleet

1953   Exercise Mariner

1954   Formosa Strait Patrol protects merchant shipping trading in and out of China

1956   Suez; first helicopter-borne amphibious assault; institution of the General List (AFO 1/56)

1957   Defence White Paper adopts massive-deterrent, short-war strategy; scrapping of reserve fleet, coastal forces

1958   Coup in Iraq; first 'cod war' with Iceland

1959   Completion of *Hermes*, the last carrier left over from wartime programmes

1960   Macmillan's 'Wind of Change' speech

1961   End of conscription

1962   Kuwait deployment to deter Iraq; Nassau Agreement to provide UK with Polaris

1963   First British nuclear-powered submarine, *Dreadnought*, commissioned

1964   Royal Marines land from *Centaur* to suppress rebellion in Tanganyika; Indonesian Confrontation: Malaysia secured by Far East Fleet; election of Labour government

1965   Defence review initiated by Healey; Rhodesia unilaterally declares independence; Beira Patrol instituted

1966   Cancellation of plans for new fixed-wing carriers

1967   Completion of *Resolution*, the first British Polaris submarine

1968	Abandonment of East of Suez policy: defence effort dedicated to NATO
1969	Institution of military salary
1970	Election of Conservative government
1971	Exercise Highwood; limitations of shore-based air support exposed; institution of the PWO structure
1973	Second 'cod war' with Iceland; Yom Kippur war between Arab states and Israel
1974	Suez Canal clearance by RN mine counter-measures craft; election of Labour government
1975	Final approval for development of Sea Harrier
1976	Third 'cod war' with Iceland
1977	Precautionary deployments in support of Falklands and Belize
1978	Counter Illegal Immigrant operations in Hong Kong
1979	Election of Conservative government; amelioration of service pay scales
1980	Completion of *Invincible*; outbreak of Gulf war (Iran–Iraq); Armilla Patrol instituted
1981	'The Way Forward' (Nott defence review) published
1982	The Falklands campaign: 40 warships, 20 RFA, and 45 ships taken up from trade (STUFT), 10 naval air squadrons, 3 RM Commandos involved
1983	Addbacks agreed to RN strength partly to offset war losses
1986	Announcement of US Maritime Strategy
1988	NATO Exercise Teamwork 88, massive demonstration of the Maritime Strategy with substantial RN participation
1990	Kuwait invaded by Iraq; WRNS go to sea in combatant ships
1991	19 RN units participate in Operation Desert Storm to liberate Kuwait
1993	Announcement of further cuts in naval force levels; abolition of Royal Naval Auxiliary Service and one-third reduction of Royal Naval Reserve; naval units operate in Adriatic during Balkan crisis
1994	Dissolution of WRNS as a separate service and absorbtion into RN

FURTHER READING

1. The Wall of England, to 1500

C. Allmand, *The Hundred Years War: England and France at War c.1300–c.1450* (Cambridge, 1989), sets the war in context with good sections on naval objectives and the forces at the rulers' disposal.

R. C. Anderson, *Oared Fighting Ships from Classical Times to the Coming of Steam* (London, 1968), gives full technical details of English medieval galleys (Chapter 5).

F. W. Brooks, *The English Naval Forces 1199–1272* (London, 1932), a detailed study of this period.

Marquis Terrier de Loray, *Jean de Vienne, Amiral de France 1341–1396* (Paris, 1877), old and possibly difficult to find, but the only work on this topic.

K. Fowler (ed.), *The Hundred Years War* (London, 1971), has a chapter by C. F. Richmond, 'The War at Sea', a very useful study.

C. Harper-Brill *et al.* (eds.), *Studies in Mediaeval History Presented to R. Allen Brown* (Woodbridge, 1989), includes Nicholas Hooper's essay, 'Some Observations on the Late Saxon Navy'.

J. Haywood, *Dark Age Naval Forces: A Re-Assessment of Frankish and Anglo-Saxon Seafaring Activities* (London, 1991), a good general survey.

H. J. Hewitt, *The Organisation of War under Edward III* (Manchester, 1966), makes clear the difficulties in organizing a successful cross-Channel expedition.

Navy Records Society, *British Naval Documents 1204–1960* (London, 1993), contains the chronicle narratives of most of the battles described.

N. H. Nicolas, *History of the Royal Navy* (London, 1847), old but a full and stirring narrative.

W. Stanford Reid, 'Sea Power in the Anglo-Scottish War 1290–1328', *Mariners' Mirror*, 46 (1960), the only reasonably accessible writing on this topic.

S. P. Rose, *The Navy of the Lancastrian Kings* (London, 1982), prints in full the accounts and inventories of Henry V's ships before their sale together with an introduction covering the use of the ships and the career of William Soper.

C. Warren-Hollister, *Anglo-Saxon Military Institutions on the Eve of the Norman Conquest* (Oxford, 1962), contains a study of the late Saxon Navy (Chapter 6).

J. F. Willard (ed.), *The English Government at Work 1327–1386* (Cambridge, Mass., 1940), includes in vol. i a very useful chapter by A. L. Prince on the army and navy.

2. From the King's Ships to the Royal Navy, 1500–1642

S. Adams, 'New light on the "Reformation" of John Hawkins: The Ellesmere naval survey of 1584', *English Historical Review*, 105 (1991).

R. C. Anderson, *Oared Fighting Ships* (London, 1962).

—— 'Henry VIII's *Great Galley*', *Mariner's Mirror*, 6 (1920).

K. R. Andrews, *Elizabethan Privateering* (Cambridge, 1964).

J. S. Corbett (ed.), *Papers Relating to the Navy during the Spanish War, 1585–1587* (London, 1898).

C. S. L. Davies, 'The Administration of the Royal Navy under Henry VIII: The Origins of the Navy Board', *English Historical Review*, 80 (1965), an excellent research article.

—— 'England and the French War, 1557–9', in J. Loach and R. Tittler (eds.), *The Mid-Tudor Polity, c.1540–1560* (London, 1980).

C. H. L'Estrange Ewan, 'Organised Piracy around England in the Sixteenth Century', *Mariner's Mirror*, 35 (1949).

Felipe Fernandez Armesto, *The Spanish Armada: The Experience of War in 1588* (London, 1988).

E. W. Fowler, *English Sea Power in the Early Tudor Period, 1485–1558* (Ithaca, NY, 1965).

T. Glasgow, 'The Navy in Philip and Mary's War', *Mariner's Mirror*, 53 (1967).

—— 'The Maturing of Naval Administration, 1556–1564', *Mariner's Mirror*, 56 (1970), with the above, two of the best articles from a prolific writer.

J. Guilmartin, *Gunpowder and Galleys* (Cambridge, 1974).

P. Kirsch, *The Galleon* (London, 1990), a technical study rather than a history, full of key information.

W. Laird Clowes, *The Royal Navy: A History from the Earliest Times to the Present* (London, 1897), a useful narrative, particularly for the usually neglected pre-Elizabethan period.

David Loades, *The Tudor Navy: An Administrative, Political and Military History* (Aldershot, 1992), the only modern work to relate the navy to the remainder of the activities of government.

M. Oppenheim (ed.), *Naval Accounts and Inventories of the Reign of Henry VII* (London, 1896).

—— *The History of the Administration of the Royal Navy, 1509–1660* (London, 1896), still valuable.

G. Parker and C. Martin, *The Spanish Armada* (London, 1988), the best and most comprehensive work on the Armada.

M.-J. Rodriguez Salgado, *Armada: An International Exhibition to Commemorate the Spanish Armada* (Greenwich, 1988), an outstanding work, written to catalogue the exhibition at the National Maritime Museum.

—— and S. Adams, *England, Spain and the Gran Armada, 1585–1604* (London, 1991).

Margaret Rule, *The Mary Rose* (Greenwich, 1982), full of information not readily obtainable elsewhere.

Alfred Spont (ed.), *Letters and Papers Relating to the War with France, 1512–13* (London, 1897).

E. G. R. Taylor, *Tudor Geography, 1485–1583* (London, 1930).

—— *Later Tudor and Early Stuart Geography, 1583–1650* (London, 1934), with the above, excellent scholarship providing essential information on contemporary knowledge and practice.

S. Usherwood and E. Usherwood, *The Counter Armada, 1596* (London, 1983).

D. W. Waters, *The Art of Navigation in England in Elizabethan and Early Stuart Times* (London, 1958), an excellent work of scholarship, giving the student of the navy a knowledge of the wider context of navigation.

J. A. Williamson, *Maritime Enterprise, 1485–1558* (Oxford, 1913).

3. A PERMANENT NATIONAL MARITIME FIGHTING FORCE, 1642–1689

M. Baumber, *General-at-Sea: Robert Blake and the Seventeenth-Century Revolution in Naval Warfare* (London, 1989), a sound biography of Blake but rather tentative on the 'revolution in naval warfare'.

B. Capp, *Cromwell's Navy: The Fleet and the English Revolution 1648–60* (Oxford, 1989), a magisterial treatment of its subject, particularly strong on the political and religious contexts.

J. D. Davies, *Gentlemen and Tarpaulins: The Officers and Men of the Restoration Navy* (Oxford, 1991), looks at both the social and professional structures of the service and many administrative and operational aspects.

F. Fox, *Great Ships: The Battlefleet of King Charles II* (London, 1980), an excellent, detailed survey of the warships of the period (including those built before 1660); exceptionally well illustrated.

S. Hornstein, *The Restoration Navy and English Foreign Trade 1674–88* (Aldershot, 1991), a comprehensive survey of operations in the Mediterranean and of the organization of the convoy system.

R. Latham (ed.), Pepys's *Navy White Book* (London, 1994).

R. Latham and W. Matthews (eds.), *The Diary of Samuel Pepys*, 11 vols. (London, 1970–83), a remarkable undertaking, with companion and index volumes almost as fascinating as Pepys's own inimitable text; essential to any understanding of the navy in the 1660s, and much else besides.

J. Miller, *James II: A Study in Kingship* (Hove, 1978).

—— *Charles II* (London, 1991), with the above, the best biographies of the two brothers in terms of the political and international contexts which affected the navy.

Navy Records Society, *Papers Relating to the First Dutch War, 1652–4*, ed. S. R. Gardiner and C. T. Atkinson, 6 vols. (1899–1930); *The Journal of Edward Montagu, First Earl of Sandwich*, ed. R. C. Anderson (1929); *The Journals of Sir Thomas Allin*, ed. R. C. Anderson, 2 vols. (1939–40); *Journals and Narratives of the Third Dutch War*, ed. R. C. Anderson (1946); *The Rupert and Monck Letterbook 1666*, ed. J. R. Powell and E. K. Timings (1969); between them, these volumes contain many of the most important documents of the three wars, together with introductions which (taken as a whole) provide detailed operational histories. *Samuel Pepys's Naval Minutes*, ed. J. R. Tanner (1925), and *The Tangier Papers of Samuel Pepys*, ed. E. Chappell (1935) provide a mass of information about the Restoration navy and Pepys's attitudes, but need to be treated with a healthy regard for Pepys's prejudices and belief in his own infallibility.

R. Ollard, *Pepys: A Biography* (London, 1974), more concise and judicious than Sir Arthur Bryant's great three-volume panegyric of the 1930s.

J. R. Powell, *The Navy in the English Civil War* (London, 1962), a straightforward and clear operational account, effectively the only detailed treatment of its subject.

C. Wilson, *Profit and Power* (London, 1957), becoming dated, but still a convincing account of the causes and course of the first two Anglo-Dutch wars.

4. THE STRUGGLE WITH FRANCE, 1689–1815

David D. Aldridge, 'Admiral Sir John Norris and the British Naval Expeditions to the Baltic

Sea 1715–1727' (Ph.D. thesis, University of London, 1971), the only detailed study of the subject.

Jeremy Black, *The Collapse of the Anglo-French Alliance 1727–1731* (London, 1987), an examination of the diplomatic background.

—— and Philip Woodfine (eds.), *The British Navy and the Use of Naval Power in the Eighteenth Century* (Leicester, 1988), includes a wide range of new research.

J. S. Bromley, *Corsairs and Navies* (London, 1987), a collection of essays providing valuable insight into privateering.

J. R. Bruijn, 'William III and His Two Navies', in *Notes Rec. R. Soc. London*, 43 (1989), 117–32, outlines the king's dual naval relationships.

G. N. Clark, 'War Trade and Trade War', *Economic History Review*, 1 (1929), 262–80, a key article on commerce warfare.

Sir Julian Corbett, *The Campaign of Trafalgar* (London, 1910), still the best study of the broad strategic aspects.

John Cresswell, *British Admirals of the Eighteenth Century: Tactics in Battle* (London, 1972), the most recent authoritative examination of tactics.

Michael Duffy, *Soldiers, Sugar and Seapower: The British Expeditions to the West Indies and the War Against Revolutionary France* (Oxford, 1987), provides valuable insight into the importance of the West Indies.

John Ehrman, *The Navy in King William's War 1689–1697* (Cambridge, 1953), a pioneering study linking administration and operations.

G. C. Gibbs, 'The Revolution in English Foreign Policy', in Geoffrey Holmes (ed.), *Britain after the Glorious Revolution, 1689–1714* (London, 1969), a key article providing the background to later policy.

Eric Grove (ed.), *Great Battles of the Royal Navy* (London, 1994), includes the latest research and judgements on the most famous battles.

Richard Harding, *Amphibious Warfare in the Eighteenth Century: The British Expedition to the West Indies, 1740–1742* (London, 1991), a detailed case study that provides a new basis on which to judge amphibious operations throughout the century, correcting Richmond on this point.

John B. Hattendorf, *England in the War of the Spanish Succession* (New York, 1989), a broad analysis showing the relationship of naval strategy to grand strategy.

—— *et al.* (eds.), *British Naval Documents, 1204–1960*, Publications of the Navy Records Society, 131 (London, 1993), a seminal work with topical introductions to the key documents illustrating the broad pattern of naval affairs.

Henry Kamen, 'The Destruction of the Spanish Silver Fleet at Vigo in 1702', *Bulletin of the Institute of Historical Research*, 39 (1966), 165–73, based on Spanish archival research and providing a case study on the limited value of these operations.

Paul Langford, *Modern British Foreign Policy: The Eighteenth Century 1688–1815* (London, 1976), a good general overview of foreign policy and strategy.

Ruddock Mackay, *Admiral Hawke* (Oxford, 1985), a key modern biography of an important leader.

Piers Mackesy, *Statesmen at War: The Strategy of Overthrow* (London, 1974), with the following, a masterful two-volume study of the War of the Second Coalition and the process by which naval operations were directed.

Piers Mackesy, *War Without Victory* (Oxford, 1984).

—— *The War in the Mediterranean, 1803–1810* (London, 1957), complements Corbett's *Trafalgar* and corrects many mistaken impressions created by the traditional overemphasis on that battle.

Richard Middleton, *The Bells of Victory: The Pitt–Newcastle Ministry and the Conduct of the Seven Years' War 1757–1762* (Cambridge, 1985), the latest interpretation correcting Corbett's earlier work.

Carola Oman, *Nelson* (London, 1948), the standard modern biography.

Richard Pares, *Colonial Blockade and Neutral Rights* (London, 1938), an important study linking naval and maritime history.

Edward B. Powley, *The Naval Side of King William's War* (Hamden, Conn., 1972), a detailed study of naval operations in the first part of the war.

Herbert Richmond, *The Navy in India, 1763–1783* (London, 1931), the standard work on these operations.

—— *The Navy in the War of 1739–1748* (Cambridge, 1920), the standard work on this war.

N. A. M. Rodger, 'The British View of the Functioning of the Anglo-Dutch Alliance, 1688–1795', in G. J. A. Raven and N. A. M. Rodger, *Navies and Armies: The Anglo-Dutch Relationship in War and Peace, 1688–1988* (Edinburgh, 1990), an outline of the elements of the naval alliance in 1688–1714.

A. N. Ryan, 'The Royal Navy and the Blockade of Brest, 1689–1805', in Martine Accera, José Merino, and Jean Meyer (eds.), *Les Marines de Guerre Européennes XVII–XVIIIe siècles* (Paris, 1985), outlines the development of this recurring operation.

Service Historique de la Marine, *Guerre et Paix 1688–1815* (Paris, 1988), includes papers representing much new research on both sides of the Channel.

David Spinney, *Rodney* (London, 1969), a valuable modern biography.

Marshal T. Smelser, *The Campaign for the Sugar Islands, 1759: A Study in Amphibious Warfare* (Chapel Hill, 1955), an important campaign study.

C. P. Stacey, *Quebec, 1759* (Toronto, 1959), the most detailed study of naval and military co-operation.

Geoffrey Symcox, *The Crisis of French Sea Power, 1688–1697: From Guerre d'escadre to Guerre de course* (The Hague, 1974), a fundamental work for understanding the context of English naval operations in this war.

Brian Tunstall, *Naval Warfare in the Age of Sail: The Evolution of Fighting Tactics, 1650–1815*, ed. Nicholas Tracy (Annapolis, 1990), long delayed in publication, this beautifully illustrated volume was written before Cresswell's work, but places the issues in a more detailed context.

J. C. M. Warnsinck, *De Vloot van de Koning-Stadhouder 1689–1690* (Amsterdam, 1934), the standard Dutch study of the early period of the allied fleet under William III.

5. The Eighteenth-century Navy as a National Institution, 1690–1815

Daniel A. Baugh, *British Naval Administration in the Age of Walpole* (Princeton, NJ, 1965), covers all aspects of the navy as an institution except shipboard life. The book focuses on the 1740s, but selectively examines prior developments.

Daniel A. Baugh, (ed.), *Naval Administration, 1715–1750* (London, 1977), the documents and introductions extend the chronological range of the 1965 study.

Henry Baynham, *From the Lower Deck: The Royal Navy, 1780–1840* (London, 1969), excerpts from seamen's memoirs of the Napoleonic era.

Christian Buchet, *La Lutte pour l'espace caraïbe et la façade atlantique de l'Amérique centrale et du sud (1672–1763)* (Paris, 1991), a closely researched study in comparative naval history offering a pioneering analysis of how victualling, strategic planning, and other factors affected West Indian operations.

Jonathan Coad, *Historic Architecture of the Royal Navy: An Introduction* (London, 1983), an authoritative study of the navy's shore facilities, featuring many photographs of structures that are still standing.

John Ehrman, *The Navy in the War of William III, 1689–1697* (Cambridge, 1953), an integrated analysis of the political, financial, and administrative foundations of England's emerging naval power.

Robert Gardiner and Brian Lavery (eds.), *The Line of Battle: The Sailing Warship, 1650–1840* (London, 1992), notwithstanding the title, the book is noteworthy for its penetrating discussions of the development of cruisers and small specialized warships.

Jan Glete, *Navies and Nations: Warships, Navies and State Building in Europe and America, 1500–1860*, 2 vols. (Stockholm, 1993), a provocative analysis based on a vast, comparative, statistical study of the number and tonnage of warships over time; an extraordinary accomplishment.

Stephen Gradish, *The Manning of the British Navy during the Seven Years War* (London, 1980), the navy's most troublesome problem is thoroughly researched in this study of a single war.

R. J. B. Knight (ed.), *Portsmouth Dockyard Papers, 1774–1783: The American War* (Portsmouth, 1987), Knight's introduction ranges over many topics and cites his important articles, most notably those on the adoption of copper sheathing and the supply of masts.

Brian Lavery, *The Ship of the Line, i: The Development of the Battlefleet, 1650–1850* (London, 1983), the authoritative account of the British ship of the line's evolution; individual ships are listed and categorized in the appendix.

—— *Nelson's Navy: The Ships, Men and Organisation, 1793–1815* (London, 1989), a comprehensive and readable volume with numerous maps and illustrations, arranged in a manner that makes it useful for reference.

Michael A. Lewis (ed.), *A Narrative of My Professional Adventures (1790–1839) by Sir William Dillon, K.C.H., Vice-Admiral of the Red*, 2 vols. (London, 1953–6), no one else's memoirs offer such a rich and comprehensive view of naval life in the Napoleonic era.

—— *A Social History of the Navy, 1793–1815* (London, 1960), an admirable study of sea officers and men in the Napoleonic era. The main topics are social origins, ranks, ratings, shipboard responsibilities, and chances of survival.

Christopher Lloyd (ed.), *The Health of Seamen: Selections from the Works of Dr. James Lind, Sir Gilbert Blane and Dr. Thomas Trotter* (London, 1965), the key source book on the subject.

—— *The British Seaman, 1200–1860: A Social Survey* (London, 1968), a short, readable survey, with an admirable chapter on 'The Law and the Press'.

—— and Jack L. S. Coulter (eds.), *Medicine and the Navy, iii: 1714–1815* (Edinburgh and

London, 1961), comprehensive and still valuable, though in many respects superseded by recent research.

G. J. Marcus, *Heart of Oak: A Survey of British Sea Power in the Georgian Era* (Oxford, 1975), not really a survey of sea power, the book assembles 20 brief and engagingly written essays on naval and maritime lore.

R. D. Merriman (ed.), *Queen Anne's Navy: Documents concerning the Administration of the Navy of Queen Anne, 1702–1714* (London, 1961), Merriman's introductions still offer the best guide to naval administrative history in Anne's reign.

Roger Morris, *The Royal Dockyards during the Revolutionary and Napoleonic Wars* (Leicester, 1983), a carefully researched enquiry into the dockyards and their organization, their work-force, and the erratic progress of dockyard reform.

Richard Pares, 'The Manning of the Royal Navy in the West Indies, 1702–1763', in Pares (ed.), *The Historian's Business and other Essays* (Oxford, 1961), still the best monograph on an important subject.

Thomas Pasley, *Private Sea Journals, 1778–1782*, ed. Rodney M. S. Pasley (London, 1931), Pasley's diary is arguably the most authentic and informative record of shipboard activity that has survived from the eighteenth century.

N. A. M. Rodger, *The Wooden World: An Anatomy of the Georgian Navy* (London, 1986), focuses on the Seven Years War. An expert, highly readable study of shipboard life and sea-service personnel. Rodger's important articles on desertion, victualling, and the delayed defeat of scurvy are listed in his bibliography.

Oliver Warner, *The Life and Letters of Vice-Admiral Lord Collingwood* (Oxford, 1968), a well-balanced biography containing many excerpts from Collingwood's letters. For full printed texts of the admiral's letters see the references in Warner's Preface.

6. THE SHIELD OF EMPIRE, 1815–1895

C. J. Bartlett, *Great Britain and Seapower 1815–1854* (Oxford, 1963).

—— 'The Mid-Victorian Re-appraisal of Naval Policy' in K. Bourne and D. C. Watt (eds.), *Studies in International History* (London, 1967).

K. Bourne, *The Foreign Policy of Victorian England* (Oxford, 1970), the standard work.

P. H. Colomb, *Naval Warfare* (London, 1891), a pioneering work of historical analysis covering some nineteenth-century campaigns.

—— *Memoirs of Sir Astley Cooper Key* (London, 1898), a major figure in the navy, 1850–85.

J. K. Daly, *Russian Seapower and the Eastern Question: 1827–1841* (London, 1991), an important monograph on the realities of Russian seapower.

G. S. Graham, *The Politics of Naval Supremacy* (Cambridge, 1965), the distilled reflections of a major naval–imperial historian on the nineteenth-century British Empire and the sea.

—— *The China Station: 1830–1865* (Oxford, 1978), diplomacy and war in China and Japan.

B. Greenhill and A. Giffard, *The British Assault on Finland* (London, 1988), detailing the Baltic and the impact of the British campaigns of 1854 and 1855.

K. Hagan, *In Peace and War* (Westport, 1984), US naval history 1774–1984.

C. I. Hamilton, *Anglo-French Naval Rivalry 1840–1870* (Oxford, 1993).

A. D. Lambert, *The Crimean War: British Grand Strategy against Russia 1853–1856* (Manchester, 1990), maritime strategy and the defeat of Russia.

A. D. Lambert, *The Last Sailing Battlefleet: Maintaining Naval Mastery 1815–1850* (London, 1991), rebuilding the battlefleet and power diplomacy.

M. Lewis, *The Navy in Transition: 1814–1865* (London, 1965), a social history.

A. J. Marder, *The Anatomy of British Seapower: 1880–1905* (London, 1940), now rather dated but still the standard work.

C. N. Parkinson, *Viscount Exmouth* (London, 1934), the standard life of the victor of Algiers.

M. S. Partridge, *Military Planning for the Defence of the United Kingdom, 1814–1870* (Westport, Conn., 1989), a major study of the military case for home defence.

B. Ranft, *Technical Change and British Naval Policy 1860–1939* (London, 1977).

N. A. M. Rodger, *The Admiralty* (London, 1979), the structure of naval administration.

T. Ropp, *The Creation of a Modern Navy* (Annapolis, 1987), the reconstruction of the French navy after 1871.

D. M. Schurman, *The Education of a Navy: The Development of British Naval Strategic Thought, 1867–1914* (London, 1965), a study of the revolution in naval scholarship and strategic thought that influenced the 1890s.

C. S. White, 'The Bombardment of Alexandria, 1882', *Mariner's Mirror*, 66 (1980), 31–50, the best account of this critical event.

7. Wood, Sail, and Cannonballs to Steel, Steam, and Shells, 1815–1895

Admiral G. A. Ballard, *The Black Battle Fleet* (Greenwich, 1980), Ballard had known many of these ships; excellent 'colour', weak on technology.

J. P. Baxter, *The Introduction of the Ironclad Warship* (Cambridge, Mass., 1933, repr. Hamden, Conn., 1968), seems generally reliable.

D. K. Brown, RCNC, *A Century of Naval Construction* (London, 1983), the centenary history of the Royal Corps of Naval Constructors.

—— *Before the Ironclad* (London, 1990), naval technology from 1815 to 1860.

R. A. Burt, *British Battleships 1889–1904* (London, 1988), very good design history of the later ships.

R. Gardiner (ed.), *All the World's Fighting Ships, 1860–1905* (London, 1970), by far the best reference book for the period.

W. Hovgaard, *Modern History of Warships* (London, 1920; repr., 1971), clear technical account but little background.

J. W. King, *The Warships of Europe* (London, 1878), an excellent snapshot of the day by the Chief Engineer, US Navy.

A. D. Lambert, *Battleships in Transition* (London, 1981), an excellent study of the generally neglected wooden steam battleship.

—— *Warrior* (London, 1987).

—— *The Last Sailing Battle Fleet* (London, 1991), the changes in wooden sailing ships between Trafalgar and the end of sail are often forgotten.

O. Parkes, *British Battleships* (London, 1956), a classic in its day; now seen as under-researched, unreferenced, and unreliable on technical matters. It does, however, describe all the ships with fine illustrations.

E. J. Reed, *Our Ironclad Ships* (London, 1869), an excellent account by a great designer who was also an accomplished writer.

E. C. Smith, *A Short History of Naval and Marine Engineering* (Cambridge, 1937), good but with some gaps; better than more recent books.

J. G. Wells, *The Immortal Warrior* (London, 1987), this and Lambert's book give a good picture of the political and technical problems associated with the procurement of *Warrior*.

8. ENDEAVOUR, DISCOVERY, AND IDEALISM, 1760–1895

G. M. Badger, *The Explorers of the Pacific* (Kenthurst, NSW, 1988), a good general outline from the early Polynesians to the 1830s.

A. W. Beaglehole, *The Life of Captain James Cook* (London, 1974), the classic biography of Cook.

M. Blewitt, *Surveys of the Seas* (London, 1957), a good general outline of surveying and charting from a cartographer's point of view.

E. H. Burrows, *Captain Owen of the African Survey, 1774–1857* (Rotterdam, 1979), a good biography, before, after and for the Great African Survey.

L. S. Dawson, *Memoirs of Hydrography* (Eastbourne, 1885; repr. London, 1969), detailed history of the RN Hydrographic service to 1885, with some earlier and foreign surveys.

Archibald Day, *The Admiralty Hydrographic Service, 1795–1919* (London, 1967), the official history, fuller from 1885 onward.

E. S. Dodge, *The Polar Rosses. John and James Clark Ross and their Explorations* (London, 1973), a sound biography, fuller on Polar than personal matters.

A. Friendly, *Beaufort of the Admiralty. The Life of Sir Francis Beaufort, 1774–1857* (London, 1977), gives full details both of his Asia Minor surveys and of his long service as Hydrographer.

R. C. Howell, *The Royal Navy and the Slave Trade* (London, 1987), deals mainly with the East Coast of Africa.

G. C. Ingleton, *Matthew Flinders, Navigator and Chartmaker* (Guildford, 1986), a very full, luxuriously presented biography.

W. Kaye Lamb (ed.), *The Voyage of George Vancouver, 1791–1795* (London, 1984), modern edition of Vancouver's own account.

E. Linklater, *The Voyage of the* Challenger (London, 1974), a popular account of the first great oceanographic voyage.

C. Lloyd, *The Navy and the Slave Trade* (London, 1949), fuller on the West Coast of Africa than the East.

—— *Mr Barrow of the Admiralty* (London, 1970), a good biography of the power at home behind much of the Naval exploration of the first half of the nineteenth century.

A. Moorehead, *Darwin and the* Beagle (London, 1973), a well-illustrated popular account of the voyage and its outcome.

L. H. Neatby, *The Search for Franklin* (London, 1970), a popular account of the mapping and charting of the Canadian Arctic which stemmed from the search.

A. Parry, *Parry of the Arctic* (London, 1963), a good popular biography by one of Parry's descendants.

A. L. Rice, *British Oceanographic Vessels, 1800–1950* (London, 1986), in tracing the history of the ships, gives much of the history of oceanography in potted form.

G. S. Ritchie, *The Admiralty Chart, British Naval Hydrography in the Nineteenth Century* (London, 1967), a good popular account, more anecdotal than Dawson or Day.

L. Withey, *Voyages of Discovery. Captain Cook and the Exploration of the Pacific* (London, 1988), a good popular account of late eighteenth- and early nineteenth-century exploration.

9. LIFE AND EDUCATION IN A TECHNICALLY EVOLVING NAVY, 1815–1925

'A Naval Nobody', 'On Naval Education', *Macmillan's Magazine*, 37 (Nov. 1877–Apr. 1878).

Henry Baynham, *From the Lower Deck: The Royal Navy 1780–1840* (London, 1969).

—— *Before the Mast: Naval Ratings of the 19th Century* (London, 1971).

—— *Men from the Dreadnoughts* (London, 1976).

J. S. Bromley (ed.), *The Manning of the Royal Navy* (Selected Public Pamphlets 1693–1873, William Clowes for the Navy Records Society, 1976).

Henry D. Capper, *Aft From the Hawsehole: Sixty-Two Years of Sailors' Evolution* (London, 1927).

P. H. Colomb, *Memoirs of Admiral The Right Honble Sir Astley Cooper Key, GCB DCL FRS* (London, 1898).

W. Hickman (ed.), *Reports and Opinions of Officers on the Acts of Parliament & Admiralty Regulations for Maintaining Discipline and Good Order in the Fleet Passed and Issued since the Year 1860* (London, 1867).

E. A. Hughes, *The Royal Naval College Dartmouth* (London, 1950).

William Laird Clowes, *The Royal Navy: A History from the Earliest Times to the Present*, vol. 6 (London, 1901) and vol. 7 (London, 1903).

Michael Lewis, *The Navy in Transition, 1814–1864: A Social History* (London, 1965).

Christopher Lloyd, 'The Origins of HMS *Excellent*', *Mariner's Mirror*, 41 (1955).

—— *The British Seaman 1200–1860: A Social Survey* (London, 1963).

—— 'The Royal Naval Colleges at Portsmouth and Greenwich', *Mariner's Mirror*, 52 (1966).

Thomas J. Spence Lyne, *Something About a Sailor: From Sailor Boy to Admiral* (London, 1940).

Arthur J. Marder, *Fear God and Dread Nought: The Correspondence of Admiral of the Fleet Lord Fisher of Kilverstone*, volume 1, *The Making of an Admiral 1854–1904* (London, 1952).

S. W. C. Pack, *Britannia at Dartmouth* (London, 1966).

Randolph Pears, *Young Sea Dogs: Some Adventures of Midshipmen of the Fleet* (London, 1960).

Geoffrey Penn, *'Up Funnel, Down Screw!' The Story of the Naval Engineer* (London, 1955).

—— *Snotty: The Story of the Midshipman* (London, 1957).

E. N. Poland, *The Torpedomen: HMS* Vernon's *Story, 1872–1986* (Emsworth, 1992).

H. Pursey, 'Lower Deck to Quarterdeck, 1818 to 1937', *Brassey's Annual* (1938), ch. 6.

RNEC 1880–1980 (Plymouth, 1980).

Charles Napier Robinson, *The British Tar in Fact and Fiction: The Poetry, Pathos, and Humour of the Sailor's Life* (London and New York, 1909).

G. B. Sayer, (ed.), *H.M.S.* Vernon: *A History* (Portsmouth, 1930).

B. B. Schofield, *Navigation and Direction: The Story of HMS* Dryad (Havant, 1977).

E. P. Statham, *The Story of the 'Britannia': The Training Ship for Naval Cadets* (London, 1904).

Edgar C. Smith, 'The Engineers' Button', *Engineering* (28 Jan. 1944).

—— 'The Executive Curl', *Engineering* (21 Apr. 1944).

—— 'From Warrant to Flag Rank', *Engineering* (12 May 1944).

D. L. Summers, *HMS* Ganges *1866–1966: One Hundred Years of Training Boys for the Royal Navy* (Shotley Gate, Suffolk, 1966).

R. Taylor, 'Manning the Royal Navy: The Reform of the Recruiting System, 1852–1862', *Mariner's Mirror*, 44 (1958), 45 (1959).

J. T. Ward, *Sir James Graham* (London, 1967).

John G. Wells, *Whaley: The Story of HMS* Excellent *1830 to 1980* (Portsmouth, 1980).

Agnes Weston, *My Life Among the Bluejackets* (London, 1911).

Walter White (ed.), *A Sailor Boy's Log Book: From Portsmouth to the Peiho* (London, 1862).

J. Winton, *Hurrah for the Life of A Sailor!: Life on the Lower-deck of the Victorian Navy* (London, 1977).

Lionel Yexley, *The Inner Life of the Navy* (London, 1908).

—— *Our Fighting Seamen* (London, 1911).

10. THE BATTLESHIP FLEET: THE TEST OF WAR, 1895–1919

R. A. Burt, *British Battleships of World War One* (Annapolis, 1986), a comprehensively illustrated guide to the design and construction of British capital ships, incorporating much new archival material and technical information.

Patrick Beesly, *Room 40: British Naval Intelligence* (London, 1982), analyses the development and operation of the British naval signals intelligence organization.

Geoffrey Bennett, *Naval Battles of the First World War* (London, 1968), a well-written popular survey of the major sea actions of the war, now somewhat dated by recent research.

John Campbell, *Jutland: An Analysis of the Fighting* (Annapolis, 1986), a highly detailed examination of the battle with particular emphasis given to hits achieved and damage analysis.

Winston Churchill, *The World Crisis*, 5 vols. (London, 1925–31), Churchill's apologia for his role in the First World War. The first volume is particularly important in detailing his period as First Lord.

Julian S. Corbett, *Some Principles of Maritime Strategy* (London, 1911), Corbett's 'primer' on naval strategy, derived from his lectures to students of the RN War College.

—— and Henry Newbolt, *History of the Great War: Naval Operations*, 5 vols. (London, 1920–31), the official British history of the war at sea. Written with access to all documents but under considerable constraints, both official and individual, the work nevertheless remains an effective and generally fair narrative.

James Goldrick, *The King's Ships Were at Sea: The War in the North Sea, August 1914–February 1915* (Annapolis, 1984), examines the opening months of the war at sea and the impact of novel and poorly understood technology upon the progress of naval operations.

Paul Halpern, *The Mediterranean Naval Situation 1908–1914* (Cambridge, Mass., 1971), *The Naval War in the Mediterranean 1914–1918* (Annapolis, 1987), and *A Naval History of the First World War* (Annapolis, 1993). The first two books present a comprehensive picture of the naval war in the Mediterranean and its background, assessing the record of all the protagonist nations and drawing on a wealth of international archival material. The third is his comprehensive survey of every facet of the war at sea.

Earl Jellicoe, *The Grand Fleet 1914–16: Its Creation, Development and Work* (London, 1919), and *The Submarine Peril* (London, 1934); Jellicoe's attempts to justify his policies as Commander-in-Chief Grand Fleet and as First Sea Lord. The two books are nevertheless straightforward accounts which provide a useful insight into the conduct of the war.

Paul Kennedy, *The Rise of the Anglo-German Naval Antagonism 1860–1914* (London, 1980), an excellent account of the Anglo-German naval rivalry.

Nicholas Lambert, 'The Influence of the Submarine Upon Naval Strategy, 1898–1914' (D.Phil. thesis, University of Oxford, 1992), a revisionist and compelling analysis of the critical part played by the submarine in the evolution of Fisher's strategic thinking.

Peter H. Liddle, *The Sailor's War 1914–1918* (Poole, 1985), a selection of photographs and accounts from the remarkable archive of the First World War assembled by Peter Liddle in Sunderland.

Ruddock F. Mackay, *Fisher of Kilverstone* (Oxford, 1973), the best extant biography of Lord Fisher, thoroughly researched and well argued.

Arthur J. Marder, *British Naval Policy 1880–1905: The Anatomy of British Sea Power* (London, 1941), and *From the Dreadnought to Scapa Flow: The Royal Navy in the Fisher Era, 1904–1919*, 5 vols. (London, 1961–70). Elegantly written and sweeping in their breadth of vision, Marder's works remain unmatched, although many of his assessments are being progressively revised in the light of new archival research.

Peter Padfield, *Aim Straight: A Biography of Admiral Sir Percy Scott* (London, 1966), a useful biography of an important figure in the shaping of the pre-1914 Royal Navy.

—— *The Battleship Era* (London, 1975), a well-constructed popular survey of the era.

Stephen Roskill, *Admiral of the Fleet Earl Beatty: The Last Naval Hero, An Intimate Biography* (London, 1980), a masterly biography of a complex figure which draws together Beatty's professional and personal lives.

Jon Tetsuro Sumida, *In Defense of Naval Supremacy: Finance, Technology and British Naval Policy 1889–1914* (Boston, Mass., 1989), the seminal work in the progressive reassessment of the driving factors behind British naval policy in the pre-war era.

11. RETRENCHMENT, RETHINKING, REVIVAL, 1919–1939

Augustus Agar, *Baltic Episode* (London, 1963), a very readable account of the Royal Navy's operations against the Russian fleet.

Martin H. Brice, *The Royal Navy and the Sino-Japanese Incident, 1937–1941* (London, 1973), important for the navy's role in China.

James Cable, *The Royal Navy and the Siege of Bilbao* (London, 1979), a scholarly account of the navy's role in the Spanish Civil War.

A. Carew, *The Lower Deck of the Royal Navy 1930–1939: Invergordon in Perspective* (Manchester, 1981), the best account of lower-deck concerns.

W. S. Chalmers, *The Life and Letters of David Earl Beatty* (London, 1951), valuable for the insights provided into Admiral Beatty's policy.

Lord Chatfield, *It Might Happen Again* (London, 1947), an invaluable first-hand analysis by one of the leading naval policy-makers of the period.

L. Gardiner, *The* Royal Oak *Courts Martial* (London, 1965), a thoroughly readable account of a major personnel scandal of the period.

R. Gardiner (ed.), *The Eclipse of the Big Gun: The Warship 1906–45* (London, 1992), an interesting collection of technical analyses of ships and submarines.

G. Gordon, *British Seapower and Procurement Between the Wars: A Reappraisal of Rearmament* (London, 1988), the best account of the Admiralty's procurement policy.

Russell Grenfell, *Main Fleet to Singapore* (London, 1951), this book is a sensible review of the influence of the Far East on British naval policy.

B. Hunt, *Sailor–Scholar* (Waterloo, Ont., 1982), a good analysis of the role and thoughts of one of the navy's leading thinkers, Admiral Sir Herbert Richmond.

A. J. Marder, *From the Dardanelles to Oran* (Oxford, 1974), a useful collection of articles.

J. Neidpath, *The Singapore Naval Base and the Defence of Britain's Eastern Empire 1919–1941* (Oxford, 1981), explores Britain's defence policy towards the Far East.

George Peden, *British Rearmament and the Treasury* (Edinburgh, 1979), shows the influence of economic constraint and the Treasury on British naval policy.

Stephen Pelz, *Race to Pearl Harbor: The Failure of the Second London Naval Conference and the Onset of World War II* (Cambridge, Mass., 1974), provides a useful background to the Pacific War.

P. Pugh, *The Costs of Seapower* (London, 1986), a thorough analysis of the economic determinants of British naval policy.

Bryan Ranft (ed.), *Technical Change and British Naval Policy 1860–1939* (London, 1977), this collection of articles shows the importance of technological developments on naval policy.

S. Roskill, *Naval Policy between the Wars*, 2 vols. (London, 1968 and 1976), the standard text on the period.

—— *Hankey—Man of Secrets*, 3 vols. (London, 1970–4), an insider's view on British defence policy.

Geoffrey Till, *Airpower and the Royal Navy* (London, 1979), explores the impact of air power on naval policy in the period.

12. A SERVICE VINDICATED, 1939–1946

Correlli Barnett, *Engage The Enemy More Closely: The Royal Navy in the Second World War* (London, 1991), the best up-to-date one-volume study of the subject. Strong on strategy and amphibious operations, less so on technological detail and inter-war background.

Patrick Beesley, *Very Special Intelligence. The Story of the Admiralty's Operational Intelligence Centre 1939–45* (London, 1977), an excellent and considered account by an insider of the Admiralty's operational intelligence organization.

John Campbell, *Naval Weapons of World War Two* (London, 1985), the most comprehensive guide.

Conway's All The World's Fighting Ships 1922–46 (London, 1980), the best reference work on the warships of 1939–46.

Lord Cunningham, *A Sailor's Odyssey* (London, 1957), although 'ghosted', this still gives an excellent flavour of the war as seen by the navy's most notable fleet commander and second wartime Chief of Staff.

Peter Ellott, *Allied Escort Ships Of World War Two* (London, 1977), a complete and detailed survey of these vital ships and their activities.

William H. Garzke and Robert O. Dulin, *Battleships: Allied Battleships in World War Two; Battleships: Axis and Neutral Battleships of World War Two* (Annapolis, 1980–6), the best sources on capital ship actions with detailed studies of damage inflicted on both sides.

W. Hackmann, *Seek and Strike: Sonar and Anti-Submarine Warfare and the Royal Navy 1914–54* (London, 1984), a useful technical study of the development and use of Asdic that also shows how the Admiralty applied science and scientists to war.

F. H. Hinsley *et al.*, *British Intelligence in the Second World War, its Influence on Strategy and Operations*, 3 vols. (London, 1979–84), the major source on the impact of intelligence (notably cryptanalysis) on the Royal Navy's activities. The necessary antidote to more sensational accounts.

Derek Howse, *Radar At Sea: The Royal Navy in World War Two* (London, 1993), a study of the operational application of a key technology that is a model of its kind. It is particularly important on the development of fighter control.

Alan Raven and John Roberts, *British Battleships of World War Two* (London, 1976), the fullest and best-informed technical account of British capital ships of the Second World War.

——*British Cruisers of World War Two* (London, 1980), an equally detailed and well-informed account of the ships that formed the backbone of the Royal Navy's surface forces.

J. Rohwer and C. Hummelchen, *Chronology of the War At Sea* (London, 1992), a unique source of material on all naval activities in the Second World War.

S. W. Roskill, *The War At Sea*, 3 vols. (London, 1954–61), probably the best of the official histories and still an irreplaceable source of fact and fair-minded analysis.

——*The Navy At War* (London, 1960), a one-volume condensation of the larger official history.

——*Churchill and the Admirals* (London, 1977), a stimulating but fair reassessment of Churchill's stormy relationship with the naval professionals.

M. Stephen and E. Grove, *Sea Battles in Close-up: World War 2*, 2 vols. (London, 1988, 1993), the most up-to-date short accounts of the main engagements of the war.

Martin Van Creveld, *Supplying War* (Cambridge, 1977), a key re-evaluation of the impact of Malta on Axis sea communications and hence an essential critique of British Mediterranean strategy.

C. H. Waddington, *Operational Research in World War II: Operational Research against the U-Boat* (London, 1973), the only full study of a vital factor in victory in the Atlantic.

John Winton, *The Forgotten Fleet* (London, 1969), a comprehensive account of the naval operations against Japan in 1944–5 not only in the Pacific but in the Indian Ocean.

——*Ultra at Sea* (London, 1988), a useful up-to-date summary of the contributions made by special intelligence.

13. The Realities of Medium Power, 1946 to the Present

D. Barker, *Ruling the Waves* (London, 1986), a journalist's, but not a journalistic, view of the training and social structure of the 1980s navy.

C. Barnett, *The Collapse of British Power* (London, 1972), a scathing but generally just analysis of economic, educational, and social decline.

British Maritime Charitable Foundation, *Why the Ships Went* (London, 1988), a cool analysis of the catastrophic decline of the UK merchant fleet.

D. Brown, *The Royal Navy and the Falklands War* (London, 1987), an account by the head of the Naval Historical Branch.

J. Cable, *Gunboat Diplomacy* (3rd edn., London, 1994), a seminal work on low-intensity naval coercion and conflict, from 1919 to 1991.

J. Crane, *Submarine* (London, 1984), snapshot but impartial view of life on board.

J. Ethell and A. Price, *Air War South Atlantic* (London, 1983), concentrates on the air aspect and is somewhat inclined to overemphasize its impact, but has many valuable insights.

M. H. Fletcher, *The WRNS* (London, 1989), the best illustrated history to date.

S. G. Gorshkov, *The Sea Power of the State* (Oxford, 1978), testament of the leader for 30 years of the Soviet navy.

E. Grove, *Vanguard to Trident* (London, 1987). The best account of post-1945 naval policy, particularly strong for the period before 1956.

N. Henderson, *The Birth of NATO* (London, 1983), written in 1949 but not published for 30 years, it has authentic freshness.

J. R. Hill, *British Sea Power in the 1980s* (Shepperton, 1984), up-to-date account of the navy of the time.

—— *Maritime Strategy for Medium Powers* (Beckenham, 1986), an attempt to place transient strategies in a more permanent context.

B. Jackson and D. Bramall, *The Chiefs* (London, 1992), first-class account of British defence policy throughout the twentieth century, from the standpoint of those in military charge—which the authors were.

P. Kennedy, *The Rise and Fall of British Naval Mastery* (London, 1976), sombre but authoritative, foreshadowing the author's even more pessimistic later work.

H. C. Leach, *Endure no Makeshifts* (London, 1993), the First Sea Lord at the time of the Falklands Campaign gives brief but telling insights into political and naval concerns.

L. W. Martin, *The Sea in Modern Strategy* (London, 1967), valuable because the subject was being looked at by a fresh mind; somewhat dated now.

M. Middlebrook, *Operation Corporate* (London, 1985), a good all-round account of the South Atlantic Campaign.

P. Nailor, *The Nassau Connection* (London, 1989), official account of the development of the British submarine-based deterrent.

P. Nitze *et al.*, *Securing the Seas* (Boulder, Colo., 1979), demonstrates the preoccupation of NATO with the problem of reinforcing Europe against a threat from the Soviet Navy.

P. Pugh, *The Cost of Seapower* (London, 1986), an important analysis of the financial and technical problems facing the modern naval planner.

B. Ranft and G. Till, *The Sea in Soviet Strategy* (London, 1972), analysis at the height of the Cold War gives a powerful indication of the preoccupations of the West.

S. Roskill, *The Strategy of Sea Power* (London, 1962), draws heavily on Second World War experience and results but its analysis is still of great interest.

G. Till, *Maritime Strategy and the Nuclear Age* (2nd edn., London, 1984), an excellent survey of theory ancient and modern.

J. D. Watkins, *The Maritime Strategy* (Annapolis, 1986), immensely influential document of the period, however flawed critics may have found it then and later.

J. Wells, *The Royal Navy: An Illustrated Social History 1870–1982* (Stroud, 1994), an excellent comprehensive treatment of the subject and the only one covering this timespan.

J. Winton, *Air Power at Sea* (London, 1986), it is hard to find an unemotional work on this subject but this is the best.

S. Woodward, *One Hundred Days* (London, 1992), the task force commander's view of the Falklands campaign.

J. R. Young, *The Royal Marines* (London, 1991), well-reported account of the corps.

P. Ziegler, *Mountbatten* (London, 1985), the official, and very full, biography.

14. THE ROYAL NAVY AND THE POST-WAR NAVAL REVOLUTION, 1946 TO THE PRESENT

D. K. Brown, *A Century of Naval Construction* (London, 1983), this history of the Royal Corps of Naval Constructors includes important insights into post-war British naval design, such as the frigate programme.

—— *The Future British Surface Fleet* (London and Annapolis, 1991), the author is a senior retired British warship designer.

N. Friedman, *Modern Warship Design and Development* (London, 1980). This is an account of trade-offs in post-war surface warship design.

—— *Submarine Design and Development* (London and Annapolis, 1984).

—— *The Postwar Naval Revolution* (London and Annapolis, 1987), largely based on British official documents released in the 1980s (mainly for the period up to 1956), this book includes sketches of numerous unrealized projects, such as the big missile cruisers of the 1950s.

—— *British Carrier Aviation* (London and Annapolis, 1988), covers both the ships and their aircraft, drawing on documents declassified through the mid-1980s; includes Harrier carriers and the abortive 'escort cruiser'.

—— *The Naval Institute Guide to World Naval Weapons Systems* (Annapolis, 1989, 1991, 1994), includes most post-war British naval weapons and associated equipment, with extensive notes on how they were used. Editions vary in coverage of specific systems.

—— (ed.), *Navies in the Nuclear Age: Warships Since 1945* (London, 1993).

R. Gardiner (ed.), *Conway's All the World's Fighting Ships 1947–1982*, i (London and Annapolis, 1983), British section by Antony Preston.

E. Grove, *Battle for the Fjords* (London and Annapolis, 1991), uses a 1988 NATO exercise to explain the evolution of post-war forward NATO naval strategy.

W. Hackmann, *Seek and Strike: Sonar and Anti-Submarine Warfare and the Royal Navy, 1914–54* (London, 1984), although in theory extending only up through 1954, it actually describes developments into the 1960s.

D. Howse, *Radar At Sea* (London and Annapolis, 1993), although nominally limited to the Second World War, Howse follows the story through post-war work begun in wartime, i.e. through the 1950s and early 1960s.

P. J. Kemp, *The T-Class Submarine* (London and Annapolis, 1990), covers post-war modernization, including new weapons and sensors.

J. Lambert and D. Hill, *The Submarine* Alliance (London and Annapolis, 1986), includes post-war modernization; the ship herself is on display at Portsmouth.

J. Lipiett, *Type 21* (London, 1990).

L. Marriott, *Royal Navy Frigates Since 1945* (London, 1983 and 1990).

—— *Type 42* (London, 1985).

—— *Type 22* (London, 1986).

—— *Royal Navy Destroyers Since 1945* (London, 1989).

C. J. Meyer, *'Leander' Class* (London, 1984).

R. Osborne, *The Leander Class* (Kendal, 1992).

E. N. Poland, *The Torpedomen* (Emsworth, 1993), the story of HMS *Vernon* (1892–1986), the Royal Navy's mine and torpedo establishment.

C. E. Preston, *Power for the Fleet: The History of British Marine Gas Turbines* (Eton, 1982).

S. Ward, *Sea Harrier in the Falklands* (London and Annapolis, 1993), includes valuable insights into the concept of the Sea Harrier and of the Harrier carrier; Ward was one of the original proponents of the aircraft.

R. Watton, *The Aircraft Carrier* Victorious (London, 1991), shows details of the ship's very elaborate post-war reconstruction.

ILLUSTRATION SOURCES

Submarine Museum, Gosport; **303** IWM (Q13262); **308 top left** IWM (Q55499); **309 top right** IWM (SP1903); **311** NMM (N4399); **314** IWM (Q18676); **315** IWM (Q20640); **317** National Portrait Gallery, London; **318** IWM (Q19288); **321** National Portrait Gallery, London; **323** Hulton Deutsch Collection; **325** Popperfoto; **328** Hulton Deutsch Collection; **332** Topham Picture Source; **333 top** NMM (P39119), **bottom** Royal Naval Museum, Portsmouth; **334** Royal Naval Museum, Portsmouth; **335, 337** Geoffrey Till; **338** Royal Naval Museum, Portsmouth; **342** IWM (HU45230); **345** IWM (HU55811); **348** IWM (A16722); **356** IWM (A14845); **358** IWM (E488E); **359** IWM (ZZZ 53199); **360** IWM (A16268); **363** IWM (HU1838); **364** IWM (A20516); **367** IWM (A8166); **370** IWM (CH 15359); **372** IWM (A12022); **373** IWM (A21988); **374** IWM (FL5851); **375** IWM (A23916); **376** IWM (A25476); **378 top left** IWM (FL6488); **380** IWM (ABS1258); **385** C.C.M.O.D.;

388 NATO; **391, 392, 394, 397** C.C.M.O.D; **402 top left** Rex Features / Geoff Moore; **402 bottom left, top right, bottom right** C.C.M.O.D; **404, 405** C. & S. Taylor; **406** Rex Features; **408** C.C.M.O.D; **410** IWM; **411** IWM (FXC 17373); **412** TRH Pictures; **413** IWM (FKD 140); **415** TRH Pictures; **419** IWM (FXL 15231); **421** TRH Pictures; **422** IWM (FL3239); **423** IWM (MH 33793); **425** TRH Pictures; **426** IWM (MH 33794); **430** TRH Pictures; **431** TRH Pictures; **433** TRH Pictures; **439** NMM (B4898); **440** NMM (BHC2755); **441** NMM (BHC 0277); **442** Bishop Museum, The State Museum of Natural and Cultural History; **443** NMM (BHC2889); **444** Hulton Deutsch Collection; **445** NMM (8979); **446** IWM (HU55811); **447** IWM (A14845)

Picture Research by Sandra Assersohn

Maps and Action Stations diagrams drawn by Russell Birkett

INDEX

Page numbers in *italics* refer to captions to illustrations or maps. Individuals are entered under the names by which they are most commonly known, and ranks are the highest known to have been achieved. Sub-entries are arranged in chronological order where this is clear, otherwise alphabetically. Names and classes of British ships are sub-entries under the appropriate heading (king's ships, Royal Naval ships, etc.); ships of other navies generally appear as individual items. Where several ships are named in the text, only the most prominent in that activity or action are indexed.